W9-AHG-877

The Learning of Mathematics

The Learning of Mathematics

Sixty-ninth Yearbook

W. Gary Martin
Sixty-ninth Yearbook Coeditor
Auburn University
Auburn, Alabama

Marilyn E. Strutchens
Sixty-ninth Yearbook Coeditor
Auburn University
Auburn, Alabama

❄

Portia C. Elliott
General Yearbook Editor
University of Massachusetts
Amherst, Massachusetts

NATIONAL COUNCIL OF
TEACHERS OF MATHEMATICS

Library of Congress Cataloging-in-Publication Data

The learning of mathematics / W. Gary Martin, sixty-ninth yearbook coeditor, Marilyn E. Strutchens, sixty-ninth yearbook coeditor, Portia C. Elliott, general yearbook editor.
p. cm. — (Yearbook ; 69th)
Includes bibliographical references.
ISBN-13: 978-0-87353-596-0
ISBN-10: 0-87353-596-0
1. Mathematics—Study and teaching. 2. Learning. I. Martin, W. Gary (Wayne Gary) II. Strutchens, Marilyn E., 1962- III. Elliott, Portia C.
QA11.2.L437 2007
510.71—dc22
2006036719

The National Council of Teachers of Mathematics is a public voice of mathematics education, providing vision, leadership, and professional development to support teachers in ensuring mathematics learning of the highest quality for all students.

Printed in the United States of America

Table of Contents

Preface

A member of the Editorial Panel, Nancy Washburn, who is also a mathematics instructional specialist in a small city school district, recently posed the following question to a group of teachers attending a summer institute: "If we consider a school to be like a business, what is the product that it produces?" The participants brainstormed for a few minutes and then posed a number of possible responses, such as good instruction, high test scores, or even students who are ready for life. However, no one seemed ready for her response, which was quite simply "learning!"

The learning of mathematics is the central goal of mathematics education. Yet, as shown above, discussions of issues in our field more frequently address means of achieving that goal, including teaching, assessment, and curriculum. This is perhaps indicative of the difficulties of producing a yearbook on the topic of learning. The Learning Principle frequently seems to be the one that attracts the least attention among the principles in NCTM's (2000) *Principles and Standards for School Mathematics*. In some sense, the other principles lead up to, and support, the Learning Principle, but in and of itself, it may be the most difficult for us to really get our minds around. Discussions of constructivist learning theory quickly evolve into discussions of "constructivist teaching practices," a topic that makes little sense apart from the possible meaning of "teaching practices that may support students' construction of knowledge." Such discussions of teaching often rapidly diverge from discussions of how students learn.

The last attempt of the National Council of Teachers of Mathematics to produce a yearbook on the topic of learning was in 1990, and much has changed since then. NCTM's first volume of standards, *Curriculum and Evaluation Standards for School Mathematics*, had just been released the preceding year. We seemed on the verge of a new era in school mathematics, and that impression was not entirely wrong. Indeed, the first section of the 1990 Yearbook (NCTM 1990) was entitled "New Perspectives on Teaching and Learning," and those articles put forth a view of learning that many may now take somewhat for granted. The second part outlined "Effective Models and Methods for Teaching and Learning." Topics included cooperative learning, mathematics as communication, problem posing, writing, and student motivation—topics that (perhaps unsurprisingly) we might today consider not that groundbreaking.

However, as we consider our current situation, things turned out to be much more complex than we might have imagined in 1990. Disagreement about what constitutes mathematics learning is, if anything, more pronounced; no one would have envisioned the prolonged "math wars" that have stretched over the past decade or more. Furthermore, teachers who are trying to change their teaching practices to engage more students in the learning process are concerned about the balance between focusing on developing concepts and fostering relational understanding and providing practice for memorizing certain facts and rules. We face additional challenges related to learning that were only beginning to emerge seventeen years ago, including an increasing population of English language learners and the inclusion of students with learning disabilities into the regular mathematics classroom; how can their needs be met within the context of a discourse-rich classroom? Thus, perhaps the time is ripe for another attempt to initiate serious discussions about students' learning.

Articles in this book fall under five major categories, each of which is presented in a section of the yearbook. The first section presents perspectives on how students learn and how that might improve our understanding of current challenges faced by the field. The second section discusses issues related to students' learning in school contexts. The third section includes articles related to measuring and interpreting students' learning. The fourth addresses teachers' learning of mathematics; and the final section concerns reflections on mathematics teaching and learning. It may be interesting to note that these final two sections had little representation in the 1990 Yearbook. Each of these five sections begins with an introduction that highlights that category of articles and offers perspectives on the articles included in that particular section.

As editors of the yearbook, we would like to thank a number of people for their contributions to the yearbook. Thanks are extended to the authors for sharing their insights with others in the field. We also express sincere thanks and gratitude to the members of the Sixty-ninth Yearbook Panel, who provided their services willingly to shape the yearbook:

Patricia F. Campbell	University of Maryland
Gladis Kersaint	University of South Florida
Joseph G. Rosenstein	Rutgers University, New Brunswick, New Jersey
Nancy S. Washburn	Alexander City Schools, Alabama

We would also like to thank Portia C. Elliott, general yearbook editor, for her help in shaping the yearbook and keeping us on track. We also extend thanks to the NCTM staff for their hard work and dedication to making this yearbook the best that it could be. We all hope that the readers will find the Sixty-ninth Yearbook thought-provoking and helpful as they strive to advance mathematics education.

Marilyn E. Strutchens
W. Gary Martin
Sixty-ninth Yearbook Coeditors

REFERENCES

National Council of Teachers of Mathematics (NCTM). *Teaching and Learning Mathematics in the 1990s: 1990 Yearbook.* Edited by Thomas J. Cooney. Reston, Va.: NCTM, 1990.

———. *Principles and Standards for School Mathematics.* Reston, Va.: NCTM, 2000.

PART ONE

Perspectives on How Students Learn

W. Gary Martin

Learning theory and research on learning can provide a foundation on which we can better understand student learning. The first section of the yearbook includes four articles that offer general perspectives on how learning theory can improve our understanding of current issues in mathematics education.

Lambdin and Walcott open this section with an update of an article written for the 1989 Yearbook. Their article gives a historical perspective on the connections between learning theory and school mathematics. As they take the reader through six historical phases in the teaching and learning of mathematics, the influence of psychological learning theory provides a powerful lens through which to view changes in classroom practice. The continuing tension between emphases on meaningful or relational understanding on the one hand and instrumental understanding on the other is characterized as a "swinging pendulum" that has been oscillating throughout the past century. The authors argue that this viewpoint can furnish important insights into our current situation.

In the second article, Beatty and Moss investigate the learning of children with learning disabilities, a crucial issue facing many teachers of mathematics. Whereas learning disabled students are often given remedial instruction that emphasizes procedural knowledge, they report on a teaching experiment designed to improve the learning of these students' conceptual understanding of the equal sign and of equations, an area that has been found to be problematic for students of all abilities. They conclude that their approach, which was built on principles of learning drawn from both theory and research, provided these students with a conceptual understanding that could well undergird their further development of algebraic thinking, building from more concrete representations to increasingly abstract representations.

In the third article, Bay-Williams and Herrera consider another important student population, English language learners. They consider how one particular learning theory, sociocultural learning theory, can offer insights into the mathematics learning of English language learners. On the basis of this theory, they argue that culture and language must be considered as integral parts of the learning process. It is not enough to argue that mathematics is a universal language, or that "good practices" such as cooperative groups or the use of manipulatives will by themselves be sufficient. Teachers must consider both the linguistic and the content needs of their students. Moreover, the social interactions among students must be carefully structured in order to promote both mathematical growth and growth in the use of language to express mathematics.

In the final article of this section, Battista explores how learning theory can help to explain students' development of meaning for geometric ideas. Building on general principles of learning along with the van Hiele theory, a description of students' development of geometric thinking, Battista explores how a pair of students' understanding of geometric shapes grew through their use of a computer microworld. He illustrates the difficulty they have in building on their current level of thinking and—through interactions with each other and their teacher—developing a deeper understanding of geometric shapes. The article ends with a powerful description of how an understanding of learning theory can enable teachers to better support their students as they negotiate their way through these difficulties.

1

Changes through the Years: Connections between Psychological Learning Theories and the School Mathematics Curriculum

Diana V. Lambdin
Crystal Walcott

DURING the twentieth century, and extending to the present day, the teaching of mathematics in American schools has experienced six identifiable phases with differing emphases: (1) drill and practice, (2) meaningful arithmetic, (3) new math, (4) back to basics, (5) problem solving, and—currently—(6) standards and accountability. Each of these phases is worthy of attention because each corresponds to a period when American education in general was going through radical and fundamental changes, and each introduced new and innovative practices to mathematics education.

In fact, the existence of these historical phases would seem to belie the old cliché that there's nothing new under the sun. The field of education seems to be continually passing through cycles of change. On closer examination, however, many educational innovations are actually recycled in slightly modified forms. Not only is it difficult for newcomers to the profession to tell which aspects of mathematics education are recent innovations and which represent established practice, but it is often difficult for veteran teachers as well to recognize old ideas when they come around again, perhaps in new clothing. Thus, another reason for looking back at major phases of mathematics education over the past century or so is to discover the extent to which certain contemporary practices have their roots in the educational changes of yesteryear.

Finally, and most important, through historical analysis we may gain some perspective on the forces and issues that contribute to change in education. Historical perspective helps us avoid tunnel vision about the uniqueness of the educational problems we face today and suggests options to be considered as we ponder their solutions.

This paper is a revision and update of an article entitled "Connections between Psychological Learning Theories and the Elementary Mathematics Curriculum," written by the first author under her former name (Diana Lambdin Kroll), that was published in NCTM's 1989 Yearbook, *New Directions for Elementary School Mathematics.*

During any era, numerous diverse factors play a part in directing and influencing educational practice. Indeed, the factors surrounding the emergence of the phases listed above were complex: a constellation of mathematical, political, psychological, and sociological elements. This article focuses attention on the influence of one significant factor—psychological learning theories—both on earlier and on contemporary phases in the teaching of mathematics, but it also highlights the influences of other important factors. Although numerous strands of psychological theory and educational practice can be discerned during each phase, major psychological trends are generally recognizable, and the work of particular theorists can usually be considered central. Table 1.1 gives a brief overview of each of the phases, including their main theorists, classroom focus, and primary teaching methods.

The Drill-and-Practice Phase

In the early years of the twentieth century, drill and practice, which had long been one component of mathematics instruction, became its primary focus. Prior to 1900, the aim of schooling had been to confront students with difficult mental exercises to build up their powers of reasoning and thought. However, implicit faith in "mental discipline" as a theory of learning had begun to wane by the early 1900s, when Edward Thorndike proposed a new theory—a theory that became known in its several forms as "connectionism," "associationism," or "S-R bond theory." Thorndike claimed that learning is the formation of connections or bonds between stimuli (events in the environment) and responses (reactions of an organism to the environment). His theory maintained that through conditioning, specific responses are linked with specific stimuli.

In 1922 Thorndike published *The Psychology of Arithmetic*, in which he demonstrated how his theory applied to the teaching of arithmetic. He explained that teachers needed to recognize and make explicit the essential bonds that constitute the subjects they teach. As an example of what he meant by bonds, Thorndike listed seven separate S-R bonds in "simple two-column addition of integers." Among the list were bonds such as "learning to keep one's place in the column as one adds," "learning to add a seen [number] to a thought-of number," and "learning to write the figure signifying units rather than the total sum of a column." The list concluded with the statement that "learning to carry also involves itself at least two distinct processes" (1922, p. 52). According to Thorndike, the teacher's aim was to arrange for students to receive the right type of drill and practice on each of the right bonds for the right amount of time.

That Thorndike's theory did, in fact, influence mathematics education can be seen by examining yearbooks of professional societies and textbooks of the time. For example, in the introduction to the 1930 Twenty-ninth Yearbook of the National Society for the Study of Education (entitled *Report of the*

Table 1.1
Relationships between Phases of Mathematics Education and Psychological Learning Theories

Phase	Main Theories and Theorists	Focus	How Achieved
Drill and practice (approx. 1920–1930)	Connectionism or Associationism (e.g., Thorndike)	Facility with computation	Rote memorization of facts and algorithms Break all work into series of small steps
Meaningful arithmetic (approx. 1930–1950s)	Gestalt Theory (Brownell, Wertheimer, van Engen, Fehr)	Understanding of arithmetic ideas and skills Applications of math to real-world problems	Emphasis on mathematical relationships Incidental learning Activity-oriented approach
New math (approx. 1960–1970s)	Developmental psychology, socio-cultural theory (e.g., Bruner, Piaget, Dienes)	Understanding the structure of the discipline	Study of mathematical structures Spiral curriculum Discovery learning
Back to basics (approx. 1970s)	(Return to) connectionism	(Return to) concern for knowledge and skill development	(Return to) learning facts by drill and practice
Problem solving (approx. 1980s)	Constructivism, cognitive psychology, and sociocultural theory (Vygotsky)	Problem solving and mathematical thinking processes	Return to discovery learning, learning *through* problem solving
Standards, assessment, and accountability (approx. 1990s to present)	Cognitive psychology, sociocultural theory vs. renewed emphasis on experimental psychology (NCLB)	Math wars: concern for individual mathematical literacy vs. concern for administration of educational systems	NSF-developed student-oriented standards-based curricula vs. focus on test preparation for state-specified expectations

Society's Committee on Arithmetic), the editor presents the overall perspective of the volume: "Theoretically, the main psychological basis is a behavioristic one, viewing skills and habits as fabrics of connections" (Knight 1930, p. 5). As another example, figure 1.1 shows a portion of a page from Thorndike's (1924) arithmetic text for third graders, a page showing drill on basic addition facts. Since Thorndike did not provide any prior instruction to encourage pupils to relate the basic facts to one another, pupils working on this page would probably view the exercise 3 + 1 as totally unrelated to the exercise 1 + 3.

8. Addition

Read these lines. Say the right numbers where the dots are:

2 and 3 are	5 and 3 are	4 and 3 are ....
1 and 3 are	6 and 3 are	7 and 3 are ....
4 and 4 are	5 and 4 are	6 and 4 are ....
3 and 1 are	6 and 1 are	2 and 1 are ....
7 and 1 are	4 and 1 are	8 and 1 are ....

9.

Add and say the sums:

2	3	4	2	1	4	1	2	4	4
3	4	2	1	6	3	5	7	4	1
1	2	2	4	1	3	3	3	3	5
8	6	2	5	4	3	1	6	2	1
3	3	1	2	4	2	1	2	1	4
5	7	2	4	6	8	3	5	1	2

Fig. 1.1. A portion of a page from a 1924 third-grade text: Thorndike's *The Thorndike Arithmetics*

One major effect of Thorndike's theories was the segmentation of the curriculum into many disjoint bits. Teachers attempted to be certain that an entire collection of individual bonds was established and exercised in order for each higher-level skill to be mastered. Since each bond was believed to exist in isolation, it was thought that mixed, unorganized drill was perhaps even more effective than practice on a systematic arrangement of facts; with mixed drill the problem of interference between similar bonds was avoided. Another effect of connectionist theory was the prescriptive teaching methods that it encouraged. Teachers tended not to permit unorthodox algorithms or novel solutions, since the most efficient way to direct pupils to form correct bonds was to carefully keep them from forming incorrect bonds. Thus, in the curriculum of the drill-and-practice era, mathematics was taught by concentrating on drill with skills that had been segmented into small, distinct, easily mastered units.

Although reformulated types of drill and practice returned during the "back to basics" era of the 1970s and continue to be used to this day, in the years immediately following the drill-and-practice era the focus of mathematics education shifted considerably. Parents and educators of the 1930s and 1940s began to question the extreme emphasis on drill for drill's sake and to wonder whether the mathematics learned in school was of any practical use. The focus of instruction during those years—the progressive era—became an attempt to ensure that the skills students had mastered constituted "meaningful" arithmetic.

The Meaningful Arithmetic Phase

With the 1930s came the Great Depression. High unemployment and other social pressures encouraged more adolescents to remain in school and to continue their education through high school (Bidwell and Classen 1970, p. 531). From approximately 1930 to 1950, when the progressive education movement was influential in the United States, there was a new emphasis on "learning for living." The major change in mathematics education was a shift from an emphasis on drill for drill's sake to a focus on attempting to develop mathematics concepts in a "meaningful" way. This vague term meant different things to different educators of the progressive era.

For some progressives, mathematics was meaningful when it was encountered in the context of practical activity. Mathematics was learned for social utility—to acquire the tools for dealing with problems that would be encountered in later life. These educators recommended an activity-oriented approach. Some believed that students (particularly at the elementary school level) would learn all the mathematics they needed—and learn it better—through incidental experience rather than by systematic teaching. As a writer in the 1935 Tenth Yearbook of the National Council of Teachers of Mathematics (NCTM) commented:

> A large part of the most efficient learning is incidental, that is, learning a special subject with reference to some broader interest or aim without realizing it: Learning number relationships in connection with telling time or making change;

learning baseball averages (without effort) through sheer interest in big league contests (Wheeler, p. 239).

However, incidental learning was often haphazard, slow, and time-consuming, and according to William Brownell—a psychologist and influential mathematics educator of the time—the arithmetic learned under such circumstances was "apt to be fragmentary, superficial, and mechanical" (Brownell 1935, p. 18). When children were left to do whatever they felt like, their experiences were often so diverse and unstructured that they were unable to relate the different bits and pieces.

As a result, many mathematics educators of the era debated the merits of incidental learning. They proposed a different interpretation of "meaningful arithmetic," claiming that "to learn arithmetic meaningfully it is necessary to understand it systematically" (McConnell 1941, p. 281). In a 1938 *Mathematics Teacher* article, Buckingham attempts to clarify this view of meaningful arithmetic by emphasizing what he saw as an important distinction between the significance and the meaning of arithmetic. For him, the significance of number was functional: "its value, its importance, its necessity in the modern social order," whereas the meaning of number was mathematical—embodied in the structure of the number system (p. 26). In general, although proponents of this type of "meaning theory" acknowledged the motivating and enriching value of number experiences that emanate from student-initiated activities, they emphasized as well the notion of "arithmetic as a closely knit system of understandable ideas, principles, and processes" (Brownell 1935, p. 19).

This emphasis on mathematical relationships is clearly illustrated in figure 1.2, a reproduction of a page from Buswell, Brownell, and Sauble's (1959) third-grade text. In introducing basic subtraction facts to pupils, these textbook authors emphasized the relationship of the subtraction facts to previously learned addition facts, explicitly pointing out the "whole story" involving two different addends, a "story" in which the student relates two addition facts and two subtraction facts into one coherent whole.

Another important factor in the emergence of the new meaning theory was the introduction to this country of the "Gestalt" or "field" theory of learning (Fehr 1953). The central idea of field psychology is expressed in the German word *Gestalt,* which means an organized whole in contrast to a collection of parts. Gestalt psychologists regarded learning as a process of recognizing relationships and of developing insights. They believed that it is only when the relationship of a part of a situation to the whole situation is perceived that insight occurs and a solution to a problem can be formulated. Rather than drill on individual skills that in sum might lead to the solution of a problem (as a connectionist would), the field psychologist would try at the start to bring all the elements of a problem together.

A writer in the 1941 Sixteenth Yearbook of the NCTM rejected the connectionist theories of the earlier era as he explained the relationship of field

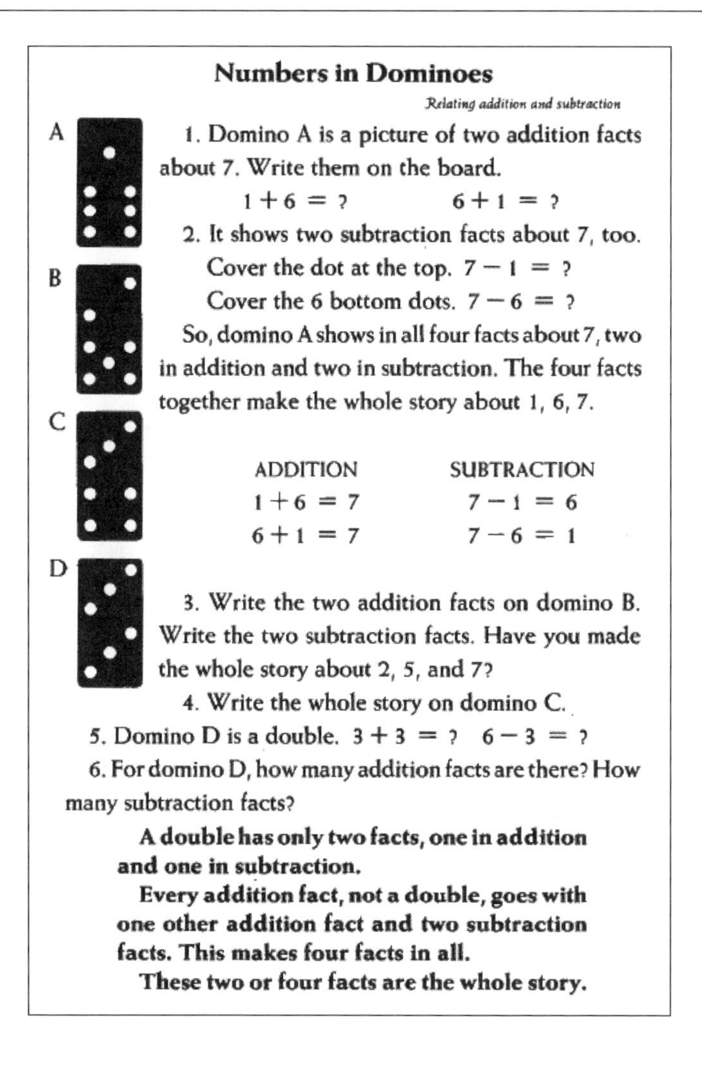

Numbers in Dominoes

Relating addition and subtraction

A B C D

1. Domino A is a picture of two addition facts about 7. Write them on the board.

$1 + 6 = ?$ $6 + 1 = ?$

2. It shows two subtraction facts about 7, too.
Cover the dot at the top. $7 - 1 = ?$
Cover the 6 bottom dots. $7 - 6 = ?$

So, domino A shows in all four facts about 7, two in addition and two in subtraction. The four facts together make the whole story about 1, 6, 7.

ADDITION	SUBTRACTION
$1 + 6 = 7$	$7 - 1 = 6$
$6 + 1 = 7$	$7 - 6 = 1$

3. Write the two addition facts on domino B. Write the two subtraction facts. Have you made the whole story about 2, 5, and 7?

4. Write the whole story on domino C.

5. Domino D is a double. $3 + 3 = ?$ $6 - 3 = ?$

6. For domino D, how many addition facts are there? How many subtraction facts?

A double has only two facts, one in addition and one in subtraction.

Every addition fact, not a double, goes with one other addition fact and two subtraction facts. This makes four facts in all.

These two or four facts are the whole story.

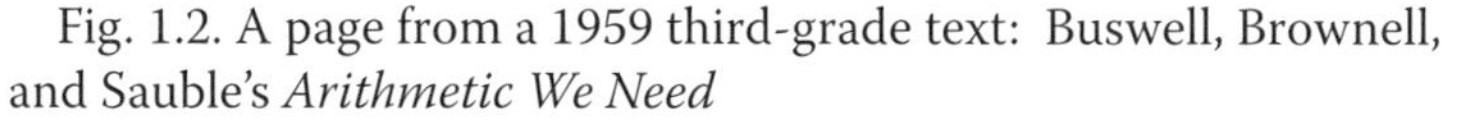

Fig. 1.2. A page from a 1959 third-grade text: Buswell, Brownell, and Sauble's *Arithmetic We Need*

psychology to the new meaning theory of arithmetic. Arguing that "meaning inheres in relationships," McConnell (1941, p. 280) claimed that the connectionists had "mutilated" arithmetic by decomposing it into numerous unrelated facts and by emphasizing "discreteness and specificity" (p. 275). The new "meaning theory," however, emphasized "understanding and relating the many specific items which are included [in our decimal number system]" (p. 280).

The meaning theory of arithmetic attempted to combine the progressive idea of activity learning with the ideas of the Gestalt psychologists. Teaching meaningful arithmetic demanded a very different kind of instruction from what had been the norm during the earlier drill-and-practice phase. Rote memorization was de-emphasized, and activity-based discovery was used to help students see connections among the many discrete skills and concepts they were learning. A new era in mathematics education was born. In fact, one important aspect of the "new math"—a focus on clarifying mathematical structures that was threaded throughout the curriculum from the earliest years right through high school—might be seen as a refinement and an extension of the meaningful arithmetic programs initiated during the progressive era.

The "New Math" Phase

The profound and far-reaching changes introduced by what became known as the "new math" during the 1960s were actually stimulated by severe criticism of American education in the years immediately following the Second World War. These changes received additional attention and support when the USSR's launch of the *Sputnik* satellite in 1957 fueled concerns that U.S. schools were not doing an adequate job of preparing graduates capable of understanding those science and mathematics concepts necessary to compete in a new, elite, technology-driven national workforce (Ferrini-Mundy and Graham 2003). In an article in the *Saturday Review*, an education writer blamed outdated teaching methods for the students' difficulties:

> In many classrooms teachers still rely exclusively on the drill-and-memory system of the past. Others teach arithmetic "incidentally," as it is needed in other subjects and school activities. (Dunbar 1956, p. 54)

In response to such criticisms, mathematicians, educators, and psychologists joined together for the first time in a massive effort to restructure the teaching of mathematics at all levels.

In September 1959, thirty-five scientists, scholars, and educators gathered at Woods Hole on Cape Cod to discuss how science and mathematics education might be improved. A major theme for the Woods Hole conference was that a knowledge of the fundamental structures of a discipline was considered crucial. The conference participants believed that "an understanding of fundamental principles and ideas ... appears to be the main road to adequate 'transfer of training'" and that the learning of "general or fundamental principles" ensures that individuals can reconstruct details when memory fails" (Bruner 1960, p. 25).

Thus, in the early 1960s, at least a dozen curriculum development projects were established to write new mathematics textbooks. Many of these efforts were funded by the recently established National Science Foundation.

The new math texts differed considerably from traditional texts both in organization and in content. Most of the new math curricula for secondary school evolved from the course of study proposed by a Commission on Mathematics appointed by the College Entrance Examination Board (CEEB 1959).

Topics and concepts were unified in a four-year sequence of high school courses (Algebra I, Geometry, Algebra II, Precalculus) that would eventually become nearly universal throughout the United States. Increased focus was placed on underlying mathematical theory, "with the apparent assumption that by including theory, both skills and understanding would greatly increase" (Lott and Souhrada 2000, p. 100).

A major innovation was the attempt to introduce abstract but fundamental ideas early in the curriculum (in modified form when necessary) and to return continually to these ideas in subsequent lessons, relating, elaborating, and extending them. In the elementary school curriculum, lessons were included on such previously unheard-of topics as sets, numeration systems, intuitive geometry, and number theory, and these new topics were related to more familiar content through a spiral organization of unifying strands. For example, figure 1.3 illustrates how the notions of subsets, take-away subtraction, and the writing of equations are all intertwined on one page from a third-grade new math text (School Mathematics Study Group 1965).

It was a bold undertaking to attempt to design materials that would help children understand the underlying structure of a subject as abstract and complex as mathematics. And the notion that young children could profit from lessons on topics (such as set theory) that had previously been studied only at the college level seemed incredible to many. Although the educators and mathematicians who cooperated to write the new texts may not have relied heavily on any specific learning theory to guide their efforts, one psychologist is closely associated with the new math movement: Jerome Bruner (Resnick and Ford 1981, p. 111).

Bruner (1960) supported two major recommendations for the new math curricula: the idea of a spiral curriculum and the idea of discovery learning. Grounding his psychological theory in part in the stage-theory work of Piaget, Bruner claimed that children move through three levels of representation as they learn. The first is the "enactive level," where the child directly manipulates objects. In the second, or "ikonic level," the child manipulates mental images of the objects rather than directly manipulating the objects themselves. The final level the child attains is the "symbolic level," where symbols are manipulated rather than objects or mental images of objects.

Just as children evolve through levels in which they use first enactive, then ikonic, and eventually symbolic representations, so the basic structures of a discipline can be represented manipulatively, visually, or in formal symbolic language. Many proponents of new math advocated using this structure in the design of mathematics curricula: teachers should first promote the discovery of mathematical concepts through the manipulation of blocks, sticks, chips, or other objects; then present these concepts pictorially; and finally introduce the appropriate mathematical symbolism. In other words, during the new math era, mathematicians and mathematics educators emphasized the notion of a

Removing a Subset

1. Look at these pictures.

How many members are in Set A? ________

2. How many members are in the subset being removed? ________

3. Draw a picture of the set that would be left when Subset B is removed from Set A.

4. How many members are in the set remaining when Subset B is removed from Set A? ________

5. Write an equation which describes the set remaining. ________

6. Look at these pictures. Ring a subset in each picture and write an equation for the set remaining.

________ ________ ________

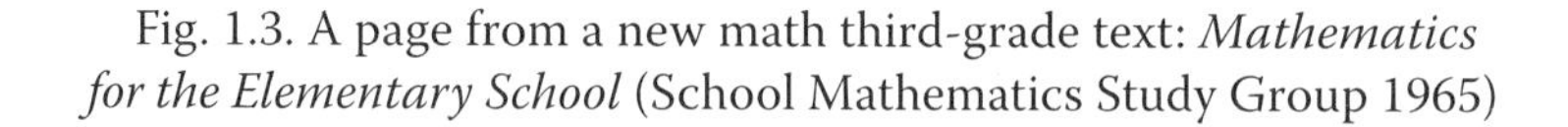

Fig. 1.3. A page from a new math third-grade text: *Mathematics for the Elementary School* (School Mathematics Study Group 1965)

spiral curriculum—a curriculum in which ideas are returned to again and again in increasingly more complex and abstract forms. Related to this approach was Bruner's oft-quoted assertion that "any subject can be taught effectively in some intellectually honest form to any child at any stage of development" (Bruner 1960, p. 33).

Bruner's second recommendation for the new math curriculum was to promote discovery learning. Of course, discovery learning was not a completely new idea, but it has been said that

> more than any one man, Bruner managed to capture its spirit, provide it with a theoretical foundation, and disseminate it. Bruner is not the discoverer of discovery; he is its prophet. (Shulman 1970, p. 29)

Mathematicians had considerable influence on mathematics education during the new math era, and they agreed that regardless of an individual's developmental stage, intellectual activity involves a continual process of discovering new ideas and relating these new ideas to previously discovered concepts. In actual practice, however, since the authors of most new math texts believed that it was inefficient to expect students to rediscover each element of the curriculum, they wrote textbooks that adopted a guided discovery approach. Texts that presented examples and explanations interspersed with leading questions, such as "Do you believe that ...?" or "Do you suppose it is true that ...?" were designed to guide students to discover the many underlying regularities and fundamental truths of mathematics. Figure 1.4 shows an example of how a new math text led junior high school students to use hands-on experiments in investigating a conjecture about geometry and measurement.

The Back-to-Basics Phase

The emphases of the new math phase were difficult for many parents, politicians, and even teachers to appreciate because both the mathematics itself and the recommended forms of instruction were so different from what many people were used to. Concerns were also raised about the utility of what students were learning in the new math classes, particularly about whether they were being adequately prepared for everyday uses of mathematics and for the mathematics needed in the workplace. Beginning in the 1970s, these concerns led to recommendations that American mathematics education should go "back to basics." The result in some schools was a rapid rejection of new math curricular materials, with a return to teaching facts and skills through demonstration, drill, and practice. In other schools, teachers continued using new math lessons but supplemented them with more traditional instruction on topics considered "basic." There seem to be no discernible new psychological learning theories evident that are due to the back-to-basics movement, though a return to connectionism (drill and practice) was common.

The Problem-Solving Phase

In the 1980s—the decade after back to basics became the rallying call—many educators thought the pendulum had, once again, swung too far. They were concerned that focusing mathematics instruction on basic facts and skills

Now let us consider the volume of an oblique prism. . . . As a beginning, let us consider a stack of rectangular cards which are congruent. You may make such a stack or use a deck of playing cards. . . .

Now push the cards a bit so that the deck will have the following appearance in cross section:

. . . it would seem that we have the basis for making the following conjecture.

Conjecture: If two prisms have congruent bases and equal heights, they have equal volumes.

To test this conjecture, look at Models 6, 11, and 12 [various oblique prisms, provided in the text]. Do they appear to have congruent bases? Do they have equal heights? For this it may help to stand them on their bases and lay a ruler across their upper bases to see if it seems level. . . . Now fill Model 6 with salt and pour into Model 11. Did you have too much salt or not enough, or did it seem to be just right? (This sounds like the three bears!) Do your results on this experiment confirm the conjecture above?

Fig. 1.4. A portion of a new math investigative lesson for junior high school (School Mathematics Study Group 1961, pp. 473–74)

did not, in fact, prepare students well for their future lives and careers. However, in contrast to the new math era's focus on preparing a technologically elite workforce, the 1980s ushered in an era of "concern about international competitiveness in commerce and finance and the need for a general workforce competent to support and advance the nation's overall economic and technological progress" (Ferrini-Mundy and Graham 2003, p. 1247). What was really important, many said, was that *all* students be able to *use* mathematics to solve problems, and thus NCTM published *An Agenda for Action*, a policy document that demanded that "problem solving must be the focus of school mathematics" (NCTM 1980, p. 2). Lessons on problem-solving strategies (such as "draw a picture" or "solve a simpler problem") soon were routinely included in nearly all elementary mathematics textbooks as a new focus on problem solving swept the American mathematics educational scene. In the latter part of the 1980s, the problem-solving movement became further refined when a distinction was drawn between teaching students how to solve problems—termed "teaching *about* problem solving" or "teaching *for* problem solving"—and using engagement in genuine problem solving to help students develop deeper understanding of mathematical concepts and skills—termed "teaching *via* problem solving" (Schroeder and Lester 1989).

During the problem-solving era, considerable attention was given to having students work in cooperative groups and having them verbalize their thinking. During the late 1980s and into the 1990s, theories of learning espoused during the new math era (e.g., Piaget's developmental psychology and Vygotsky's sociocultural theory) seemed to morph into enthusiasm for constructivism, a learning theory compatible with similar teaching strategies. Piaget asserted that children's sense-making processes and their modes of thinking change qualitatively with age. By contrast, Vygotsky focused attention on how individuals learn as a result of social interactions, particularly with more capable others. Although their perspectives were quite different, both Piaget's and Vygotsky's early research sparked an interest in modern-day constructivist researchers, such as Cobb, Glasersfeld, Davis, Maher, Steffe, and others (Davis, Maher, and Noddings 1990). Constructivist theorists believe that much can be learned from investigating an individual's sense-making strategies as he or she attempts to understand new mathematical ideas.

> The traditionally accepted hypothesis was that the teacher teaches the child and that children understand and assimilate more or less correctly what they have been taught. The new point of view is that children invent their own methods of counting, adding, and so forth; that these methods may be totally different from those usually presented by the teacher; and that children's methods may be more adequate to their own way of thinking than those proposed by the teacher. (Fischbein 1990, p. 7)

In short, constructivist educators pay attention to student-designed strategies and strive to create a learning environment where cognitive conflict (as presented by problem situations) necessitates a restructuring of the students' thinking (Booker 1996). Thus, teaching through problem solving and constructivist learning theory were plausibly aligned through the belief that students learn best when they themselves are engaged in figuring things out and that working out their own approaches to mathematical problems is the best way for students to become mathematically literate and proficient. For example, figure 1.5 shows an example—from the teacher materials for an elementary school curriculum developed in the early 1990s—of different ways that primary school students might think about 29 + 12 prior to learning any formal algorithm for addition with regrouping.

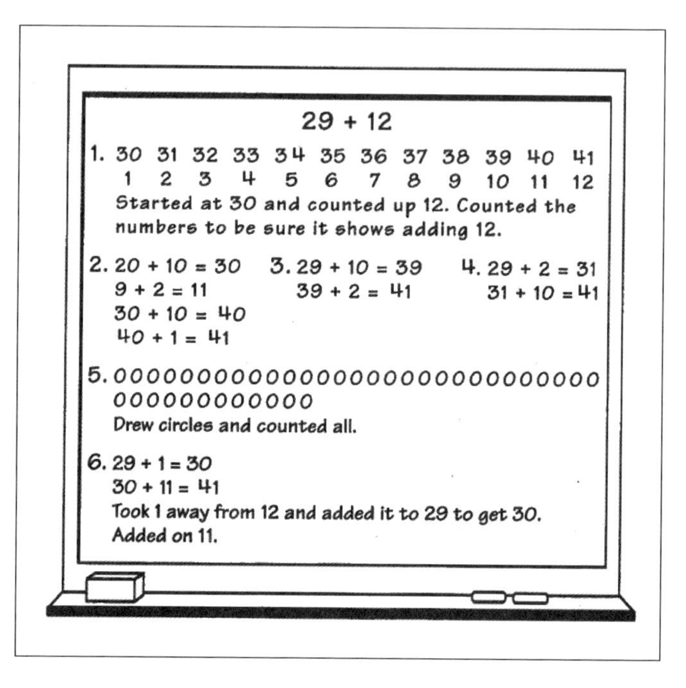

Fig. 1.5. An example of possible student approaches, taken from one of the elementary school curricula funded by the National Science Foundation during the "math wars" (Economoupoulos and Russell 1998, p. 7)

The Standards and Accountability Phase

The phase we are labeling "standards and accountability" was initiated by the unveiling in 1989 of NCTM's groundbreaking *Curriculum and Evaluation Standards for School Mathematics.* We discuss this phase in two parts, focusing first on the development of standards and standards-based curricula, and then on pressures for accountability.

Developing Mathematics Education Standards and Standards-Based Curricula

After NCTM's 1989 *Standards* document outlined a vision for curricular innovations, subsequent NCTM *Standards* documents focused on teaching

(NCTM 1991) and assessment (NCTM 1995). Soon after, the recommendations of all three documents were updated and consolidated in *Principles and Standards for School Mathematics* (NCTM 2000). All the NCTM *Standards* documents stress the importance of considering not only which mathematical concepts are important for students to understand and which skills are important for students to master, but also—perhaps most important—how students learn. Indeed, the NCTM *Standards* documents quite clearly reflect the influence of constructivist theory. Students are not viewed as "passively absorb[ing] information, storing it in easily retrievable fragments as a result of repeated practice and reinforcement"; rather, according to the NCTM *Curriculum and Evaluation Standards*, "individuals approach a new task with prior knowledge, assimilate new information, and construct their own meanings" (NCTM 1989, p. 10). Additionally, NCTM's *Standards* documents echo the influence of the sociocultural theory of learning: "Learning with understanding can be further enhanced by classroom interactions, as students propose mathematical ideas and conjectures, learn to evaluate their own thinking and that of others, and develop mathematical reasoning skills" (NCTM 2000, p. 21).

As during the new math era, the National Science Foundation funded a number of important curriculum development projects beginning in the mid-1980s and continuing into the new millennium in an effort to support the development of curricula to make the vision of the NCTM *Standards* a reality. These curricula were eventually adopted by school districts across the country, and numerous research and evaluation projects gathered data on their efficacy (Senk and Thompson 2003).

Research to document the effectiveness of the standards-based curricula was particularly important because, throughout the 1990s, opinions about how best to teach mathematics varied widely. Many mathematicians and mathematics educators placed themselves at different positions along a continuum of support or disdain for the standards-based mathematics education reform, sparking debates that came to be known as the "math wars." The *Standards* documents brought issues highlighted by the math wars to the forefront, which was arguably beneficial to the community as a whole. Reform-minded educators, whose primary focus was on learners of mathematics, interpreted the *Standards* as a call for a restructuring of the mathematics curriculum based in large part on constructivist and sociocultural views. Antireformists, who tended to focus more on the mathematics and less on the learners, argued for a more traditional curriculum devoid of any "fuzzy new-new math" (Schoenfeld 2004, p. 278). As the math wars heated up, debates about what sorts of research and evaluation were appropriate, convincing, and trustworthy also became heated (Slavin 2002). Some scientists expressed confidence in nothing other than "scientifically based research"—defined as "rigorous, systematic and objective procedures to obtain valid knowledge," which primarily includes research that "is evaluated using experimental or quasi-experimental designs" preferably with random assignment (U.S. Department of Education 2001). Yet, many educators were concerned that such research is often neither feasible nor informative in the complex environments of students, classrooms, and schools.

Accountability

The decade of the 1990s included considerable agitation for educators to recognize that U.S. students were not internationally competitive in performance, particularly on mathematics assessments. Along with calls for action to redress this situation through changes in teachers' qualifications, state-identified standards, and classroom instruction, the federal government unveiled plans for increased accountability through mandated assessments.

On January 13, 1990, President George H. W. Bush addressed the Joint Session of the Congress on the State of the Union. Sparked by the Education Summit held at the National Governors Conference the preceding fall (August-September 1989), President Bush proclaimed, "By the year 2000, U.S. students must be first in the world in math and science achievement" (Bush 1990). In addition, the president called for mandatory national assessments at grades 4, 8, and 12.

President William J. Clinton echoed similar sentiments in his 1997 State of the Union Address (Clinton 1997), offering a ten-point call to action for American public education in the twenty-first century. Along with calling for an increase in the number of Americans entering college, improving safety in schools, increasing teacher quality, and other important issues in the national education landscape, Clinton issued calls for setting national standards (along with developing accompanying national tests) in fourth- and eighth-grade reading and mathematics and expanding accountability in public education.

In 2001, following the lead of his immediate predecessors, President George W. Bush unveiled the No Child Left Behind Act (NCLB) (U.S. Congress 2001; U.S. Department of Education 2001) aimed at increased accountability in U.S. public schools. NCLB required that schools receiving federal funding make "adequate yearly progress" (AYP) toward the goal of having all students reach minimal state proficiency levels by 2014 or face a variety of sanctions. Additionally, the law mandated that all students be taught by "highly qualified" teachers and designated federal funding to promote research-based, effective classroom practices.

NCLB was extremely controversial. Proponents argued that the law increased accountability for the performance of all students by rewarding achievements and imposing santions on failures, focused attention on teaching strategies that work, and empowered the public (and parents, in particular) by providing information about school performance and offering choices when schools are failing. Opponents argued (1) that frequent high-stakes testing does nothing to improve the lives of children but can lead to an unhealthy narrowing of both the curriculum and teaching strategies (teaching to the test), and (2) that an undue emphasis on punishment, negative labels, and threats can do great damage to schools and children.

Two documents offer helpful insights into the learning theories influencing mathematics instruction at the dawn of the twenty-first century—Paul Cobb's (forthcoming) chapter from the *Second Handbook on Research in Mathematics Education* ("Putting Philosophy to Work: Coping with Multiple Theoretical Perspectives") and an influential volume from the National Research Council (NRC): *How Students Learn: History, Mathematics, and Science in the Classroom* (NRC 2005).

Cobb offered a detailed analysis of four theoretical perspectives on learning (experimental psychology, cognitive psychology, sociocultural theory, and distributed cognition). He compared and contrasted how these four perspectives characterize individuals and discussed the usefulness and limitations of each perspective, demonstrating compellingly that different theories of learning can be more or less helpful for education professionals depending on what the professionals' responsibilities are. For example, Cobb noted that contemporary accountability demands reflect an allegiance to theories of learning espoused by experimental psychology—a "research tradition whose primary contributions to mathematics education have involved the development of assessment instruments, particularly norm-referenced tests, and the findings of studies that have assessed the relative effectiveness of alternative curricular and instructional approaches" (Cobb forthcoming, p. 35). Thus, turn-of-the-century trends in mathematics education—as prompted by federal mandates—tended to focus on the learning of an "abstract, collective individual" who was actually "a statistical aggregate," rather than on how to deal with "individual differences and the challenges involved in accounting for the reasoning and learning of specific students" (Cobb forthcoming, p. 36). In other words, experimental psychology is a useful theoretical perspective for those many school administrators with responsibility for figuring out how to raise aggregate test scores to meet NCLB mandates, although it has limited relevance for making decisions at the classroom level. By contrast, Cobb observed that teachers—concerned as they must be with classroom-based decision making—are more likely to focus on the "individual as participant in cultural practices" (sociocultural theory) or the individual as an "element in a reasoning system" (distributed cognition) because it is their responsibility to design an appropriate learning environment for their individual students.

The National Research Council's *How Students Learn* (NRC 2005)—recognizing that teachers must shoulder responsibility for providing instructional environments conducive to learning—offered a framework helpful in analyzing and describing effective classroom environments and theories of learning. The book discussed four design characteristics—drawn from the implications of contemporary research—that can be used as lenses to evaluate the effectiveness of teaching and learning environments. The learner-centered lens focuses attention on the students—in particular, on instruction that helps them build new knowledge on a foundation of the knowledge, skills, and ideas they already possess. The knowledge-centered lens focuses on the subject matter—why it is taught, how it should be organized (curriculum), and what learning goals (competence or mastery) should be established. The focus of the assessment-centered lens is assessment "that supports learning by providing students with opportunities to revise and improve their thinking" (NRC 2005, p. 16). And the community-centered lens focuses on classroom or school norms, encouraging "a culture of questioning, respect, and risk taking" (NRC 2005, p. 13).

Reflecting on the Past and Looking into the Future

A visitor to contemporary American classrooms is certain to see lessons designed in response to early twenty-first-century governmental pressures for test preparation and adherence to state-mandated curricular standards. But the visitor is also likely to see evidence of most of the major phases through which mathematics education passed during the twentieth century, as well as the influences of the theories of learning undergirding these curricular phases. In contemporary classrooms, the influence of Thorndike's S-R theory and the back-to-basics movement can often be seen, for example, in state or district curricular frameworks that include long lists of discrete learning objectives or expectations, in individualized learning packages or test-preparation materials, and in many computer-based instructional systems. Continuing concern for problem solving in mathematics can be seen as an extension of the influence of the Gestalt psychologists on the formulation of a meaning theory of arithmetic during the progressive era and the reprise of that emphasis during the problem-solving phase. We continue to see high interest in helping students to develop insights and to consolidate the skills they have learned into a meaningful whole, as emphasized by constructivists' focus on problem solving during the 1980s and on reform-oriented, standards-based curricula during the 1990s. Moreover, an important component of much instruction today is the opportunity it gives students to develop skills in mathematics through engagement in real-life situations, including using electronic technologies in increasingly innovative and challenging ways. In these aspects of the contemporary curriculum we see the continued influence of the activity approach that dominated both the progressive era and recent standards-based reforms.

In spite of the fact that the new math curriculum projects of the 1960s failed to fulfill the optimistic dreams of their originators for universal and massive curricular reform, the projects did, in fact, effect certain fundamental changes both in the content and in the organization of today's mathematics curriculum. One of the major aims of mathematics instruction today is—as during the new math era—to help students understand the structure of mathematics. For example, "algebraic thinking" has become a focus of instruction starting in the earliest grades, with the aim of spiraling higher and higher during each succeeding year.

Though curricula, assessments, teacher education, and many other aspects of mathematics education have been consistently modified, extended, and improved during the past hundred years, we continue to hear clarion calls of concern similar to those that have echoed repeatedly throughout recent history. Global competition for scientific and technological work—due to our increasingly "flat world"—is the source of many of these concerns (Friedman 2005). In 2005, the authors of an influential governmental report, *Rising above the Gathering Storm*, described the situation, while calling for government-sponsored efforts to improve American mathematics and science education and to sustain and strengthen research and innovation:

> Today, Americans are feeling the gradual and subtle effects of globalization that challenge the economic and strategic leadership that the United States has enjoyed since World War II. A substantial portion of our workforce finds itself in direct competition for jobs with lower-wage workers around the globe, and leading-edge scientific and engineering work is being accomplished in many parts of the world. (Committee on Prospering in the Global Economy of the 21st Century 2006, p. 3)

Our nation is also facing a sea change in population, since former "minority" groups such as Hispanics and African Americans are growing so rapidly that the traditionally white majority population will soon be the new minority. Although "mathematics for all" has been a rallying call for decades, we have a long way to go before policies and practices that have supported differential expectations and inequitable opportunities give way to universal access and support for high-quality mathematics education for all.

Perhaps the ultimate challenge for mathematics education is to find answers for questions that were troublesome during the twentieth century and that still plague us today. For example, what goals should be paramount in school mathematics? Preparing students for an increasingly technological world of work? Educating them as mathematically literate citizens? Helping them to appreciate the power and beauty of mathematics? Increasing the pool of our nation's scientists and engineers?

In a similar vein, which learning theories seem most relevant for the challenges we face today and in the foreseeable future? Anna Sfard identified ten needs that are "the driving force behind human learning and must be fulfilled if this learning is to be successful" and analyzed how different theories of learning either address these needs or fall short. Her conclusions included this sobering thought:

> In our attempts to improve the learning of mathematics, we will always remain

> torn between two concerns: our concern about the learner and our concern about the quality of the mathematics being learned. Because of this ever-present tension, we are repeatedly thrown from one extreme solution to another. It appears impossible to bring the swinging pendulum to a stop. (Sfard 2003, p. 386)

Indeed, the pendulum is always swinging, as our overview of the influences of learning theories on mathematics education over the past century should make clear. Many educational practices and catchphrases that today seem quite current actually have their roots in bygone eras. Too often, however, we use such practices and phrases without perspective, without appreciation for the contexts that gave them birth. According to the historian Morris Cohen, this can lead to an "exaggerated idea of our own originality and of the uniqueness of our own age and problems" (1961, p. 277). In addition, Cohen continues, without historical perspective we may "fail to see all that is involved in the issues of the day" (p. 277). History gives us clues to factors that may still be in operation today, to currents and forces that move our discipline, and to motives and conflicts that shape it. Most important, a knowledge of the past empowers us to speak with more authority and to make better-informed decisions about contemporary educational practices. In particular, when we become aware of their historical roots, it is clear not only that numerous psychological theories are very influential in twenty-first-century school mathematics curricula but also that we must think more critically about the relevance of those theories—and the practices they engender—for future problems in mathematics education.

REFERENCES

Bidwell, James K., and Robert G. Clason. "Comment." In *Readings in the History of Mathematics Education,* edited by James K. Bidwell and Robert G. Clason, pp. 531–33. Washington, D.C.: National Council of Teachers of Mathematics, 1970.

Booker, George. "Constructing Mathematical Conventions Formed by the Abstraction and Generalization of Earlier Ideas: The Development of Initial Fraction Ideas." In *Theories of Mathematical Learning,* edited by Leslie P. Steffe, Pearla Nesher, Paul Cobb, Gerald A. Goldin, and Brian Greer. Mahwah, N.J: Lawrence Erlbaum Associates, 1996.

Brownell, William A. "Psychological Considerations in the Learning and the Teaching of Arithmetic." In *The Teaching of Arithmetic,* Tenth Yearbook of the National Council of Teachers of Mathematics, edited by W. D. Reeve, pp. 1–31. New York: Teachers College, Columbia University, 1935.

Bruner, Jerome S. *The Process of Education.* Cambridge, Mass.: Harvard University Press, 1960.

Buckingham, B. R. "Significance, Meaning, Insight—These Three." *Mathematics Teacher* 31 (January 1938): 24–30.

Bush, George H. W. *Address before a Joint Session of the Congress on the State of the Union*. Speech presented at the House Chamber of the Capitol, Washington, D.C., January 1990.

Buswell, Guy T., William A. Brownell, and Irene Sauble. *Arithmetic We Need: Book 3.* Boston: Ginn & Co., 1959.

Clinton, William J. *Call to Action for American Education in the Twenty-first Century.* Speech presented at the 1997 State of the Union address, Washington, D.C., January 1997.

Cobb, Paul. "Putting Philosophy to Work: Coping with Multiple Theoretical Perspectives." In *Second Handbook of Research on Mathematics Teaching and Learning,* edited by Frank K. Lester. Greenwich, Conn.: Information Age Publishers, forthcoming.

Cohen, Morris. *The Meaning of Human History*. 2nd ed. LaSalle, Ill.: Open Court Publishing Co., 1961.

College Entrance Examination Board, Commission on Mathematics. *Program for College Preparatory Mathematics.* New York: College Entrance Examination Board, 1959.

Committee on Prospering in the Global Economy of the 21st Century. *Rising above the Gathering Storm: Energizing and Employing America for a Brighter Economic Future*. Executive Summary. Washington, D.C.: National Academies Press, 2006.

Davis, Robert B., Carolyn A. Maher, and Nel Noddings. *Constructivist Views on the Teaching and Learning of Mathematics. Journal for Research in Mathematics Education* Monograph Series No. 4. Reston, Va.: National Council of Teachers of Mathematics, 1990.

Dunbar, Ruth. "Why Johnny Can't Add." *Saturday Review,* 8 September 1956, pp. 28, 54.

Economopoulos, Karen, and Susan Jo Russell. *Putting Together and Taking Apart: Addition and Subtraction*. Second-grade curriculum unit from *Investigations in Number, Data, and Space*. White Plains, N.Y.: Dale Seymour Publications, 1998.

Fehr, Howard F. "Theories of Learning Related to the Field of Mathematics." In *The Learning of Mathematics: Its Theory and Practice,* Twenty-first Yearbook of the National Council of Teachers of Mathematics (NCTM), edited by Howard F. Fehr, pp. 1–41. Washington, D.C.: NCTM, 1953.

Ferrini-Mundy, Joan, and Karen J. Graham. "The Education of Mathematics Teachers in the United States after World War II: Goals, Programs, and Practices." In *A History of School Mathematics,* Vol. 2, edited by George Stanic and Jeremy Kilpatrick, pp. 1193–1308. Reston, Va.: National Council of Teachers of Mathematics, 2003.

Fischbein, Efraim. "Introduction." In *Mathematics and Cognition: A Research Synthesis by the International Group for the Psychology of Mathematics Education,* edited by Pearla Nesher and Jeremy Kilpatrick, pp. 1–13. Cambridge: Cambridge University Press, 1990.

Friedman, Thomas. T*he World Is Flat: A Brief History of the Twenty-first Century.* New York: Farrar, Straus, & Giroux, 2005.

Knight, F. B. "Introduction." In *Report of the Society's Committee on Arithmetic,* Twenty-ninth Yearbook of the National Society for the Study of Education, edited by F. B. Knight, pp. 1–8. Bloomington, Ill.: Public School Publishing Co., 1930.

Lott, Johnny W., and Terry A. Souhrada. "As the Century Unfolds: A Perspective on Secondary School Mathematics Content." In *Learning Mathematics for a New Century,* 2000 Yearbook of the National Council of Teachers of Mathematics (NCTM), edited by Maurice J. Burke, pp. 96–111. Reston, Va.: NCTM, 2000.

McConnell, T. R. "Recent Trends in Learning Theory: Their Application to the Psychology of Arithmetic." In *Arithmetic in General Education,* Sixteenth Yearbook of the National Council of Teachers of Mathematics, edited by W. D. Reeve, pp. 268–89. New York: Teachers College, Columbia University, 1941.

National Council of Teachers of Mathematics (NCTM). *An Agenda for Action.* Reston, Va.: NCTM, 1980.

———. *Curriculum and Evaluation Standards for School Mathematics.* Reston, Va.: NCTM, 1989.

———. *Professional Standards for Teaching Mathematics.* Reston, Va.: NCTM, 1991.

———. *Assessment Standards for School Mathematics.* Reston, Va.: NCTM, 1995.

———. *Principles and Standards for School Mathematics.* Reston, Va.: NCTM, 2000.

National Research Council. *How Students Learn: History, Mathematics, and Science in the Classroom.* Washington, D.C.: National Academies Press, 2005.

Resnick, Lauren B., and Wendy W. Ford. *The Psychology of Mathematics for Instruction.* Hillsdale, N.J.: Lawrence Erlbaum Associates, 1981.

Schoenfeld, Alan H. "The Math Wars." *Educational Policy* 18 (January 2004): 253–86.

School Mathematics Study Group (SMSG). *Mathematics for Junior High School, Volume II, Part II.* Stanford, Calif.: SMSG, 1961.

———. *Mathematics for the Elementary School, Book 3.* Stanford, Calif.: SMSG, 1965.

Schroeder, Thomas L., and Frank K. Lester, Jr. "Developing Understanding in Mathematics via Problem Solving." In *New Directions for Elementary School Math-*

ematics, 1989 Yearbook of the National Council of Teachers of Mathematics (NCTM), edited by Paul R. Trafton, pp. 31–42. Reston, Va.: NCTM, 1989.

Senk, Sharon L., and Denisse R. Thompson, eds. *Standards-Based School Mathematics Curricula: What Are They? What Do Students Learn?* Mahwah, N.J.: Lawrence Erlbaum Associates, 2003.

Sfard, Anna. "Balancing the Unbalanceable: The NCTM Standards in Light of Theories of Learning Mathematics." In A *Research Companion to "Principles and Standards for School Mathematics,"* edited by Jeremy Kilpatrick, W. Gary Martin, and Deborah Schifter, pp. 353–92. Reston, Va.: National Council of Teachers of Mathematics, 2003.

Shulman, Lee S. "Psychology and Mathematics Education." In *Mathematics Education,* Sixty-ninth Yearbook of the National Society for the Study of Education, edited by Edward G. Begle, pp. 23–71. Chicago: University of Chicago Press, 1970.

Slavin, Robert E. "Evidence-Based Educational Policies: Transforming Educational Practice and Research." *Educational Researcher* 31(7) (2002): 15–21.

Thorndike, Edward L. *The Psychology of Arithmetic.* New York: Macmillan Co., 1922.

———. *The Thorndike Arithmetics: Book One.* New York: Rand McNally, 1924.

U.S. Congress. *No Child Left Behind Act of 2001.* Washington, D.C: U.S. Congress, 2001.

U.S. Department of Education. *No Child Left Behind.* Retrieved January 10, 2006, from www.ed.gov/nclb/, 2001.

Wheeler, Raymond H. "The New Psychology of Learning." In *The Teaching of Arithmetic,* Tenth Yearbook of the National Council of Teachers of Mathematics, edited by W. D. Reeve, pp. 233–50. New York: Teachers College, Columbia University, 1935.

2

Teaching the Meaning of the Equal Sign to Children with Learning Disabilities: Moving from Concrete to Abstractions

Ruth Beatty
Joan Moss

A CURRENT focus of mathematics instruction centers on the push for algebra reform and the resulting recommendation from the National Council of Teachers of Mathematics (NCTM 2000) that algebra become an essential strand in the elementary school curriculum. The lack of algebraic understanding still remains a barrier for participation in high levels of mathematics. In response, researchers are investigating how to design curricula that incorporate instruction in algebraic reasoning for all grades, allowing elementary school students to develop early algebraic thinking that is fundamental for participating in algebra at later grades (Kaput 1998; Schliemann et al. 2003; Warren 2003).

An understanding of equality is central to the development of algebraic thinking. This understanding refers to the relationship between two mathematical expressions that hold the same value and is expressed symbolically in a mathematical sentence or equation. Within an equation, the equal sign indicates that the quantities on either side of it are the same. For instance, in the equation $2 + 3 = 5$, the equal sign means that $2 + 3$ and 5 are different mathematical expressions for the same value. Although this idea seems evident to adults, it is a difficult concept for children to develop.

Given the importance of algebra, a number of researchers from the mathematics education community have been specifically focusing on children's understanding of the equal sign as part of the "algebra for all" movement (Blanton and Kaput 2003; Herscovics and Kieran 1980; Baroody 2000; Carpenter and Levi 2000; Falkner, Levi, and Carpenter 1999; MacGregor and Price 1999; Saenz-Ludlow and Walgamuth 1998; Stacey and MacGregor 1997). What all these studies reveal is that instead of conceiving of the equal sign as a symbol that denotes a relationship between two mathematical expressions that hold the same value, children misconceive the equal sign as a signal to perform a computation and record the answer. For the majority of elementary school children, this misconception is robust and long lasting (Falkner, Levi, and Carpenter 1999; Warren 2003), which makes even more difficult the already abrupt move from arithmetic to algebraic instruction (Matz 1982; Kieran, Battista, and Clements 1991; Swafford and Langrall 2000).

The prevailing procedural or operational interpretation of the equal sign has been demonstrated in studies in which children were unable to understand equations such as 5 = 4 + 1 or 9 = 9 (Falkner, Levi, and Carpenter 1999; Osborne and Wilson 1988; Behr, Erlwanger, and Nichols 1980). Additionally, children struggled with solving for missing addends. When presented with 5 + 7 = ____ + 8, most children place a 12 in the blank, indicating they interpreted the = as a signal to compute the operation on the left side and record it on the right.

Although no research to date has been done with children with learning disabilities, it would seem predictable that these children would also find the conceptual underpinnings of the equal sign as elusive as typically developing children. The kinds of instructional approaches that might support the development of a more sophisticated and mathematically useful understanding of the equal sign for this population of children have not been investigated. Research on learning disabilities in the domain of mathematics is in its infancy (Gersten, Jordan, and Flojo 2005). A principal area of consideration is the divide between procedural and conceptual instructional practice and whether explicit and inquiry-based instruction can and should be integrated for students with learning disabilities (Pedrotty Bryant 2005).

Students with learning disabilities often have difficulty retaining facts (Geary 1993), and so the instructional approach for these students tends toward memorization through repetition rather than the development of conceptual knowledge (Cawley and Parmar 1992). For instance, when learning addition facts, children with learning disabilities are very often drilled to respond to "2 + 3 =" by writing "5" as their response without understanding that "2 + 3" and "5" are different ways of writing the same value. This rote drill approach may seem a successful strategy in offering students a means of producing the correct answer, but it is an extremely limited way of understanding either the operation of addition or the relationship of equivalence. It also does not promote the inferential meaning-making of symbols. Therefore, the misconception of the equal sign may be particularly entrenched in these children.

The National Council of Teachers of Mathematics emphasizes equitable instruction for all students. Given the proportion of students with learning disabilities (5% to 8%, according to Shalev et al. 2000) and the current move to integrate students with learning disabilities into general classroom environments, there is a need to examine remedial mathematics education in general and specifically how these students can be helped to be part of "algebra for all."

In this article we report on a research study we conducted to investigate how learning disabled students conceive of the equal sign and the implementation of a teaching approach specifically designed for this special population of children to broaden their understanding of the equal sign and equations. We had four specific questions:

1. How do children with learning disabilities interpret the equal sign?
2. How would children with learning disabilities deal with an instructional approach in which concepts rather than procedures were highlighted?
3. Could a short-term intervention make a difference to the children's conceptual understanding of the equal sign and equations?
4. Would any gains in conceptual understanding be robust enough to persist over the long term?

The approach for the design of this teaching intervention came from two different sources. Although other researchers who work in this area have documented the propensity for children to view the equal sign as an operator rather than as a symbol of relationship between values, it is our conjecture that this misconception may also be due to syntactic and semantic constraints. Children are usually given only one form of equation to solve (Falkner, Levi, and Carpenter 1999), and their interpretation from such a narrow exposure of equations makes sense. They form a hypothesis that is based on their experiences with solving mathematical problems, and so they come to the conclusion that the equal sign is a signal to carry out the operation, generate the "right" answer, and record the answer to the right of the sign. This is verbally reinforced by teachers, who use statements such as "3 + 4 makes 7" (Stacey and MacGregor 1997).

Children learn to read equations the way they learn to read text, from left to right, with certain grammatical symbols always appearing in a specific order. It would seem, then, that when the mathematical sentence is changed (e.g., $9 = 9$, $7 = 3 + 4$), they correctly identify that it no longer corresponds to the mathematical sentences of their experience. In symbolic classification and higher cognitive linguistic interpretation, the equal sign plays a specific role in children's thinking about a mathematical problem according to its grammar, its directionality, and the fact that it requires a solution.

The mathematical syntactical constraint—the exposure to only one form of equation—directly relates to a semantic constraint. Children are never explicitly taught that the equal sign means "is the same as," which leads them to formulate their own interpretations. Given that this is a universal misconception for primary school children lends credence to our conjecture that this "misconception" may in fact be a sophisticated interpretation springing from children's active construction of their knowledge both of mathematical syntax (the order in which the symbols make sense) and of English language semantics (that equals means "makes" or "results in"). With this in mind, a central component of our instructional design was the focus on specific language in order to explicitly map the concept of equivalence onto the symbolic representation. We took the phrase that children spontaneously use when considering equivalence in regard to quantity—"is the same as"—and used it to refer to the equal sign when reading equations, for example, the equation $4 + 5 = 9$ was read, "Four plus five is the same as nine."

Another impetus for our teaching approach came from a consideration of the developmental literature of children's understanding of equivalence in general. Very young children have no problem in determining whether groups of like objects—for example, stacks of Unifix cubes—are equivalent (Baroody 2000; Falkner, Levi, and Carpenter 1999). Children are able to tell when one group of cubes has more than another group, and they are able to show how to make both groups of cubes equivalent. Although this understanding of equality has not been assessed in children with learning disabilities, it has been observed in children as young as kindergarten age. For this intervention, we wondered if we could access children's protoquantitative intuitions about sameness to foster an understanding of number relationships.

To guide the instructional design, we borrowed principles from Case's design for mathematical instruction in order to develop a curriculum that fosters the integration of children's analogic (qualitative or nonnumeric) sameness with their developing quantitative or numeric understanding (Kalchman, Moss, and Case 2000). Initial tasks were designed to provide conceptual bridges from one level of understanding (qualitative concrete equivalence) to a higher level of understanding (symbolic equivalence) in the hope that this higher understanding would then allow students to expand their conceptions of equations from procedural to relational, and from concrete to abstract.

The Students

The six third-grade students in this study were from a small self-contained learning disabilities classroom in a school that offers a special education stream for students from grades 1 to 8. All the children (five girls and one boy) had been identified as severely learning disabled in both reading and mathematics. Their specific challenges ranged from having problems with sequencing and decoding text to difficulty accessing long-term and working memory, central auditory processing disorders, and attentional deficits with and without hyperactivity.

The mathematics instruction in this classroom exclusively involved the repetitive presentation of "fact family" equations for which the students memorized the correct answer. The goal was to foster fluency in addition and subtraction. By the time this study took place, all the children were familiar with, and able to solve, addition and subtraction equations when presented in typical form. The use of manipulatives was not encouraged, nor were class discussions of alternative strategies and solutions.

Initial Assessment of Children's Understanding

During our first visit to the classroom, we began with a discussion to discover what the children knew about the equal sign. We wrote $2 + 2 = 4$ on the whiteboard, circled the equal sign, and asked the students, "What does the equal sign mean?" Not surprisingly, as illustrated by the comments below, all the children answered in a way that suggested they viewed the equal sign as an operator, a signal to do something:

- "If you have a math question, it will help you know the answer. Kind of like it gives you the answer."
- "If you are doing a math problem, it kind of tells you, like, that you have to add."
- "Equals means that's the answer."
- "It tells you what you have to do ... it tells you to add or subtract or whatever and then write down the answer."
- "It means add!"
- "It tells you that you plus the numbers."
- "It's how you get the answer, the right answer, how you do it. And you write it down."

Next, we asked the children to judge whether each of a set of five equations was true or false and why. The equations included one standard form and four nonstandard forms, one of which was untrue.

Are these math sentences true (T) or false (F)? Why?

$4 + 5 = 9$
$7 = 3 + 4$
$8 = 8$
$10 - 5 = 2 + 3$
$8 + 2 = 10 + 4$

The responses to the five equations presented were unanimous and consistent with previous research findings. In other words, the only equation that the students asserted was true was $4 + 5 = 9$, and their response to that was "because it just is." None of the children articulated that it was true because the values on either side of the equal sign were the same.

As for the second equation, $7 = 3 + 4$, they reasoned that the equation was "backwards" with "the equal sign in the wrong place." Some students stated that it would be true if it was written the other way around, which indicates that they saw that the values were the same, but they believed that the signal to compute needed to come after the operation, not after the answer.

In response to the $8 = 8$ item, all the children stated that was false because it was not asking for a solution. "There is no plus sign and it looks weird, so it's not true."

For the final two questions involving separate operations on either side of the equation, the students claimed that both of these equations were not only false but also ridiculous, since there were two operations written with no space to write down the answers. For $10 - 5 = 2 + 3$ they stated, "The equals and the plus are in the wrong place." Similarly, $8 + 2 = 10 + 4$ was judged to be false, not because the two numerical values were unequal, but because there were two operations listed and no room for the answer.

Finally, the students were asked to solve for unknowns on two additional items. One of the items, $6 - 2 =$ ____, was given so that the children would understand that the purpose of the blank was to act as a placeholder for the missing number. The second equation, $5 + 7 =$ _____ $+ 8$ (Falkner, Levi, and Carpenter 1999), was given in order to determine whether children viewed the equal sign as a signal to compute or as a symbol indicating that the numerical expressions on either side had to have the same value.

What number goes in the box?

$6 - 2 = \square$
$5 + 7 = \square + 8$

Again, the responses corresponded to the research literature, and all the children answered the equation $5 + 7 =$ ___ $+ 8$ by putting a 12 in the blank, indicating (as in Falkner, Levi, and Carpenter's 1999 study) that the children were recording the answer to the computation to the left of the equal sign as opposed to balancing the two numerical values. Following this initial assessment, the first author taught three forty-five-minute lessons in the classroom during the students' regularly scheduled math class—one lesson a week for three weeks. The lessons are described in the sections that follow.

Lesson Outline

Lesson 1: M&Ms

We began by folding a piece of paper in half and placing eight M&Ms on one side of the paper and none on the other side. This concrete model was also simultaneously represented as an equation, $8 = 0$. We asked students whether the amounts of M&Ms on either side of the fold in the paper were the same. Of course they said no, that the two amounts were definitely not the same. Then, pointing to the equation and using the specific language of our lessons, we asked if the equation "eight is the same as zero" was true or false. The students immediately responded that this was false and said that eight M&Ms should be placed on the empty half of the paper. We added eight M&Ms, wrote $8 = 8$ (see fig. 2.1), and again, using specific language, asked the students whether "eight is the same as eight" was a true statement. All the children agreed that this was a true statement. Recall that during our initial discussion, this very same equation had been considered false. Clearly this was a change of the students' consideration of representation when evaluating the equation.

Fig. 2.1

We continued to work on the two sides of the paper with quantities of M&Ms and illustrated these relationships using atypical forms of equations. As the lesson progressed, students were challenged to judge equations written either with the same numeral on either side of the equation or with the operation on the right side and the answer on the left (see fig. 2.2). Although the students had initially found these kinds of equations to make no sense, they seemed to accept the new forms of representations, and in every instance they were able to base their evaluation of the equation on the numerical values of the candies they represented, not on the format used.

Fig. 2.2

Lesson 2: Numeric Atypical Equations

In this lesson, the children were asked to discuss and make judgments on a variety of different forms of equations, but this time the students were not presented with the concrete modeling using M&Ms. Again, we used the whiteboard to present a series of equations. The equations were all in numeric symbolic form, and in this lesson we included equations containing operations on either side of the equal sign. The exchanges between researcher and students below illustrate the development of students' reasoning.

Researcher:	[*Writes*] $8 + 2 = 5 + 5$
Students:	[*Read aloud*] "Eight plus two is the same as five plus five."
R:	Is that true?
Students:	Yes!
R:	G, do you think this is true?
G:	Yes!
R:	Why is this true?
G:	[*G walks up to the board*] Because this [*pointing to* $8 + 2$] and this [*pointing to* $5 + 5$] both is ten!
R:	OK, K, this one's for you [*writes* $9 = 3 + 3 + 2 + 2$].
K:	Nine is the same as three plus three plus two plus two.
R:	True or false?
K:	False!
R:	Why?
K:	Because ... three plus three is six, two plus two is four, so six, seven, eight, nine, ten. Ten isn't the same as nine.

As with their evaluation of equations using M&Ms the previous week, it appeared that the students evaluated the equations on the basis of the equivalence of the numerical values represented, as opposed to the form of the equation. However, it was interesting that one student, P, reverted to her initial way of thinking. In responding to $7 = 3 + 4$, she insisted that this was false "because you can't have equal before the thing that you're actually doing, or else it's backwards. And you can't even add them together if you don't know what you're adding. How can you say 7, when you don't know what the other two numbers are?" To help her understand why that particular number sentence was true (semantics) and correct (syntax), we went back to working with piles of M&Ms while emphasizing that an equation is true if one value is the same as the other value.

To conclude this lesson, the children generated their own atypical equations with an emphasis on treating both sides symmetrically, as opposed to an "input-output" conception. All the children enjoyed this activity immensely and made up elaborate equations, for example:

$$3 + 2 + 1 + 4 = 2 + 3 + 4 + 1;$$
$$8 = 2 + 2 + 2 + 2;$$
$$9 = 1 + 1 + 1 + 1 + 1 + 1 + 1 + 1 + 1.$$

Lesson 3: Solving for Unknowns

In the third and final lesson, the children were asked to apply their developing knowledge of equations to solve for unknown variables. This increased the challenge significantly, and, indeed, dealing with unknown variables can be difficult even for typically developing children. Again, there was a specific focus on the language that the children were to use. In keeping with our modeling of verbal expression, we began by writing two equations on the whiteboard and modeled how to read them, this time including the word *something* to refer to the missing variable represented by the box.

> $6 + \square = 10$ was read as "six plus something is the same as ten."
> $3 + 3 = \square + 1$ was read as "three plus three is the same as something plus one."

Following this introduction, the students were presented with a number of other equations. The students appeared to be successful in applying their knowledge of balanced equations, as the dialogue below illustrates:

Researcher: [*Writes 8 + 4 = ____ + 5.*]
C: Seven!
R: Why seven?
C: Because eight plus four is twelve and five to get to twelve is seven, it goes 5, 6, 7, 8, 9, 10, 11, 12 [*counting on fingers*].
R: OK, go through it again because that was a really nice explanation.
C: Eight plus four is twelve and you need to add as much more numbers as you can to get to twelve.

Although in this vignette C, a student with many learning difficulties, explains her reasoning using counting, the underlying logic behind her counting reveals a real understanding of unknown variables.

Following this challenge, the researcher wrote $8 + 2 =$ ____ $+ 5$.

Students: Five!
R: Why is it five?
P: Because eight plus two is ten, and five plus five is ten.

It was surprising and gratifying to see how well the students coped with this task.

For the final activity, as in the lesson before, the children created their own equations with unknown variables for their classmates to solve. Some examples of their equations included $3 + \_\_\_ = 2 + 2$, $10 + 5 = \_\_\_ + 3 + 2$, and $8 = \_\_\_ + 4$. The students appeared to have little trouble either inventing or responding to their classmates' challenges.

Posttest

At the end of this lesson, we gave the children the same set of assessment items they had seen prior to the teaching sequence in order to determine how much they had learned.

First, when evaluating atypical equations as true or false, in contrast with the preassessment where they made judgments that were based solely on the form of the equation, at the postassessment the children based their evaluation on the relationship between the numerical values and were able to articulate why the equations were true or false.

The equation $8 = 8$ was true "because eight is the same as eight. Because they are the same." The two equations with operations on either side of the equal sign were judged on the basis of numerical equivalence, and the children correctly responded that $10 - 5 = 2 + 3$ was true because "two plus three is five and ten minus five is five" and that $8 + 2 = 10 + 4$ was false because "eight plus two isn't the same as ten plus four."

For the final question, all the children were able to correctly use their knowledge of balanced equations to solve $5 + 7 = \_\_\_ + 8$.

Three-Month Follow-Up

Although we were very gratified at how much the children seemed to have grasped during the course of our brief intervention, we wondered how much of this developing understanding of equations would be retained. Three months later we returned to the classroom in order to give the students a follow-up assessment.

As we did during our initial visit, we began by asking the children what the equal sign represents by writing an equation on the board and circling the equal sign. All the children responded that it means "is the same as." As J explains, "Like two plus two is the same as three plus one, because they're both four."

Then we asked the children to judge whether a variety of atypical equations, written on the whiteboard, were true or false. The students were once again successful at determining whether equations were true or false on the basis of numerical equivalence, as the dialogues below illustrate.

Researcher: [*Writes* $10 - 5 = 2 + 3$.]
Students: Ten minus five is the same as two plus three.

R:	Now what do we think. Is this true?
Students:	Yes!
R:	Why?
J:	Because 10 – 5 is five and 2 + 3 is five.

Researcher:	[*Writes 8 + 2 = 10 + 4.]*
Students:	Eight plus two is the same as ten plus four.
R:	Is that true?
G:	No.
R:	How come?
G:	Because it doesn't make the same number.

In order to probe their understanding further, we decided to see if the students would be able to represent M&Ms placed on either side of a folded piece of paper. For this activity we deliberately chose different colors of candies to represent different quantities and asked the students to write down an equation that would represent the M&Ms. For instance, a paper with ten green M&Ms on the left side and four yellow and six blue on the right side of the paper would be represented as 10 = 4 + 6. The students were able to generate equations that were based on the candy quantities; however, the response of one student, K, illustrates that there were challenges:

Researcher:	[*Places seven blue M&Ms on one side of a folded piece of paper and four green and three yellow on the other side.*] K, can you write an equation to show how many M&Ms are on each side?
K:	[*Writes 7 = 7.*]
R:	That's great! Can you write it another way showing the way I put the different colors down? Think of how many I put down first and then how many I put down after that....
K:	[*Writes 7 + 3 = 4 and looks at it dubiously. Then looks at the researcher and frowns.]*
R:	Say it out loud, K; what do you want to write down?
K:	Seven is the same as three plus four. [*K then erases the + and = of her equation and replaces them with = and + to make the equation 7 = 3 + 4.*]

Her recognition of a gap between her understanding of the numerical values she wanted to represent and the equation she had written indicates a conceptual understanding of equations that goes well beyond the procedural understanding she had had three months previously. The fact that she could not precisely identify her mistake may indicate a fluctuation between her deeply entrenched knowledge of what a "correct" equation looks like and her understanding of the numerical values she wanted to represent. Through reading the equation out loud and using the specific language of the intervention, K was able to recognize the source of her cognitive dissonance and make the appropriate corrections.

Finally, the children were asked to solve for unknown variables.

R:	[*Writes* $5 + 7 = \_\_\_\_ + 8$.]
P:	Five plus seven is the same as something plus eight.
R:	What do you need to do to work it out?
P:	Figure out 5 + 7—that's twelve....
R:	Right—and what's on this side?
P:	Eight—so it would be ... four.

However, the same sort of interference from prior procedural learning, as seen in the example above of K's confusion about the order of symbols, also caused some confusion when solving for unknown quantities:

R:	[*Writes* $3 + 3 = \_\_\_\_ + 1$.]
J:	Seven?
R:	Read it out ...
J:	Three plus three is the same as something plus one—oh! Five.
R:	Can you explain?
J:	I only had one more to make six, so I put in five.

Conclusion

One important finding from this study is the indication that an instructional approach based on intuitive knowledge of concepts may at times be appropriate for students with learning disabilities. The lessons of this study, typical of most remediation, were very structured. However, instead of requiring students to engage in rote learning through the memorization of number facts, this instruction allowed students to link their understanding of concrete equivalence to a more abstract understanding of symbolic equivalence. The ability to make this link was evidenced by their increasing competence in judging the truth of a variety of forms of equations. It was also demonstrated by their enjoyment of generating and solving atypical equations. The children were able to play with numerical values in order to write balanced equations, which indicated a shift from a procedural to a relational understanding.

This more sophisticated understanding of symbolic equality seemed to persist over the long term. Following the intervention, the classroom teacher had continued to teach using only standard forms of equations and algorithms, and so it was particularly noteworthy (1) that months later the children were still able to articulate their understanding of equations and judge different forms of equations according to numerical equivalence and (2) that this understanding was robust even in the face of three months of procedural instruction.

In their 1999 article on children's misconception of the equal sign, Falkner, Levi, and Carpenter state that a "concerted effort over an extended period of time is required to establish appropriate notions of equality" (p.233). Although more research is needed, we found that a combination of explicit instruction that focused on semantic structures, the use of concrete examples directly related to written equations, and the use of specific language (e.g., *true, false, same, different, is the same as*) achieved at least an initial breakthrough in these particular students' understanding of what the equal sign means and initiated a deeper understanding of equations.

This study suggests some important implications for further investigations into mathematical remediation. It indicates the importance of instruction that explicitly links learning disabled students' informal understandings of equality involving concrete manipulations to more abstract numeric manipulations and symbolic expressions, coupled with opportunities to discuss and articulate this understanding using specific semantically linked language.

This study also indicates that children with learning disabilities can be taught what the equal sign and equations represent early in their mathematics instruction. This may be a first step to developing emergent algebraic thinking in these children and is an area that requires a great deal of further investigation.

REFERENCES

Baroody, Art J. "Does Mathematics Instruction for Three- to Five-Year-Olds Really Make Sense?" *Young Children* 55 (July 2000): 61–68.

Behr, Merlyn, Stanley H. Erlwanger, and Eugene Nichols. "How Children View the Equals Sign." *Mathematics Teaching* 92 (1980): 13–15.

Blanton, Maria L., and James J. Kaput. "Developing Elementary Teachers' Algebra 'Eyes and Ears.'" *Teaching Children Mathematics* 10 (October 2003): 70–77.

Carpenter, Thomas P., and Linda Levi. "Developing Conceptions of Algebraic Reasoning in the Primary Grades." Research Report 00-2. Madison, Wis.: National Center for Improving Student Learning and Achievement in Mathematics and Science, 2000. Available at www.wcer.wisc.edu/ncisla/publications/reports/RR-002.pdf.

Cawley, John F., and Rene S. Parmar. "Arithmetic Programming for Students with Disabilities: An Alternative." *Remedial and Special Education* 13 (1992): 6–18.

Falkner, Karen P., Linda Levi, and Thomas P. Carpenter. "Children's Understanding of Equality: A Foundation for Algebra." *Teaching Children Mathematics* 6 (December 1999): 232–36.

Geary, David C. "Mathematical Disabilities: Cognitive, Neuropsychological, and Genetic Components." *Psychological Bulletin* 114 (1993): 345–62.

Gersten, Russel, Nancy C. Jordan, and Jonathan R. Flojo. "Early Identification and Interventions for Students with Mathematics Difficulties." *Journal of Learning Disabilities* 38 (2005): 293–304.

Herscovics, Nicolas, and Carolyn Kieran. "Constructing Meaning for the Concept of Equation." *Mathematics Teacher* 73 (November 1980): 572–80.

Kalchman, Mindy, Joan K. Moss, and Robbie Case. "Psychological Models for the Development of Mathematical Understanding: Rational Numbers and Functions." In *Cognition and Instruction: 25 Years of Progress*, edited by Sharon M. Carver and David Klahr, pp. 1–39. Mahwah, N.J.: Lawrence Erlbaum Associates, 2000.

Kaput, James. "Transforming Algebra from an Engine of Inequity to an Engine of Mathematical Power by 'Algebrafying' the K–12 Curriculum." In *The Nature and Role of Algebra in the K–14 Curriculum: Proceedings of a National Symposium*, edited by the National Council of Teachers of Mathematics and the Mathematical Sciences Education Board, pp. 25–26. Washington, D.C.: National Research Council, National Academy Press, 1998.

Kieran, Carolyn, Michael T. Battista, and Douglas H. Clements. "Research into Practice: Helping to Make the Transition to Algebra." *Arithmetic Teacher* 38 (March 1991): 49–51.

MacGregor, Mollie, and Elizabeth Price. "An Exploration of Aspects of Language Proficiency and Algebra Learning." *Journal for Research in Mathematics Education* 30 (July 1999): 449–67.

Matz, Marilyn. "Towards a Process Model for High School Algebra Errors." In *Intelligent Tutoring Systems*, edited by Derick Sleeman and John Seeley Brown, pp. 25–50. New York: Academic Press, 1982.

National Council of Teachers of Mathematics (NCTM). *Principles and Standards for School Mathematics.* Reston, Va.: NCTM, 2000.

Osborne, Alan, and P. S. Wilson. "Moving to Algebraic Thought." In *Teaching Mathematics in Grades K–8: Research Based Methods*, edited by Thomas R. Post. Boston: Allyn & Bacon, 1988.

Pedrotty Bryant, Diane. "Commentary on 'Early Identification and Interventions for Students with Mathematics Difficulties.'" *Journal of Learning Disabilities* 38 (2005): 340–45.

Saenz-Ludlow, Adalira, and Catherine Walgamuth. "Third Graders' Interpretations of Equality and the Equal Symbol." *Educational Studies in Mathematics* 35 (1998): 153–87.

Schliemann, Analucia D., David W. Carraher, Barbara M. Brizuela, Darrell Earnest, Anne Goodrow, Susanna Lara-Roth, and Irit Peled. "Algebra in Elementary

School." In *Proceedings of the 2003 Joint Meeting of PME and PME-NA*, edited by Neil A. Pateman, Barbara J. Dougherty, and Joseph T. Zilliox, Vol. 4, pp. 127–34. Honolulu, Hawaii: University of Hawaii, 2003.

Shalev, Ruth S., Judith Auerbach, Orly Manor, and Varda Gross-Tsur. "Developmental Dyscalculia: Prevalence and Prognosis." *European Adolescent Psychiatry* 9 (2000): 58–64.

Stacey, Kaye, and Mollie MacGregor. "Ideas about Symbolism That Students Bring to Algebra." *Mathematics Teacher* 90 (February 1997): 110–14.

Swafford, Jane O., and Cynthia W. Langrall. "Grade 6 Students' Preinstructional Use of Equations to Describe and Represent Problem Situations." *Journal for Research in Mathematics Education* 31 (January 2000): 89–112.

Warren, Elizabeth. "Young Children's Understanding of Equals: A Longitudinal Study." In *Proceedings of the 2003 Joint Meeting of PME and PME-NA*, edited by Neil A. Pateman, Barbara J. Dougherty, and Joseph T. Zilliox, Vol. 4, pp. 379–86. Honolulu, Hawaii: University of Hawaii, 2003.

3

Is "Just Good Teaching" Enough to Support the Learning of English Language Learners? Insights from Sociocultural Learning Theory

Jennifer M. Bay-Williams
Socorro Herrera

IN A CLASSROOM with predominantly English language learners (ELLs) varying in their proficiency of English, the teacher has asked students to compare two linear graphs (one "steeper" than the other), noting the similarities and differences and justifying whether or not those two lines could be a representation of the same function. Students are working in cooperative groups discussing and preparing their solutions. One group includes two non-English speakers and one bilingual student. That group has been asked, instead, to find coordinates of the lines on each of the two graphs. When asked why she had modified the lesson in this way, the teacher reported that the language required for the original task was too demanding for this group and that finding coordinates would be a meaningful task for which they could be successful. This brief vignette illustrates a teacher planning a high-quality lesson and creating modifications that consider students' language background. As we will discuss later, she has incorporated some specific strategies that have potential to increase student learning for language learners, but she has focused on what the learners can do and not necessarily moved them *to the next level* in regard to their mathematics understanding and their linguistic growth. "The provision of assistance constitutes the heart of the teaching and learning enterprise as learners are guided by more experienced others to participate in the complex, meaningful sociocultural practices of the community" (Brown, Stein, and Forman 1996, p. 65). The purpose of this article is to use a sociocultural framework to articulate means for making content and language comprehensible to English language learners.

How to guide English language learners so that they can participate and learn in a mathematics classroom is a priority in both the mathematics education and the English as a second language (ESL) arenas (Teachers of English to Speakers of Other Languages [TESOL] 1997; National Council of Teachers of Mathematics [NCTM] 2000). (*Note:* "ELL" is commonly used to refer to students, whereas "ESL" is used to refer to the academic field of study.) Efforts in both fields have targeted those practices that support the learning of all students. In mathematics education, research and national standards have highlighted the value of communication and discourse in the learning of mathematics (Lampert and Cobb 2003; NCTM 2000; Silver and Smith 1996). The ESL reform has emphasized an emergent perspective that language must be learned simultaneously with content development (Diaz-Rico and Weed 2002; TESOL 1997). Not surprisingly, this resounds with the NCTM Principles, which include the following statements (NCTM 2000):

- Excellence in mathematics education requires equity—high expectations and strong support for all students (Equity Principle, p. 12)
- Effective mathematics teaching requires understanding what students know and need to learn and then challenging and supporting them to learn it well (Teaching Principle, p. 16)
- Students must learn mathematics with understanding, actively building new knowledge from experience and prior knowledge. (Learning Principle, p. 20)

The question arises, if one is implementing *Standards*-based mathematics practices (i.e., those practices articulated in *Principles and Standards* [NCTM 2000]), is she or he already supporting the learning of all students, including those who are acquiring English? Before pursuing the answer in regard to sociocultural learning theory, we first discuss three pervasive philosophical views on supporting the mathematics learning of ELLs.

Views on Supporting the English Language Learner

Paramount to the issue of how ELLs can best learn mathematics is the issue of the quantity of language they should encounter in their learning. One common view is that mathematics, which uses a variety of symbols that are common across cultures, is universal and therefore easily accessible to language learners. Limiting language and focusing primarily on symbols enable the learner to participate in the mathematics classroom. For example, in a third-grade lesson introducing multiplication, a lesson that is modified to offer access to language learners may not have situations or contexts but just have students look at statements of repeated addition or pictures of rectangles as they find different products.

There are numerous reasons why this view, although well intentioned, limits students' access to important mathematics and language development. First, symbols are only one of many representations used to illustrate mathematical concepts. A deep understanding of the use of language, in both written and oral forms, is essential if students are to develop a deep understanding of mathematics. As articulated in *Adding It Up* (Kilpatrick, Swafford, and Findell 2001), a student's understanding of a mathematics topic spans five strands (conceptual understanding, procedural proficiency, strategic competence, adaptive reasoning, and productive disposition). A focus on symbols alone can deny a student access to full mathematical proficiency. Second, given such a viewpoint, teachers may move too quickly to symbolic representation before conceptual understanding is developed. Third, symbols do not exist separate from language but are language-dependent. Students must be able to relate symbols such as $\geq$ and parentheses to mathematics concepts and processes (Dale and Cuevas 1995). Fourth, even the use of symbols to convey mathematical concepts may prove linguistically problematic for language learners. For example, addition, even as basic as $5 + 7$, can be described with different linguistic synonyms (*add, plus, combine, increased by, join,* and *sum*). One readily apparent linguistic challenge for an ELL is the subtle difference between the homonyms *sum* and *some.* English language learners may experience few difficulties with the problem $10 \div 2 =$ _____? However, if spoken orally as "ten divided by two is _____?" students may translate the problem as $10\overline{)2}$. With no context to ground their thinking, it may be a toss-up about how the phrase is translated into symbols. Such language-based errors are common because students tend to write mathematical expressions in the same manner in which they learn to read and write English (i.e., left to right, sequentially). Thus, a language-impoverished classroom fosters a notable variety of challenges to ELLs by limiting their access to contextual and linguistic clues that are essential to conceptual understandings. This perspective may effectively limit a student's access to higher mathematics (Garrison and Mora 1999). The teacher in the opening vignette, who may subscribe to this perspective, has reduced the language demands of the lesson, but in so doing she has also limited her students' access to the important mathematics and the related language of the lesson.

On the contrary, literacy-rich classrooms foster ELLs' opportunities for the simultaneous development of academic language proficiency and content-area (mathematics) knowledge, skills, and capacities. A second viewpoint on teaching ELLs is to embrace the recommendations outlined in *Principles and Standards for School Mathematics* (NCTM 2000), in particular the Process Standards (problem solving, reasoning and proof, communication, connections, and representations). An ELL needs to use such things as physical models, small-group discussions, writing, and applications to learn important mathematics. The strategies listed here (e.g., those that maximize interaction, provide context and visuals, and incorporate the use of language) are considered essential experiences for the ELL to succeed (Echevarria, Vogt, and Short 2004; Herrera and Murry 2005; Santiago and Spanos 1993; Flores 1997). Flores (1997), for example, describes a teacher, "Isabel," who models "just good teaching," which results in higher achievement for her ELLs (p. 84). Her strategies include accommodating different learning styles, focusing on conceptual development, using meaningful contexts to develop concepts, working in small groups, and using tools such as manipulatives and calculators. This list of strategies maximizes participation and language use in the classroom. Working in small groups, for example, enables a language learner to seek assistance, including having content translated, having instructions clarified, observing what peers are doing, as well as practicing language as it applies to mathematics concepts. In this regard, a *Standards*-based classroom is more effective because strategies such as the ones used by Isabel enable a language learner to acquire a second language while also learning content. Offering opportunities to talk in a nonthreatening environment, using contexts so that students can glean meaning from new words, and using visuals are a few of the major learning strategies developed for ELLs (Echevarria, Vogt, and Short 2004).

Nevertheless, the argument that "just good teaching" (i.e., *Standards*-based practices) is adequate and sufficient for ELLs to learn mathematics neglects the specific sociocultural, cognitive, academic, and linguistic assets and needs that these students bring to mathematics learning—not to mention the linguistic challenges that the acquisition of a second language presents for these students. Table 3.1 outlines the stages that are most commonly used when referring to the progression of English language acquisition (Krashen 1982). A knowledge of which stage a language learner is in can inform a teacher on how a learner will participate in discussions, group activities, and solving mathematics tasks.

Again, let us consider the opening vignette. If the teacher had given the same task to this group, a quick review of the task can uncover several constraints that language learners will encounter. Depending on their stage of language acquisition, they may not know what the terms *similarity* or *difference* mean. Moreover, they must understand the term and concept for *function*. Although they may be able to do the task in their native language, they may not be able to communicate their understanding in English. Even though the students have been given a task that has potential to increase their knowledge of mathematics and improve their language skills, without assistance specific to their language needs, they may not be able to learn what is intended. Therefore, although this viewpoint may go further in providing access, it alone is not enough. Communicating the message that "just good teaching" is all that it takes could mean that those teachers implementing *Standards*-based practices (e.g., including manipulatives, using groups) may believe that they do not need to make specific accommodations or modifications for ELLs.

The third viewpoint on making mathematics accessible to language learners is to integrate *Standards*-based practices with intentional language instruction and support. This is not an additional strategy to add to the collection of other instructional strategies but rather a paradigm shift. In the same way that learners construct their knowledge of mathematics based on the experiences

Table 3.1
Overview of the Stages of Second Language Acquisition (Adapted from Krashen 1982)

Stage of Second Language Acquisition	Student Characteristics
Preproduction	The student at this stage may progress through a silent period or communicate nonverbally in the classroom.
Early Production	The student at this stage will often begin to speak, using isolated words of the second language in conversations with peers.
Speech Emergent	The student's English proficiency progresses to BICS (Basic Interpersonal Communication Skills).
Intermediate Fluency	The student exhibits growth in the accuracy of listening, speaking, reading, and writing.
Advanced Fluency	The student develops highly accurate language and grammatical structures in the second language.

Note: Caution must be taken in determining advanced fluency. The student's level of academic language proficiency must be appropriately assessed.

planned by the teacher, learners must also construct their understanding of the English language. Khisty (1997) argues that creating effective learning for language-minority students involves integrating principles of bilingual education with content instruction. Among those bilingual principles are the use of students' home language and small-group work that takes into consideration language proficiency. Vocabulary development is crucial and complex. To understand mathematics, students must understand and use words that are used outside of mathematics and have the same meaning (e.g., *average, graph, probability*), those that are usually only encountered in mathematics (e.g., *quadrilateral, y-axis, solve*), and the large list of terms and concepts that are used in mathematics differently from the way they are used in the world (e.g., *function, mean, product, similar*). Some teachers do this naturally, intuitively identifying those words or phrases that can be stumbling blocks for students and explicitly instructing students on the meaning of such words. However, integrating language instruction with mathematics instruction is more comprehensive than identifying, reviewing, and emphasizing key words. Moschkovich (1999a) argues that beyond vocabulary development, mathematics teachers must engage students in discourse focused on making meaning of the mathematical task. These discussions can support ELLs' learning when they include instructional strategies, such as using several expressions for the same concept, that nurture students' ability to explain mathematical concepts. A teacher must reconceptualize his or her role in learning, identifying both the content and the language needs for a given topic and determining the assistance that needs to be in place for all learners, including English language learners. A teacher operating from this paradigm has her eye on language *and* mathematics as she plans for instruction, identifying language objectives as well as mathematics objectives. In the opening vignette, for example, the teacher's mathematics objective was for students to analyze two linear graphs, determining how the slope of the line may depend on the scale of the graph. A language objective could have been for students to use the terms *independent variable* and *dependent variable* to compare each graph. Placing a bilingual student in the group is an effective strategy because this student can provide assistance in language development. These students could, for example, discuss the important vocabulary and concepts in their native language, discuss the task that was posed in their native language and in English, and then use English both in writing and orally to describe their comparison of the two graphs to the rest of the class.

This third perspective on what can best support the ELL in learning mathematics is inherent in *Principles and Standards for School Mathematics* (NCTM 2000). The Equity and Teaching Principles focus on *support* for all students. Too often we as teachers focus on what the ELL lacks (in language, in speed of learning, in classroom expectations) rather than on the rich variety of social, cultural, and transnational experiences and language the student may bring to classroom learning (Moschkovich 1999b; Herrera and Murry 2005). Language learners require support in their language development while having access to important and challenging mathematics. It is through sociocultural learning theory that we can better understand the kinds of assistance needed to support the learning of ELLs.

Sociocultural Theory and the English Language Learner

Sociocultural theory, grounded in the work of Lev Vygotsky, and research conducted from a sociocultural perspective can furnish insights into how to support the English language learner. Vygotsky's theory has implications for educational reform because of its focus on the process of learning, the connections from theory to practice, and its focus on the importance of social and cultural factors in a student's learning (Forman 2003, p. 335):

> The 1991 NCTM Standards were intended to improve mathematics instruction for all students, not just one segment of society.... Teachers are supposed to recognize these students' distinct learning histories, attitudes, motivations, and beliefs about mathematics and to encourage all of them to engage in age-appropriate but also high-level mathematical reasoning and problem solving.

Two ideas of sociocultural theory, the zone of proximal development and semiotic mediation, furnish insights into processes that can support the learning of ELLs.

The Zone of Proximal Development

One well-known aspect of sociocultural learning theory is Vygotsky's zone of proximal development (ZPD), which is the difference between a learner's assisted and unassisted performance on a task (Vygotsky 1978). In the context of a mathematics task involving new content for a learner, the zone can represent the level of success a student can achieve on that task with and without adult guidance or in collaboration with more capable peers. The learner's level of success on that task can be conceptualized on a continuum. According to Vygotsky, instruction should be aimed at the upper end of the continuum in order to tap into a learner's potential instead of concentrating on a learner's existing level of competence. Although the term *zone of proximal development* may not be familiar to practitioners, instruction strategies based on this concept (such as guided practice, situated learning, reciprocal teaching, and scaffolding) are prevalent in many classrooms. Each of these strategies involves a form of assistance that enables a learner to develop new understanding. For an ELL, the targets are (*a*) to stretch the student toward conceptual learning of the content at the next level and (*b*) to bring her or his language development forward at the same time. This will not happen without the intervention of a teacher, adult, or more capable peer who is able to support that student's language acquisition and content-area learning.

Let's revisit the classroom discussed earlier, with the group of three students who have varying levels of English proficiency. The teacher identified an alternative task (identifying coordinates) for the ELLs that they could do independently and unassisted. This task, however, was aimed at what the students already knew and not at advancing their conceptual understanding. To learn new material, the teacher must reconceptualize her alternative task by considering the students' potential to learn with assistance. The question becomes, "What support structures should be in place so that the ELLs can achieve the academic learning goals?" Or, if the lesson is too far beyond what they are able to do, then the question is, "What modification of this task can move these students forward and what instructional strategies (e.g., scaffolding, group configurations) must be in place for that learning to occur?" In this classroom there were numerous possibilities. Among the possible instructional strategies would be these four:

1. A structured preview of the lesson, including a focus on important vocabulary for graphing and for the context of the situation represented in the graph (without taking away the discovery in the lesson)

2. Launching the lesson with a fully labeled graph that has a meaningful (and culturally relevant) context and asking questions about the meaning of the graph
3. Providing a large graph on the wall that labels the essential vocabulary in English and the languages represented in the classroom
4. Scaffolding the task by first asking students to identify coordinates, then guiding them more explicitly to the higher-level task of determining if the two graphs could describe the same context

For the English language learner, a teacher's planned assistance must consider not only the mathematics but also the academic language development. In this lesson, the teacher's original goal was to have the ELLs identify points, a task for which the academic language was minimal and already known. Even though she placed a more knowledgeable peer (in English language proficiency) in the group, she did not create a task that enabled the students to engage in mathematical discourse (in English or Spanish). She must conceptualize what they could accomplish with assistance and determine an appropriate language goal, such as using the appropriate mathematics language for describing a line on the graph. As noted earlier, this could be scaffolded by first having students use their native language and focus only on the terminology of "independent and dependent variable."

The teacher's decision regarding the ELLs' ability to participate at a higher level is limited by the belief that the students may not be able to move to their potential level of "linguistic competence" and can perform only at their existing level. In classrooms where the goal is to increase proficiency in academic language while learning important mathematical content, a student's learning is limited by neither his or her academic background nor his or her current level of English proficiency. Instead, assistance is provided to ensure higher levels of content learning and growth in academic language proficiency.

Semiotic Mediation and Communication

A second tenet of sociocultural theory is the way in which information moves from the social plane to the individual plane. This is referred to as *semiotic mediation*. Semiotic mediation is defined as the "mechanism by which individual beliefs, attitudes, and goals are simultaneously affected by and affect sociocultural practices and institutions" (Forman and McPhail 1993, p. 134). Semiotic mediation includes language, but in a broader sense it also includes diagrams, mathematical symbols, and body language—all of which are referred to as *tools* that enable mediation to occur. In a summary of research findings on ELLs, Dale and Cuevas (1995, p. 31) state:

> Language works as a mediator for mathematical thinking and metacognition. Whether the thinking defines the language or the language defines the thinking remains to be answered. Probably both occur. The important point is that

mathematical thinking, mediated by linguistic processes, is a prerequisite for mathematics achievement.

In fact, the specialized speech for mathematics, or mathematics register, develops along with everyday language as learners engage in discussing mathematics (Forman 1996). Vygotsky's focus on the social nature of learning leads to a focus on learning within communities of practice (Lave and Wenger 1991). This approach takes into consideration the "individual's and group's goal-directed activities as constituting and constituted by institutionally, culturally, and historically situated communities of practice" (Forman 2003, p. 336). Individual learning in a community of learners must take into consideration more than just the mental operations of the learner; it must also include the norms established in the classroom. Failure to learn may be due to a mismatch of norms, goals, beliefs, or cultural background rather than a lack of ability of the learner (Goodnow 1990). In mathematics classrooms, language learners may not understand the interaction patterns of the classroom, the context of the problem being explored, or the purpose of doing the tasks. This lack of understanding may inhibit a learner from participating in the classroom community and therefore limit learning. Building background by connecting to prior learning and culturally relevant experiences can forge a link whereby language learners can be full participants in the classroom community.

Traditionally, mathematics instruction has followed a pattern where the teacher poses a question, a student or students respond, and the teacher gives feedback. Building on Rotman's (1988) analysis of classroom discourse, Forman (2003) argues that students' learning is enhanced when students and teachers play different roles in the discussion of mathematics. According to Rotman, role playing includes being the one who explains (Mathematician), the one who responds to questions (Agent), and the one who mediates informally between the two other voices (Person). Such mathematical discussions are particularly important for language learners, who may struggle with the formal mathematics register (academic language).

Students' learning can be facilitated by furnishing opportunities for (1) native language use with more capable peers, (2) explicit focus on learning strategies, and (3) meaningful activities that include listening, speaking, reading, and writing about the mathematics concept. Often a student's linguistic abilities will not yet be at the level where the processes for finding a solution can be considered automatic. For example, factoring a trinomial, even though it is a symbolic manipulation task, may require additional time for a language learner because he or she may need to think through the task in his or her native language or move between the two languages in social interactions or in self talk. If a student is stronger in his or her first language than in English, he or she may move between languages, trying to internalize the information in his or her first language while using English in the social plane. It is essential in supporting ELLs that connections be made between academic language and associated meanings through pictures, diagrams, and cognates (words that have Latin roots and have similar meanings in both languages). If classroom instruction limits the use of language and tools, academic performance may suffer, especially among ELLs (Moll 1991).

Implications for Students' Learning

Sociocultural learning theory takes into account the relevance of culture and language in the learning process. In culturally and linguistically responsive learning environments, specific considerations must be addressed. In this section, we discuss four main themes that address the learning needs of students who are learning academic content as they learn English.

Linguistics

Language is the means by which learners encounter, make sense of, and internalize concepts. Linguistics must be a priority in planning, teaching, and assessment. The following questions can guide the process:

1. Is the language of the lesson or curriculum within a student's academic and linguistic profile?
2. Have mathematical content objectives and language objectives been identified, and are they appropriately challenging?
3. Is the learner explicitly involved in meaningful activities that allow all four linguistic domains (reading, writing, speaking, and listening) to be used?
4. Are vocabulary words identified and targeted throughout the lesson? Is the targeted vocabulary critical to learning the mathematics concept?

Although these questions might sound like the outline of a language arts lesson, it is part of the planning process that is essential in assisting a language learner in learning both content and language. Here, an introduction of a fourth-grade class is shared to illustrate that the teacher is focused on the mathematics goals but has a constant eye on the use of language:

> The class is about to begin a day of exploring rectangles in order to see patterns that can be used in connecting repeated addition with multiplication. The classroom has five language learners, varying in their proficiency, but all speaking some English. In considering their language needs, the teacher wants to be sure that the students connect the symbolic representation to the appropriate language and to the visual image of a rectangle. She identifies a content goal—

> students will be able to create rectangles of different dimensions and record the number of total tiles symbolically using repeated addition and multiplication. She considers all her learners and her language learners and decides that linguistically, students will need to understand the meaning for the symbol × and to read and explain the meaning of multiplication number sentences. She posts her objectives on the board, using kid-friendly language.
>
> As she begins the lesson, she places students with partners, placing her language learners with partners who can help them with the academic language, explicitly stating that they may use English or their native language as they work. She models an example, showing the tiles and pointing at the tiles, rows, and columns as she adds that vocabulary to the overhead next to the tiles. The students speak aloud that the example (4 by 5) would be stated as four rows of five tiles. She asks them, "How many tiles?" They share an answer with a partner. She asks students to explain to their partners how they would write four rows of five using repeated addition and using the mathematical phrases she has written on the overhead. A similar process proceeds for recording the multiplication number sentence. The group reads each sentence together. Reminding students of the vocabulary they need to use, she introduces a group task involving finding areas of rectangles and recording the related number sentences.

This may appear to some as "just good teaching" (e.g., using manipulatives, engaging students in communication, and making sure the task is clear before proceeding). However, this lesson explicitly focuses on both the linguistic and the content needs of the students. The use of language in the lesson introduction allows students to be able to use the language that they will need during their exploration, where they will be able to clarify concepts with more capable peers. When students complete their exploration, the teacher will return to a classroom discussion where she maximizes both academic and conversational language opportunities, revisits important vocabulary, and, most important, engages the students in using language to describe the mathematics they have explored.

Many strategies can be incorporated to help students with their academic language development. Analyzing lessons for words that may carry different meaning outside of mathematics, for instance, needs explicit attention. A simple focus on boldfaced words in the text is not sufficient for the appropriate development of the concepts. For example, in a lesson on the *mean*, the term *mean* may be boldfaced and defined, yet phrases critical to a conceptual understanding of *mean*, such as *evening out* or *equal distribution*, must be addressed in classroom discussions. Similarly, students will benefit from describing the word (in their own vernacular), not just focusing on the definitions. This is the assistance that will enable language learners to learn both mathematics and language.

Comprehensible Input and Conceptual Understanding

For many years, ESL and bilingual programs have encouraged a focus on making content comprehensible to ELLs. This focus has been referred to as *comprehensible input* (Krashen 1982). In a similar manner, mathematics standards suggest that a variety of strategies should be used to ensure learning by *all* students (NCTM 2000). Visuals, representations, explanations, and appropriate use of language have emerged as effective strategies in both domains (NCTM 2000; Diaz-Rico and Weed 2002). Simultaneous second language acquisition among ELLs is essential to their development of a conceptual understanding of mathematics. The following questions can guide differentiated classroom practices for these learners:

1. Is learning contextualized to include the type of mathematics instruction the ELLs received prior to coming to the United States?
2. Is the student learning the vocabulary that has been selected for the development of the mathematics register (i.e., the terms that are specific to the field of mathematics and not part of everyday conversation)?
3. Have the curriculum and text been reviewed to target the "critical concepts" that will maximize learning for an ELL's academic success?

From a sociocultural framework, meaning is developed first on the social plane and then on the individual plane. The student's culture, the classroom culture, and the student's knowledge all have an impact on what is learned. Therefore, comprehensible input must take into consideration all these aspects. The key to making content comprehensible is being able to identify what is essential to the learning of the mathematics concept first and then shifting to a language lens that focuses on how to overcome the barriers a student with limited English skills might encounter. Because the learner may have a strong or weak mathematics background, these decisions are highly dependent on the student's background. For example, a student may be strong in conversational English but struggle with academic English. With effective vocabulary development, language learners are equipped with words they can use to engage in mathematical conversations. Teachers need to explicitly address synonyms (e.g., *combine* and *add, greater* and *more than*) and homophones (e.g., *plane,* which can be an aircraft, surface area, level, or tool; also, *right* and *write* or *weight* and *wait*), providing the student with the tools to use terminology correctly (see Adams, Thangata, and King 2005 for a more thorough discussion of mathematical homophones). Mathematics uses many phrases to identify a single concept or process, such as "least common multiple" and "simultaneous equations." Translating them word by word will result in an incomplete or inaccurate understanding of these phrases. A compounding issue in mathematics is that the same word can designate a different mathematics process given a new context (Dale and Cuevas 1999). For example, students may learn that *by* implies multiplication, as in "5 is multiplied by 3," but students may encounter tasks that ask them to "increase 25 by 3," signifying addition. Students must learn vocabulary not as isolated words but in the context of situations in which the focus is on understanding.

Connecting words to pictures, diagrams, and real contexts makes language accessible to students. Students can keep a mathematics journal in which they record pictures, diagrams, and translations so that they have a reference for looking up concepts and terms that have been previously taught. Consistent with *Standards*-based practices, making content comprehensible means using meaningful contexts for students, using visuals, realia, and tools to develop understanding, using multiple representations, and valuing different solution strategies.

Interaction, Practice, and the Application of Learning Strategies

Vygotsky has argued that social interactions among students are valuable tools with which to enhance the learning of mathematical and other concepts and processes (Forman and McPhail 1993). Moreover, such interactions, especially those that may occur between and among ELLs and their more language-capable peers, concomitantly promote English language acquisition. The following questions can guide teachers' best practice with ELLs:

1. Are grouping configurations structured in such a way that they will promote the learning of mathematical content as well as English language acquisition?
2. Are there frequent opportunities to engage students in elaborate responses about lesson concepts?
3. Do interactions between the teacher and the students encourage higher-order thinking and academic language development?
4. Do instructional activities and groupings in the classroom promote interactions that encourage students to elaborate on their conceptual understandings?

Configuring groups according to the students' linguistic as well as academic backgrounds offers students opportunities to build learning relationships, practice and learn in a new language, and build mathematical skills that may accelerate learning. If a language learner is placed in a group without careful structures in place, his or her peers may not operate as more knowledgeable peers but instead ignore the needs of the language learner.

All students should have access to cognitively demanding tasks. If a task is cognitively demanding, it is more likely to show connections within mathematics, involve reasoning, and engage students in discussions. Unfortunately, some students, such as the ones in the classroom revisited throughout this paper, are denied access to cognitively demanding tasks because the level of linguistic demand of such tasks is judged to be too great. Language learners can think at a higher level if the content is made comprehensible. So, the text may need to be modified to reduce the unfamiliar vocabulary or the context may need to be adapted to one that is familiar. Compare the two texts in figure 3.1, the original offered as background to two questions posed in an NSF-supported middle school curriculum and the modified version an adaptation of the text that removes unfamiliar and nonessential vocabulary.

Even with words such as *cartridges* and *users* removed, a student with limited English proficiency may still struggle with the meaning of the text. Other strategies can be used. Translation into the first language may be possible. Visuals and realia can be used. For example, in this instance having available an ad on 8 ½″ by 11″ paper and the two other sizes of paper to show the students would be helpful. Pointing and using gestures can help students understand. A simpler, related task can be used as a tool for scaffolding (or gradually building the difficulty level of) the activity, without taking away the way to solve the focus task. In the task in figure 3.1, this is already included in the text by providing an example of the scale factor of 1.5. In general, a teacher must first decide if the task is within a student's ZPD from a mathematics perspective and, if so, determine what can be done to make the language and context comprehensible to the student so that he or she can engage in the task with assistance and learn what is intended.

Review and Assessment

Given the complexity of mathematics teaching and learning, ongoing review and assessment—along with the appropriate refinement of instruction to meet students' needs—are essential to incremental progress among ELLs. These reviews and assessments must encompass gains both in academic knowledge and skills as well as in academic English language acquisition pertinent to mathematics. The following questions can guide practice:

1. Have students had the opportunity to communicate their prior knowledge, skills, and experiences using the language with which they are currently most comfortable?
2. Have students been given consistent opportunities to articulate their understanding of important mathematical concepts?

Original Text

Raphael wants to make posters for his sale by enlarging his 8 ½″ by 11″ ad. Raphael thinks big posters will get more attention, so he wants to enlarge his ad as much as possible.

The copy machines at the copy shop have cartridges for three paper sizes:
8 ½″ by 11″, 11″ by 14″, 11″ by 17″. The machines allow users to enlarge or reduce documents by specifying a percent between 50% and 200%. For example, to enlarge a document by a scale factor of 1.5, a user would enter 150%. This tells the machine to enlarge the document to 150% of its current size.

(Taken from Lappan et al. 2004, p. 44)

Modified Text

Raphael is having a sale. He has an ad that is on 8 ½″ by 11″ paper. He wants to make it as big as possible.

There are three sizes of paper: 8 ½″ by 11″, 11″ by 14″, or 11″ by 17″. He can make the copy machine change the size of the paper by choosing a percent between 50% and 200%. For example, to make the ad bigger by a scale factor of 1.5, Raphael would choose 150%. This will make the ad 150% bigger than it is now.

Fig. 3.1. An example of modifying text to remove unfamiliar and nonessential vocabulary

Assessing and reviewing concepts are part of any classroom that focuses on student learning. Here we share some of the additional considerations for classrooms containing language learners. First, is the strategy of "revoicing" (O'Conner and Michaels 1993; Moschkovich 1999a), a technique used in classroom discourse. Revoicing involves restating, rephrasing, or expanding on what a student has said, allowing the language learner to have concepts clarified by hearing minor differences in the way the concept is usually stated in the discipline. In revoicing, a teacher can make the content of the statement more comprehensible to others in the classroom while helping clarify appropriate mathematics concepts. Moschkovich (1999a) says that revoicing supports students' participation in discussions in two ways—first, by allowing the student to participate in general, and second, by keeping the discussion mathematical. The following is a brief excerpt from a fifth-grade discussion on circumference:

Teacher: I would like to hear your ideas on how the length of the diameter compares to the circumference of the circles you measured.

Student 1: The part around is bigger than across.

T: The circumference is bigger than the diameter? [*Motions with fingers at a circle as she uses the mathematical terms.*] Is it always bigger?

S2: If you put the diameters around the circle, it takes about three.

T: It takes three of the diameters to go around the circle?

S2: Yes, three diameters make a circumference.

T: So, you are saying that the circumference is about three times the diameter; what do the rest of you think about that? Look at your measurements in your tables and see if you agree or disagree.

This dialogue illustrates how the teacher is able to insert the appropriate mathematical phrases, but the purpose of her responses is not to correct students but to continue to probe into their understanding of the relationship between the diameter and the circumference.

Underlying the effective use of revoicing (and other strategies to engage students in discourse) is the creation of an environment in which students' participation is expected. Language learners must understand the norms of the classroom in which they are participating. Khisty offers three norms that serve to equalize participation by members of a group: "(1) everyone in the group *ought* to have a chance to talk, (2) everyone's contribution should receive a fair hearing, and (3) not everyone has the same abilities, but all are worthwhile" (Khisty 1997, p. 98). Furnishing explicit and specific feedback to all students in the classroom in regard to appropriate participation in mathematics conversations is important in helping a language learner understand the dynamics of engagement in the conversation.

Summary

The importance of social interaction in learning underscores the need for ELLs to have access to a language-rich learning environment. Accordingly, the effective mathematics teacher creates and nurtures those learning environments and situations by designing experiences that maximize opportunities to communicate and explore mathematics concepts and language. Cognition is enhanced for language learners through social interactions with more capable peers. In such interactive, low-risk environments, academic learning is also bolstered as ELLs negotiate content-area learning with peers and become more willing to engage in mathematical conversations.

Ultimately, it falls on the mathematics teacher to best orchestrate those interactive, highly contextualized, and constructive environments that will best support both the acquisition of academic language among ELLs and content-area learning in the social plane. For the teacher of diverse student populations, this orchestration necessarily requires careful consideration of those grouping configurations that will maximize both language acquisition and concept learning among ELL students. Table 3.2 presents a brief overview of the academic and linguistic considerations in making decisions about the most appropriate groupings for targeting language acquisition and concept learning among ELLs in mathematics classrooms. For an elaboration of strategies for students falling in the four categories in table 3.2, see Garrison and Mora (1999).

Table 3.2
ELL Characteristics and Implications for Professional Practice

ELL Characteristics	What to Anticipate and Guidelines for Instructional Practice
Limited Academic Background **and** Limited Language Proficiency	For an ELL who exhibits these characteristics, a modification of the content may be necessary to build background. Begin with less technical words, but build toward increasing word complexity over time. For example, ELLs may understand *put together*, but they may need to be taught that it means the same thing as *add* and *sum*. Conduct informal preassessments with ELLs before teaching a new concept. Place a student with these characteristics in a group with a peer who is proficient in both language and the relevant content-area knowledge.
Proficient Academic Background **and** Limited Language Proficiency	An ELL with these characteristics tends to understand important academic concepts. For example, the student may already know how to solve a one-step equation. Yet the student often does not have the language skills necessary to understand the concept verbally or to explain to someone else how to solve an equation. If feasible, use the student's first language. The student may already possess the academic language for the new concept in his or her first language. Use visuals and hands-on activities to introduce concepts. Then introduce new vocabulary, such as *variable* and *equation*. Pair the student with a peer who understands the target language in order to encourage the student to discuss the concept and use the new vocabulary. Group such students with a peer who is proficient in the target language.

(Continued on p. 60)

Table 3.2—*Continued*

ELL Characteristics	What to Anticipate and Guidelines for Instructional Practice
Limited Academic Background **and** Language Proficiency	An ELL with these characteristics tends to understand the second language, but the academic concept being taught is new. This means it should be easier for the student to make the connection to English. However, ensure that the student has a CALP (Cognitive Academic Language Proficiency)-based and not a BICS (Basic Interpersonal Communication Skills)-based understanding before proceeding (Cummins 1984). Use concrete examples and manipulatives to teach the new concept. Group this student with a peer who possesses the relevant prior content-area knowledge.
Proficient Academic Background **and** Language Proficiency	An ELL with these characteristics often possesses both the academic and the language background necessary to fully understand what is being taught. In a classroom with a mix of students at different levels of academic and language proficiency, put these students in groups where they can help other students understand the content and language. Learning occurs through application and articulation. Encourage such students to enhance their knowledge, skills, and capacities by serving as translators and mentors for their peers who are not as proficient in the English language or content-area background knowledge.

Language is the vehicle through which learning occurs. If ELLs are to succeed in a mathematics classroom, specific tools must be incorporated to enable their full participation in learning mathematics. Classroom instruction for ELLs must be adapted to appropriately accommodate their different needs. However, instruction should also constructively maximize the assets these students already bring to the learning environment. Resources available in bilingual or ESL education, such as the Sheltered Instruction Observation Protocol (Echevarria, Vogt, and Short 2004), can provide guidance. However, the utility of these tools is bounded by the teacher's willingness to reconceptualize his or her role in the classroom as a teacher of language as well as a teacher of mathematics.

REFERENCES

Adams, Thomasenia L., Fiona Thangata, and Cindy King. " 'Weigh' to Go! Exploring Mathematical Language." *Mathematics Teaching in the Middle School* 10 (May 2005): 444–48.

Brown, Catherine A., Mary Kay Stein, and Ellice A. Forman. "Assisting Teachers and Students to Reform the Mathematics Classroom." *Educational Studies in Mathematics* 31 (1996): 63–93.

Cummins, Jim. *Bilingualism and Special Education: Issues in Assessment and Pedagogy.* San Diego, Calif.: College-Hill Press, 1984.

Dale, Theresa C., and Gilbert J. Cuevas. "Integrating Language and Mathematics Learning." In *ESL through Content Area Instruction—Language in Education: Theory and Practice,* edited by JoAnn Crandall, pp. 9–54. Washington, D.C.: Office of Education Research and Improvement, 1995.

Diaz-Rico, Lynne T., and Kathryn Z. Weed. *The Crosscultural, Language, and Academic Development Handbook: A Complete K–12 Reference Guide.* 2nd ed. Boston: Allyn & Bacon, 2002.

Echevarria, Jana, MaryEllen Vogt, and Deborah J. Short. *Making Content Comprehensible for English Language Learners: The SIOP Model.* Boston: Allyn & Bacon, 2004.

Flores, Alfinio. "*Sí Se Puede,* 'It Can Be Done': Quality Mathematics in More than One Language." In *Multicultural and Gender Equity in the Mathematics Classroom—the Gift of Diversity: 1997 Yearbook,* edited by Janet Trentacosta, pp. 81–91. Reston, Va.: National Council of Teachers of Mathematics, 1997.

Forman, Ellice A. "Learning Mathematics as Participation in Classroom Practice: Implications of Sociocultural Theory for Educational Reform." In *Theories of Mathematical Learning,* edited by Leslie P. Steffe, Pearla Nesher, Paul Cobb, Bharath Sriraman, and Brian Greer, pp. 115–30. Hillsdale, N.J.: Lawrence Erlbaum Associates, 1996.

———. "A Sociocultural Approach to Mathematics Reform: Speaking, Inscribing, and Doing Mathematics within Communities of Practice." In *A Research Companion to "Principles and Standards for School Mathematics,"* edited by Jeremy Kilpatrick, W. Gary Martin, and Deborah Schifter, pp. 333–52. Reston, Va.: National Council of Teachers of Mathematics, 2003.

Forman, Ellice Ann, and Jean McPhail. "Vygotskian Perspective on Children's Collaborative Problem-Solving Activities." In *Contexts for Learning: Sociocultural Dynamics in Children's Development,* edited by Ellice Ann Forman, Norris Minick, and C. Addison Stone, pp. 213–29. New York: Oxford University Press, 1993.

Garrison, Leslie, and Jill Kerper Mora. "Adapting Mathematics Instruction for English-Language Learners: The Language-Concept Connection." In *Changing the Faces of Mathematics: Perspectives on Latinos,* edited by Luis Ortiz-Franco, Norma G. Hernandez, and Yolanda De La Cruz, pp. 35–47. Reston, Va.: National Council of Teachers of Mathematics, 1999.

Goodnow, Jacqueline J. "The Socialization of Cognition: What's Involved?" In *Cultural Psychology: Essays on Comparative Human Development,* edited by James

W. Stigler, Richard A. Shweder, and Gilbert Herdt, pp. 259–86. Cambridge: Cambridge University Press, 1990.

Herrera, Socorro G., and Kevin G. Murry. *Mastering ESL and Bilingual Methods: Differentiated Instruction for Culturally and Linguistically Diverse (CLD) Students.* Boston: Allyn & Bacon, 2005.

Khisty, Lena Licon. "Making Mathematics Accessible to Latino Students: Rethinking Instructional Practice." In *Multicultural and Gender Equity in the Mathematics Classroom—the Gift of Diversity: 1997 Yearbook,* edited by Janet Trentacosta, pp. 92–101. Reston, Va.: National Council of Teachers of Mathematics, 1997.

Kilpatrick, Jeremy, Jane Swafford, and Bradford Findell, eds. *Adding It Up: Helping Children Learn Mathematics.* Washington, D.C.: National Academy Press, 2001.

Krashen, Stephen D. *Principles and Practice in Second Language Acquisition.* Oxford, England: Pergamon Press, 1982.

Lampert, Magdalene, and Paul Cobb. "Communication and Language." In *A Research Companion to "Principles and Standards for School Mathematics,"* edited by Jeremy Kilpatrick, W. Gary Martin, and Deborah Schifter, pp. 237–49. Reston, Va.: National Council of Teachers of Mathematics, 2003.

Lappan, Glenda, James T. Fey, William M. Fitzgerald, Susan N. Friel, and Elizabeth D. Phillips. *Stretching and Shrinking: Similarity.* Connected Mathematics Project. Needham, Mass.: Pearson Prentice Hall, 2004.

Lave, Jean E., and Etienne Wenger. *Situated Learning: Legitimate Peripheral Participation.* New York: Cambridge University Press, 1991.

Moll, Luis C. "Social and Instructional Issues in Literacy Instruction for 'Disadvantaged' Students." In *Better Schooling for the Children of Poverty: Alternatives to Conventional Wisdom,* edited by Michael S. Knapp and Patrick M. Shields, pp. 61–84. Berkeley, Calif.: McCutchen Publishing, 1991.

Moschkovich, Judit N. "Supporting the Participation of English Language Learners in Mathematical Discussions." *For the Learning of Mathematics* 19 (1999a): 11–19.

————. "Understanding the Needs of Latino Students in Reform-Oriented Mathematics Classrooms." In *Changing the Faces of Mathematics: Perspectives on Latinos,* edited by Luis Ortiz-Franco, Norma G. Hernandez, and Yolanda De La Cruz, pp. 5–12. Reston, Va.: National Council of Teachers of Mathematics, 1999b.

National Council of Teachers of Mathematics (NCTM). *Principles and Standards for School Mathematics.* Reston, Va.: NCTM, 2000.

O'Conner, Mary C., and Sarah Michaels. "Aligning Academic Task and Participation Status through Revoicing: Analysis of a Classroom Discourse Strategy." *Anthropology and Education Quarterly* 24 (1993): 318–35.

Rotman, Brian. "Toward a Semiotics of Mathematics." *Semiotica* 72 (1988): 1–35.

Santiago, Felicita, and George Spanos. "Meeting the NCTM Communication Standards for All Students." In *Reaching All Students with Mathematics,* edited by Gilbert Cuevas and Mark Driscoll, pp. 133–45. Reston, Va.: National Council of Teachers of Mathematics, 1993.

Silver, Edward A., and Margaret S. Smith. "Building Discourse Communities in Mathematics Classrooms: A Worthwhile but Challenging Journey." In *Communication in Mathematics, K–12 and Beyond: 1996 Yearbook,* edited by Portia C. Elliott, pp. 20–28. Reston, Va.: National Council of Teachers of Mathematics, 1996.

Teachers of English to Speakers of Other Languages (TESOL). *ESL Standards for Pre-K–12 Students.* Alexandria, Va.: TESOL, 2001.

Vygotsky, Lev S. *Mind in Society: The Development of Higher Psychological Processes.* Cambridge, Mass.: Harvard University Press, 1978.

4

Learning with Understanding: Principles and Processes in the Construction of Meaning for Geometric Ideas

Michael T. Battista

BEFORE analyzing the nature of students' learning of geometry, consider two principles of human thought offered by cognitive science.

> *Principle 1:* The human mind *constructs* rather than *receives* meaning.

According to Nobel Prize winner Francis Crick, "Seeing is a constructive process, meaning that the brain does not passively record the incoming visual information. It actively seeks to interpret it" (1994, pp. 30–31). To illustrate, consider figure 4.1. Most people see an equilateral triangle partially obscured by a right triangle. However, the right triangle is not really there; instead, our minds *construct* it.

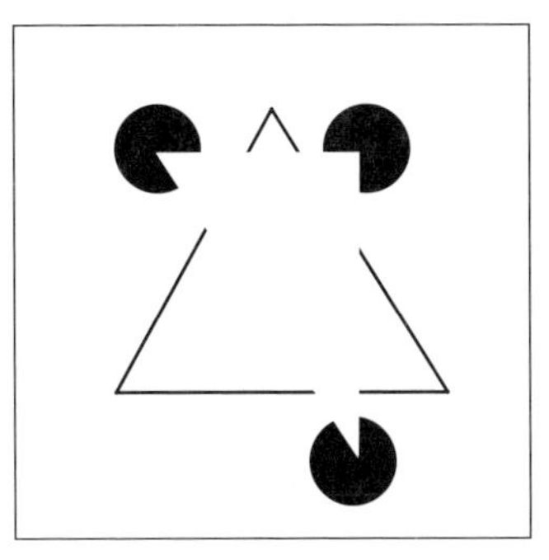

Fig. 4.1. A right triangle that is not really there

The construction of meaning, however, is not restricted to seeing; it also applies to conscious reasoning. As psychologist Robert Ornstein so eloquently asserts, "Our experiences, percepts, memories are not of the world directly but are our own creation, a dream of the world, one that evolved to produce just enough information for us to adapt to local circumstances" (1991, p. 160).

In fact, to successfully act on, and reason about, our environment, our minds must construct appropriate nonverbal, recall-of-experience-like mental versions of situations—mental models (Battista 2001).

> *Principle 2:* Individuals construct new knowledge and understanding based on what they already know and think (Bransford, Brown, and Cocking 1999).

Our interpretations of the world depend on our current mental models. For instance, examine the two dark vertical segments in figure 4.2 (Coren, Ward, and Enns 1994, p. 497). Amazingly, these segments are the exact same length. But the mental models that we already possess for interpreting the real-world context in which the segments are shown cause us to construct a perception in which the lengths of the segments appear very different.

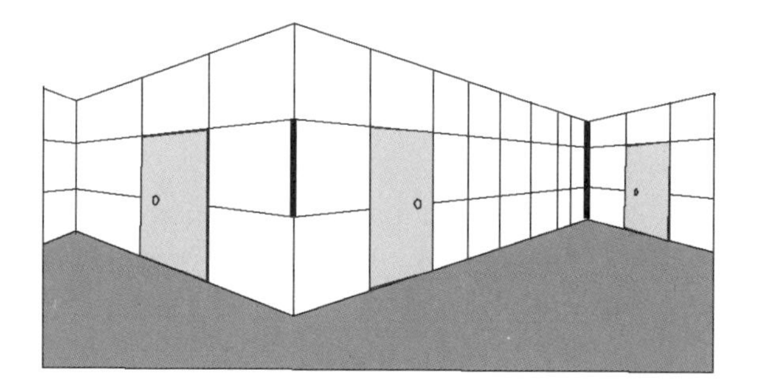

Fig. 4.2. The two dark vertical segments are the same length.

Implications for Mathematics Learning and Teaching: Attending to Students' Mathematical Constructions

Taking these two principles seriously has major implications for understanding and supporting students' mathematics learning. As teachers, we must attend to the current understandings and ways of reasoning that students bring with them as they attempt to learn new mathematical topics. No matter how incomplete, naïve, or informal students' current ideas happen to be, these ideas cannot be ignored because they form the foundation for students' construction of new mathematical meanings. To be successful in supporting students' learning, we must understand and investigate how students progress from their "starting" ideas to the formal mathematical concepts they are expected to learn in school. "There is a good deal of evidence that learning is enhanced when teachers pay attention to the knowledge and beliefs that learners bring to a learning task, use this knowledge as a starting point for new instruction, and monitor students' changing conceptions as instruction proceeds" (Bransford, Brown, and Cocking 1999, p. 11).

Investigating Students' Mathematical Constructions in Detail—the Case of Elementary School Geometry

To understand how students learn mathematics, we must carefully examine how they construct mathematical meaning. To illustrate, I examine in depth how two fifth graders constructed the concept of perpendicularity as they were involved with a special computer microworld in an inquiry-based classroom. I describe, first, the theoretical framework for examining students' mathematical thinking, second, the computer microworld, and third, the development of students' thinking.

The van Hiele Theory

An analysis of students' mathematical learning can be significantly aided by examining relevant research. For geometry learning, the van Hiele theory and its associated line of research provide major insights by describing several levels of sophistication in the development of students' geometric thinking. I summarize the levels below, giving an expanded description of Level 2 because that level is most helpful for understanding the student work described later in this article (see Clements and Battista 1992 and Battista forthcoming for more detail).

Level 0 (Prerecognition)

Because students at this level attend to only part of a shape's visual characteristics, they are unable to identify many common shapes.

Level 1 (Visual)

Students identify, describe, and reason about shapes and other geometric configurations according to their appearance as visual wholes. Their reasoning is dominated by imagery and visual perception rather than an analysis of geometric properties. When identifying shapes, students often use visual prototypes, for instance, saying that a figure is a rectangle because "it looks like a door."

Level 2 (Descriptive/Analytic)

Students recognize and characterize shapes by their geometric properties, that is, by explicitly focusing on and describing spatial relationships between the parts of a shape. At first, in the transition from Level 1 to Level 2, students describe parts and properties of shapes informally, imprecisely, and often incompletely; they do not possess the formal conceptualizations that enable precise property specifications. For instance, students might describe a rectangle as a shape having two long sides and two short sides. As students begin to acquire formal conceptualizations that can be used to make sense of and describe spatial relationships between parts of shapes, they use a combination of informal and formal descriptions of shapes. For instance, students might describe a rectangle as a four-sided shape with opposite sides equal and square corners. Finally, in a complete attainment of Level 2 reasoning, students explicitly and exclusively use formal geometric concepts and language to describe and conceptualize shapes in a way that attends to a sufficient set of properties to specify the shapes. For instance, students might think of a rectangle as a figure that has opposite sides equal and parallel and four right angles.

Level 3 (Abstract/Relational)

Students interrelate geometric properties, form abstract definitions, distinguish among necessary and sufficient sets of properties for a class of shapes, and understand and sometimes even supply logical arguments in the geometric domain. They meaningfully classify shapes hierarchically and give arguments to justify their classifications (e.g., a square is identified as a rhombus because it has the defining property of a rhombus, all sides congruent).

Level 4 (Formal Axiomatic)

Students formally prove theorems within an axiomatic system. That is, they produce a sequence of statements that logically justifies a conclusion as a consequence of the "givens." By necessity, thinking at Level 4 is required for meaningful and full participation in a proof-oriented high school geometry course.

A Computer Microworld for Investigating Geometric Shape

The Shape Makers computer microworld, a special add-on to the Dynamic Geometry program The Geometer's Sketchpad, provides students with various screen manipulable shape-making objects (Battista 1998). For instance, the dynamically interactive Parallelogram Maker is a Geometer's Sketchpad construction that can be used to make any parallelogram that fits on the computer screen, no matter what its shape, size, or orientation. Dragging the Parallelogram Maker's vertices changes the angles and side lengths of the shapes it makes, but all these shapes are parallelograms (see fig. 4.3). One of the major goals in the design of the Shape Makers program was to help students move from van Hiele Levels 0 and 1 to Level 2.

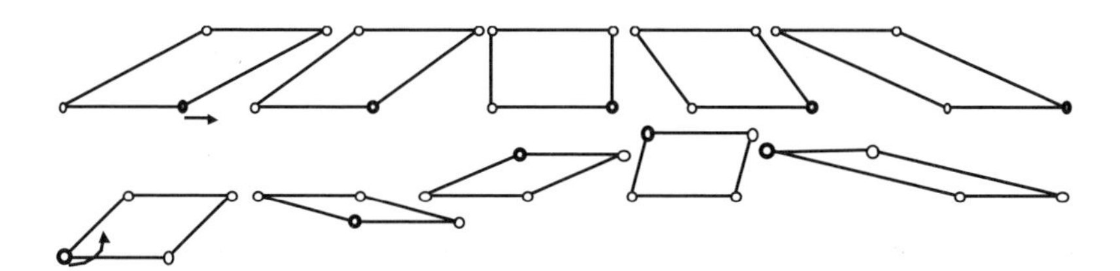

Fig. 4.3. Dragging the vertices of the Parallelogram Maker changes its angles and side lengths.

The Rectangle Maker Can't Make a *Slant*: Students' Progress and Difficulties in Constructing Meaning for a Spatial Property

The Task: What Shapes Can a Rectangle Maker Make?

After several days of working with different quadrilateral Shape Makers, a class of fifth graders, working in pairs on computers, was given the instructional task of determining which of Shapes 1–6 can be made by the Rectangle Maker (see fig. 4.4). Students had to explain and justify their conclusions. One pair of students, Matt and Tom, predicted that the Rectangle Maker could make Shapes 1–3 but not Shapes 4–6. Matt and Tom then began checking and discussing their results. (The entire episode took about twenty-five minutes.)

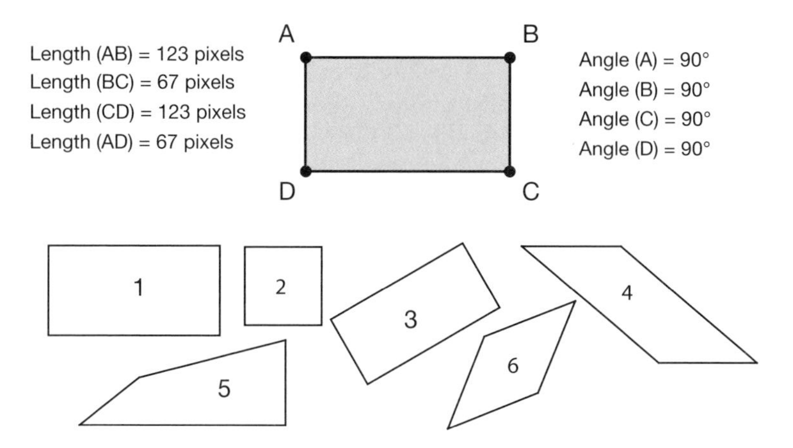

Fig. 4.4. Determining which of Shapes 1–6 can be made by the Rectangle Maker

Using Old Ideas in a New Situation

After using the Rectangle Maker to check Shapes 1 and 2, Matt and Tom had some initial difficulty manipulating the Rectangle Maker to make Shape 3 and wondered if it could make Shape 3:

Tom: Maybe it can't because it can't be slanted.
Matt: It's not at a slant....

But after trying again, the boys successfully made Shape 3, and then they moved on to checking whether Shape 4 could be made by the Rectangle Maker.

Tom: I'm positive it can't do this one.
Matt: It [the Rectangle Maker] has no slants.... It can't make a slant.... The Rectangle Maker can't make something like that slanted there [*motioning along the bottom then left side of Shape 4*].
Tchr: You mean the angle there?
Matt: Yeah, it has an angle. [Shape] 4 is no because the Rectangle Maker can't make a slant.
Tom: Cause the Rectangle Maker can only make a square and a rectangle, and that's not a square or a rectangle....

As evidenced by their correct predictions, Tom and Matt had explored the Rectangle Maker enough to learn a great deal about which kinds of shapes it could make. They had constructed sufficiently accurate mental models for rectangles and how the Rectangle Maker moves.

However, the boys were struggling to conceptualize *why* the Rectangle Maker could not make parallelograms without right angles. Furthermore, as illuminated by the van Hiele theory, Matt and Tom were reasoning about this issue in fundamentally different ways. Tom's last comment indicates Level 1 reasoning—he was thinking about shapes strictly as wholes. In contrast, as Matt attempted to describe spatial relationships between parts of shapes, he was starting to move beyond Level 1 reasoning toward Level 2. His comments that the Rectangle Maker has no slants and that Shape 4 "has an angle" were attempts to formulate a distinction that mathematicians describe using the concepts of right angle or perpendicularity. But Matt's reasoning was still extremely imprecise. And, consistent with Principle 2, Matt was building new ideas out of his current informal ideas, which were based on familiar common-language terms *slant* and *angle*. For example, the popular novelist Ken Follett employs both common-language usages in the same sentence: "A few stunted trees grew at angles, blown slantwise by the tireless wind" (2004, p. 103). An essential component of learning mathematics is developing an understanding of the formal and precise concepts mathematicians use to describe the world. These concepts, when learned with genuine understanding, furnish the lenses through which we see and analyze the world. In an attempt to get Matt and Tom to move toward formal concepts, the teacher pushed the boys to sharpen their ideas.

Tchr:	What do you mean by *slant?*
Matt:	Like this. See how this is shaped like a parallelogram [*pointing at Shape 4*]....
Tom:	This is in a slant right now [*Tom points at the Rectangle Maker, which is rotated from the horizontal*].

In this episode, matters were further complicated because "slant" meant not-perpendicular for Matt but tilted from the vertical for Tom, which raises an extremely important issue. In school geometry, shapes are described by referring *to relationships between their parts.* For example, saying that a rectangle has four right angles expresses a relationship between the sides of a rectangle; it means that adjacent sides of the rectangle meet at right angles or are perpendicular. This is the property that Matt was struggling to express with the term *slant.* However, common-language use of the terms *slant* and *at an angle* refer to the relationship of lines and segments to *the up-down or vertical frame of reference.* Thus, Matt's use of the word slant was evoking a totally different, common-language meaning for Tom. Shifting to a relationships-between-parts meaning for *slant* would have required Tom to disregard the up-down common-language perspective, a way of looking at things that is essential in real life.

In the next episode, the teacher tried again to get the students to sharpen their thinking, explicitly attempting to focus the boys' attention on parts of the shape—the angles—not the whole shape.

Tchr:	What is different about the angle here [*motioning to the upper left angle on the Rectangle Maker*] and the angle here [*motioning to the upper left angle on Shape 4*]?
Tom:	This isn't a rectangle [*referring to Shape 4*].
Matt:	The one on the Rectangle Maker has to have straight lines....
Tom:	Matt, they are straight lines. All of these are straight lines.
Matt:	Yeah, they are straight, but on a slant.

Matt and Tom's struggle to analyze the situation was further complicated by their introduction of still another common-language idea, "straight," which each boy used differently. For Tom, *straight* meant "without turning"; for Matt, *straight* meant without turning and, more important, perpendicular. Thus, consistent with Principle 2, the boys were still analyzing the situation with previously learned *informal* concepts.

Tom:	This is at a slant right now [*points at the Rectangle Maker, which is rotated from the horizontal*].
Matt:	It can't make that kind of a shape [*pointing to Shape 4 on the screen*].... It can't make something that has a slant at the top and stuff.
Tom:	Do you mean it has to have a straight line right here, like coming across [*pointing horizontally across the top of Shape 6, as shown in figure 4.5*]? [*Tom now uses* straight *to mean* horizontal.]
Matt:	I know they are straight, but they are at a slant and it [the Rectangle Maker] always has lines that aren't at a slant.

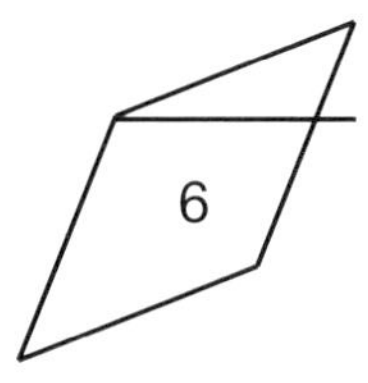

Fig. 4.5. "Do you mean it has to have a straight line here [pointing horizontally]?"

Tom's first utterance was visual-holistic (Level 1), about a shape as a whole. In contrast, Matt was struggling to move to the componential reasoning characteristic of Level 2. Being questioned by both his partner and the teacher caused Matt to start to sharpen his thinking. The phrase *lines that aren't at a slant* expresses a relationship between the adjacent sides of a rectangle. But Matt's ideas are still informal and imprecise. He does not yet possess the conceptual tools—the formal geometric concepts of perpendicularity or right angles—necessary for full attainment of Level 2 reasoning.

Difficulties in Using a Formal Concept for Geometric Reasoning

In the next episode, the teacher continued to focus the boys' attention on parts of shapes, hoping, of course, that the boys would use the concept of right angle to reason about these parts.

Tchr:	In this picture [Shape 5] are there things that it [the Rectangle Maker] could have done parts of?
Tom:	Yeah! It can make a right angle [*i.e., the Rectangle Maker could make the bottom right "corner" of Shape 5*].
Tchr:	What do you wish we could change in this picture [*for the Rectangle Maker to make Shape 5*]?

Matt: I wish this side here, this one, would go over to here [*motioning as indicated on the left side of Shape 5; see figure 4.6a*]. That [*the left side of Shape 5*] would be equal to that [*the right side of Shape 5*], and they would be parallel. We wanted this line [*the top*] *to go straight across.*

Tchr: Why isn't it coming straight across here, why is it coming down?

Tom: Well, you would have to move this handle [*the upper left vertex*] up here [*see figure 4.6b*].

Tchr: OK. So if we moved that handle, then what did we create?

Tom: A right angle, because there would be a straight line right here and a straight line right here [*motioning for segments as indicated; see figure 4.6c*]....

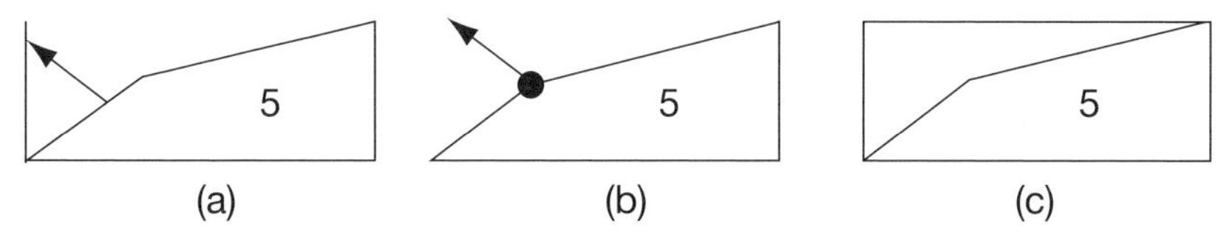

Fig. 4.6. How should Shape 5 be changed so that the Rectangle Maker could make it?

Matt and Tom clearly saw how Shape 5 must be changed so that it could be made with the Rectangle Maker. Also, the teacher was able to get the students to use the formal concept of right angle to describe a spatial relationship. In fact, the concept of right angle had been used several times in previous class discussions to describe squares and rectangles. It had also been used earlier in the year and in the previous school year. So the concept was not new to students. However, Matt and Tom had not sufficiently abstracted the concept so that they could use it in the current situation. That is, the concept of right angle was being used by the boys to describe a particular kind of spatial configuration that they recognized. It was not being used explicitly to describe a special instance of a general angular relationship between two intersecting segments.

Tchr: When you put two sides together, what do they make?

Tom: A right angle, if you put them right.

Tchr: OK. You get a right angle, and what effect does that have on the lines?

Matt: It makes them all be, like, none of them will be at a slant.

In this last episode, and mindful of Principle 2, the teacher attempted to get the boys to connect the formal relational concept of right angle to their informal concept of slant, seemingly succeeding with Matt.

Tchr: So do you think, Matt, every time you have a right angle, you have lines that are not slanted?

Matt: Yes.

Tom: Yeah, you do. Well, slanted. What do you mean by *slanted*? Do you mean like at a diagonal?...

But Tom was still having difficulty understanding what Matt and the teacher were talking about because he was holding on to his informal use of the term *slant*—an informal use that was different from Matt's. Consistent with Principle 2, Tom was having great difficulty moving beyond his previously learned idea of *slant* as referring to a deviation from vertical, which uses an up-down frame of reference. Indeed, a major difficulty in coming to grips with the concept of right angle or perpendicularity is shifting perspective—from thinking about how sides are oriented in an external vertical-horizontal frame of reference to how they are related to each other, irrespective of external reference frames.

Tchr: What if I could take this [shape] Number 4 and put it right next to [shape] Number 3. What's different about 3 and 4 [*draws Shapes 3 and 4 next to each other*]?

Matt: The top lines on these are not exactly on top of each other [*motioning to the top and bottom sides on Shape 4 as indicated by the dark ovals in figure 4.7*]; these ones are [*motioning almost vertically to the top and bottom sides on Shape 3 as indicated in figure 4.7*].

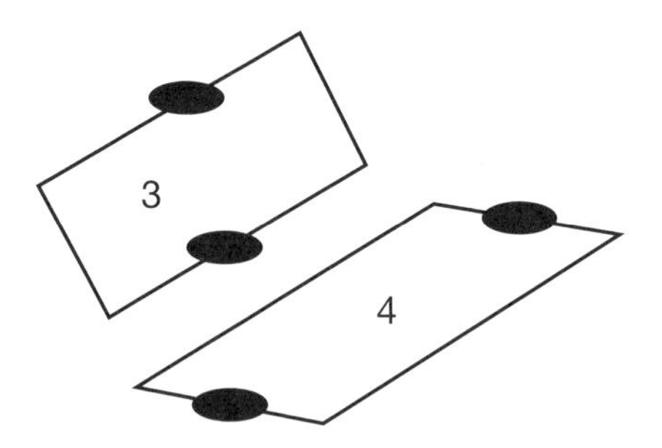

Fig. 4.7. According to Matt, the indicated sides are not "on top of each other" on Shape 4, but they are on Shape 3.

The teacher's question makes perfect sense because she saw the difference between Shapes 3 and 4 formally as the lack of right angles in Shape 4. However, Matt and Tom did not see what the teacher saw. Instead, although they attended to relationships between parts of the shapes—another indicator that they were moving toward van Hiele Level 2 reasoning—they described the relationships with respect to orientation in the plane—one side being "on top of" or above another. They still could not use the concept of right angle to encapsulate the essential *intrinsic* spatial difference between Shapes 3 and 4. The boys did not conceptualize a rectangle as a shape defined as having four right angles, and thus had not reached van Hiele Level 2 reasoning.

The "Aha!" Moment

At this time, the teacher thought that the boys needed more time to use the Shape Makers to explore their ideas, without minute-by-minute teacher intervention. But as indicated below, before leaving, she pointed them in what she thought would be a fruitful direction.

Tchr: Keep on really looking at what makes these different.... Watch the numbers up there and see if that will help you answer some of the questions [*referring to the length and angle measurements on the computer screen that are continuously updated as the Rectangle Maker is manipulated—see figure 4.4*].

Tom: [*Manipulating the Rectangle Maker*] Oh, this [the Rectangle Maker] ... I just noticed, always has to have four 90° angles. And that one [Shape 4] does not have 90° angles. And so this one [*indicating Shape 3 and moving the Rectangle Maker next to it*] has to have a 90° angle, because we made this [Shape 3] with that [Rectangle Maker]. So there is one thing different. A 90° angle is a right angle, and this [Shape 4] does not have any right angles.... Oh yeah! Because this [*pointing to the Rectangle Maker and its angle measurements*] always has to have right angles. And that [*pointing to Shape 4*] doesn't have 90° angles, so therefore you cannot make it. [*Manipulating the Rectangle Maker, then pointing to its measurements.*] See, look at the angles.

Matt: Oh, now I get what [the teacher's] question is, the angles. The angles control the shape. [*Referring to the Rectangle Maker*] See, the angles are always 90°. If [angle] A was at a 60° angle ... then it would cause it to look something like that [Shape 4]....Those two [the top and right side of the Rectangle Maker] would slant....

Finally, the "Aha!" moment had occurred! Matt and Tom used the concept of right angle to construct a formal property-based understanding that enabled them to see *why* the Rectangle Maker could not make Shape 4. And in making this realization, the boys learned some new mathematics. Indeed, from this moment on, the boys used the concept of right angle as a powerful conceptual tool to analyze properties of shapes.

What enabled this conceptual construction? It is likely that Matt and Tom's renewed manipulation of the Rectangle Maker caused them to integrate a whole set of previously disconnected fragments of knowledge. How? First, earlier that day, the class had discussed degree measure as indicating the amount of turn between an angle's sides. Students, other than Matt and Tom, said that right angles measure 90° and that right angles are "straight" and "do not slant." Second, as we observed in the episode above, Matt and Tom spent considerable time reflecting on and discussing why the Rectangle Maker could not make Shape 4, using terms like *slant, straight,* and *right angle*—but not degree measure—in their struggle to conceptualize the situation. Third, as Matt and Tom returned to manipulating the Rectangle Maker, perhaps because of the teacher's suggestion to look "at the numbers" and because, as the boys manipulated the Rectangle Maker, the length measures changed but the angle measures did not, Matt and Tom quickly noticed that the angle measures stayed constant at 90°—their attention was drawn to the constancy of the angle measures.

This shift in attention to see the Rectangle Maker's angles in relation to their measures—90°—seemed crucial. We can conjecture that thinking about right angles in relation to degree measure helped Matt and Tom see that a right angle is not just an identifiable spatial object—like a square corner. Rather, a right angle possesses one possible configuration, out of many, for the sides of an angle. This idea interrelates the angles of Shape 4 with the angles of the Rectangle Maker—they are all angles, but the Rectangle Maker's are special because they measure 90°. So the fact that every pair of sides of a rectangle forms a 90° angle is what distinguishes rectangles from nonrectangular parallelograms. Of course, I am not claiming that Matt and Tom saw this line of reasoning in an explicit, logical way. Rather, I am suggesting that the elements of this line of reasoning were implicitly recognized in a way that produced the boys' Aha! experience.

Final Reflections on Matt and Tom's Learning

For Matt and Tom, learning why the Rectangle Maker could not make Shape 4 was an extremely complex mental activity. The boys had to find a way to conceptualize and explain the spatial relationship that sophisticated users of geometry describe by saying that the sides of shapes produced by the Rectangle Maker meet at right angles. In so doing, and typical of students at van Hiele Level 1, Matt and Tom initially used holistic, vague, or incomplete ideas, all of which were inadequate conceptualizations for the task.

According to Principle 1, for learning to occur, Matt and Tom had to mentally *construct* a new way of thinking about shapes. According to Principle 2, the boys had to construct this new knowledge from their current knowledge—everyday, nonmathematical ideas of slanted, straight, and up-down. Furthermore, Matt and Tom had to reconcile two fundamentally different perspectives on shape—one critical for everyday functioning, and one critical for formal geometric analyses. That is, when the boys discussed "slanted from the vertical" and "one side on top of the other," they were using the vertical-horizontal frame of reference that is critically "privileged" in real-world experience. Because the up-down frame of reference is crucial in dealing with the physical world, students' initial mental representation of a shape, "does not just reflect its Euclidean geometry [specifying relationships among its parts], which remains unchanged as a shape is turned. It reflects the geometry relative to our up-down reference frame" (Pinker 1997, p. 266). In fact, it is natural and necessary that many of the spatial relationships that elementary school students learn outside of school deal with the up-down frame of reference (e.g., seeing a line as tilted from the vertical) rather than the internal structure of shapes (e.g., seeing two lines as perpendicular). However, in most geometric treatments of shape and space in schools, students must learn to ignore the external frame of reference and focus on relationships internal to shapes. It is not surprising, then, that according to Principle 2, elementary school students will use the already acquired up-down reference frame when initially attempting to conceptualize internal-structure spatial relationships—even when, as in the present situation, that reference frame is inadequate. Thus, Principle 2 explains a primary difficulty experienced by Matt and Tom—the formal geometric knowledge they were attempting to construct used an internal-structure frame of reference, but they were trying to build this knowledge from their current ideas, which were based on an up-down frame of reference.

The Role of the Teacher

It is important to note that the boys' learning was critically guided and supported by the instructional activities, the computer microworld, the inquiry-based culture of the classroom, and the teacher's questioning. Indeed, the teacher promoted the students' reflection on their ideas by constantly asking for clarification—What exactly do you mean by this? Guided by her knowledge of the van Hiele theory, she tried to get the boys to move from holistic to property-based thinking. Furthermore, she attempted to get the boys to use the formal concept of right angle to analyze differences in shapes. One way she did this was to implicitly encourage the boys to see that moving the handles changed the angles. But the boys were not comfortable thinking of the handles as controlling the angles. The teacher also tried to get the boys to analyze the parts of shapes (e.g., asking if there are parts of Shape 5 that the Rectangle Maker could make, why a side isn't coming straight across, and what is made when two sides are put together). Finally, recognizing that the boys needed more time to manipulate and reflect on the Shape Makers, she left them, realizing that learning with understanding requires sufficient time for productive reflection. But before leaving the boys, the teacher suggested what she thought would be a fruitful direction for further reflection—looking at the measurements of the Shape Makers.

After the teacher left the boys, Matt and Tom's exploration and thinking were quite productive. When Matt and Tom returned to manipulating the Rectangle Maker, they focused on its changing measurements, which enabled them to construct the idea that the Rectangle Maker always has four 90°, or right, angles. This new idea was sufficiently powerful for them to use it in (*a*) analyzing the differences between Shapes 3 and 4, and (*b*) seeing why the Rectangle Maker could not make Shapes 4, 5, and 6. Thus, the boys had constructed a formal, property-based geometric conceptualization (which they used consistently in subsequent activities), raising their reasoning about this situation to van Hiele Level 2. This was a major accomplishment in the boys' learning of geometry.

Learning: Reprise

Examining the episodes discussed in this article confirms everyday and research-based observations that learning mathematics with understanding is difficult. It is not clean and straightforward; it is messy and circuitous. The principles of learning explain the difficulty. According to Principle 1, learning mathematics with genuine understanding requires students to *construct* personal meaning for formal mathematical ideas—such construction requires hard, intellectual work. And according to Principle 2, the construction of new meaning is based on students' current knowledge and ways of reasoning, despite the fact that this knowledge can be inconsistent with formal mathematical ideas and inadequate for many problem-solving situations. Consequently, it can be very difficult for students to build appropriate new ideas, given their current state of knowledge.

However, Principles 1 and 2 do more than help us understand students' learning difficulties. A knowledge of these principles, if taken seriously, enables us, as teachers, to help students successfully negotiate their learning difficulties. By understanding the ways that students construct meaning for specific mathematical topics, we can instructionally guide and support their construction of meaning. For instance, the Shape Makers environment and instructional activities were designed to support and guide students' construction of knowledge about properties of shapes. Although these activities allow students to start with holistic reasoning, they constantly encourage and support students' development of ever-sophisticated knowledge of the properties of shapes. These activities also support incremental growth of knowledge, in contrast to many traditional approaches that demand that students make almost instantaneous, substantive leaps in sophistication from their current informal ideas to formal, abstract mathematical concepts. Indeed, because the teacher described in this article had a knowledge of several fundamental principles of thinking and learning, along with the van Hiele theory, she was able to appropriately guide and support her students' learning. This knowledge also gave her the confidence to let her students struggle; she understood that through such struggles (but with proper guidance and support), they would construct new and powerful mathematical ideas.

REFERENCES

Battista, Michael T. *Shape Makers: Developing Geometric Reasoning with the Geometer's Sketchpad.* Berkeley, Calif.: Key Curriculum Press, 1998.

———. "How Do Children Learn Mathematics? Research and Reform in Mathematics Education." In *The Great Curriculum Debate: How Should We Teach Reading and Math?* edited by Thomas Loveless, pp. 42–84. Washington, D.C.: Brookings Press, 2001.

———. "The Development of Geometric and Spatial Thinking." In *Second Handbook of Research on Mathematics Teaching and Learning,* edited by Frank K. Lester, Jr. Greenwich, Conn.: Information Age Publishing, forthcoming.

Bransford, John D., Ann L. Brown, and Rodney R. Cocking. *How People Learn: Brain, Mind, Experience, and School.* Washington, D.C.: National Research Council, 1999.

Clements, Douglas H., and Michael T. Battista. "Geometry and Spatial Reasoning." In *Handbook of Research on Mathematics Teaching and Learning,* edited by Douglas A. Grouws, pp. 420–64. New York: Macmillan Publishing Co., 1992.

Coren, Stanley, Lawrence M. Ward, and James T. Enns. *Sensation and Perception.* Fort Worth, Tex.: Harcourt Brace College, 1994.

Crick, Francis. *The Astonishing Hypothesis.* New York: Simon & Schuster, 1994.

Follett, Ken. *Whiteout.* New York: Dutton/Penguin, 2004.

Ornstein, Robert. *The Evolution of Consciousness.* New York: Touchstone, 1991.

Pinker, Steven. *How the Mind Works.* New York: W. W. Norton & Co., 1997.

PART TWO

Issues Related to Students' Learning in School Contexts

W. Gary Martin

Students' success in learning mathematics is affected by the context within which they learn. In order to be successful, students must be actively engaged in their own learning. The National Assessment of Educational Progress and other research projects have investigated the contextual factors of the learning environment. A number of factors have been identified, including teachers' expectations, curriculum materials, instructional practices, the use of technology, and students' opportunities to learn. The papers in this section discuss a number of these factors.

Kersaint discusses the influence of the learning environment on what is learned, including the context for learning, opportunity to learn, and the mathematics instructional program. All these issues directly influence the mathematics programs that students receive and ultimately their learning outcomes. The article outlines important findings in each of these areas and argues that ultimately teachers must take an active role in ensuring that students have a learning environment that will promote, not hinder, their mathematical development.

Rousseau offers a perspective on students' learning using a "critical equity lens" to highlight issues that must be taken seriously by those seeking to achieve the vision of NCTM's *Principles and Standards for School Mathematics*—that "*all students* ... have the opportunity and support necessary to learn significant mathematics with depth and understanding" (2000, p. 5, emphasis added). To view mathematics classrooms through the lens of learning without considering the lens of equity risks the danger of a distorted view. Moreover, this equity lens must be applied at both the classroom level and at the broader policy level. She argues that teachers, researchers, and policymakers must consider the impact of classroom practices on the learning of *all* students, whether norms for classroom interactions are understood by all students. And we must also consider the broader policy context within which students' learning is occurring, so that we do not attribute this learning strictly to the immediate context.

Chapin and O'Connor take a closer look at one aspect of the learning environment, how a teacher's skillful use of classroom talk can promote students' learning, within the context of a project carried out in a low-income urban school district. The article includes a discussion of ways in which the project found that effective classroom discourse can be "academically effective" in promoting learning; it also outlines five specific "talk moves" that can help teachers create productive discourse. However, teachers must first create a respectful learning environment in order for these talk moves to be effective.

Choppin also considers the role of the teacher in developing effective classroom discourse that promotes students' learning. Examples are furnished that highlight both the challenges and the opportunities teachers face in collaborative discussion. He suggests that a lengthy learning process will be required for teachers to develop the expertise needed to fully support students' learning of mathematics in collaborative discussions. Teachers need to learn both to initiate and sustain such discussions.

Davis explores what may result when teachers ask their students to explore mathematics in a context that involves creating mathematical models and developing new mathematical ideas in the process. A meaningful context may yield opportunities for deeper learning or it may present surprising barriers to understanding. Aspects of a context may actually obscure the mathematical concepts that students are intended to learn. To be successful, teachers must understand the importance of context, be open to learning more about modeling and the role of context, and be prepared to seize the unexpected opportunities that may arise.

Dosemagen explores a particular type of classroom environment—the use of an online discussion board as an adjunct to mathematics instruction. On the basis of an action research project with calculus students, the author outlines how having students respond to activities in their mathematics class in a public, Web-based forum can increase their reflection and metacognition and hence ultimately deepen their understanding of the mathematics. Their personal beliefs about mathematics may also be affected. In addition, such a forum can provide the teacher with a productive window into students' learning and understanding that can help shape future instruction and furnish insights into his or her teaching practices.

Finally, Hart, Keller, and Hirsch look at the potential impact of technology on students' learning, a promise that has all too rarely been realized over the years. They discuss how technology can be used to amplify learning through Java-based, interactive environments that allow students to investigate and solve problems that might be beyond their reach without such tools. Furthermore, they argue that these tools must be embedded in a curricular context in order to realize their potential. Finally, they assert that the wide availability of Java-based resources over the Internet enhances their potential to become an integral part of learning.

REFERENCE

National Council of Teachers of Mathematics (NCTM). *Principles and Standards for School Mathematics.* Reston, Va.: NCTM, 2000.

5

The Learning Environment: Its Influence on What Is Learned

Gladis Kersaint

The teacher is responsible for creating an intellectual environment where serious mathematical thinking is the norm. More that just a physical setting with desks, bulletin boards, and posters, the classroom environment communicates subtle messages about what is valued in learning and doing mathematics. Are students' discussion and collaboration encouraged? Are students expected to justify their thinking? If students are to learn to make conjectures, experiment with various approaches to solving problems, construct mathematical arguments and respond to others' arguments, then creating an environment that fosters these kinds of activities is essential.

—Principles and Standards for School Mathematics

As mentioned in the quote above, the learning environment plays a significant role in fostering the development of particular learning outcomes. What one is able to learn and how one learns within any environment depend on essential features of the environment and the ways in which those features are used to encourage learning. Accordingly, instruction should focus on the learners and on the development of learning opportunities. How are those opportunities made available in educational settings? What environmental factors influence those learning opportunities? How does the learning environment influence what one is able to learn? To address those questions and achieve the ambitious goals outlined in the National Council of Teachers of Mathematics *Standards* documents (NCTM 1991, 2000), we must examine the learning environment as a fundamental variable in student learning to determine what features of the learning environment are most conducive to enhancing students' learning opportunities and to address other issues related to those environments that may affect student learning outcomes.

In this article, I discuss three features of the learning environments that may affect students' learning opportunities: (1) the context for learning, (2) the students' opportunity to learn (OTL), and (3) the mathematics instructional program. Rather than complete a comprehensive review of all issues related to those topics, I present a brief overview to illustrate their role in supporting student learning. Those particular features were chosen because they have direct impact on the quality of mathematics programs and consequently on students' learning outcomes.

The context-for-learning discussion focuses on the type and quality of mathematics programs offered while providing a structure for examining the goals and expectations that are part of those programs. The opportunity-to-learn discussion focuses on the issues of equitable access to an education in general and equitable access to high-level or robust mathematical understandings in particular. The discussion of the mathematics instructional program is limited to three aspects of an instructional environment that influence learning: (1) the learning community, (2) the development of mathematics literacy, and (3) the need to meet the needs of a diverse student population. Those areas highlight global features of the mathematics instructional environment that influence how and what students learn. The characteristics of the established learning community communicate directly or indirectly the goals and expectations for learning. The development of mathematics literacy refers to the overarching aim of mathematics education. This discussion addresses instructional features recommended to enhance students' mathematical or quantitative development. Finally, to address learning issues for *all* suggests that the needs of a diverse student population must be considered. The article concludes with recommendations for actions that teachers can take to remain positive influences in the learning environment.

Context for Learning

The context for learning has gotten much attention in the research literature because the evidence indicates that what is learned and how it is learned may be features of the context in which it is learned. In fact, researchers have examined the differential effects of various contexts on students' achievement from multiple perspectives: rural versus urban or suburban (Fan and Chen 1999; Lee and McIntire 2000), high poverty versus low poverty (Kozol 1991; Tajalli and Opheim 2005), geographic location (Cooney 1998), and class size (Finn and Achilles 1999; Nye, Hedges, and Konstantopoulous 2000). Results from those studies suggest that the characteristics of the school and its instructional program influence students' learning opportunities and outcomes. In addition, results from those studies suggest that equitable access to education is not simply a matter of expectations and standards; instead, it may be a matter of how those standards and expectations are addressed in particular settings.

Several researchers have identified notable features of high-performing schools in general and those that have been found to be particularly effective for "disadvantaged" student populations in particular. High-performing schools were characterized by a positive learning culture that supports high expectations for both students and teachers, the use of challenging curricula, and the use of appropriate instructional approaches. Specifically, schools that were considered high performing[1]—

- held high expectations for students and teachers,
- provided a professional atmosphere for teachers and staff,
- exposed students to challenging curriculum,
- maintained a learner-centered environment,
- encouraged students to be responsible for their own learning and promoted intellectual curiosity,
- supported teacher collaboration,
- used a variety of teaching strategies,
- employed highly qualified professional teachers who kept current in their field,
- embraced collaborative governance and leadership, and
- valued and embraced school-family-community connections.

Those findings suggest that students' learning outcomes are not solely dependent on occurrences in individual classrooms. Instead, they are dependent on the collective goals, visions, and actions of a school faculty that is working toward enhancing students' learning experiences.

Opportunity to Learn

Instructional programs are typically guided by a set of standards (school district, state, national, or international) that identify what students should know and be able to do. Students' achievement of those outcomes is typically measured as part of an assessment (classroom, high-stakes, national, or international). Within any learning environment, what one learns—the link between the standards and the assessment—is a feature of the opportunity one is given to learn. Typically, *opportunity to learn* (OTL) is defined as the amount of time the learner is afforded to learn specific tasks (McDonnel 1995). However, as a measure of equity, OTL also refers to equitable access to high-quality educational opportunities (Guiton and Oakes 1995; O'Day and Smith 1992; Schmidt and McKnight 1995). In mathematics, for example, any expectation that students be proficient at writing mathematical proofs would be inappropriate if they had not had the opportunity to learn the proof-writing process or had not enrolled in courses in which proof writing was emphasized. At issue here are whether the learners are given ample opportunities to learn what is expected in the intended curriculum and whether they are provided the experiences that would allow them to use that knowledge in productive ways.

1. The items reported in the list were taken from the following sources: Bell 2001; Craig et al. 2005; Henchy 2001; Shannon and Bylsma 2003.

Tate (2005) argues that students' OTL is a factor of *time*, *quality*, and *design*. *Time* refers to actual instructional time that is allotted to learning. A related construct is time-on-task—the amount of time that the learner is engaged in the learning process. The amount of time given to the subject matter influences both the amount of content covered and the time students have to engage with the subject matter. Regarding mathematics, the number and type of mathematics courses taken and the assessment practices or policies used also influence students' OTL. Opportunities to take high-level mathematics courses allow students to engage with and learn a broader range of mathematics content. However, simply assigning them to such courses is not enough. Systems must be in place that acknowledge students' prior knowledge, provide support to fill gaps in their knowledge, and continue to consider ways to enhance their learning opportunities.

Another factor that influences time is assessment. Assessments, particularly the quantity of assessments, may result in a limited curriculum that focuses only on those topics that will be tested or may reduce instructional time to allow more time for taking tests.

Quality refers to the characteristics of the instructional program that students experience. It is defined as the "classroom pedagogical strategies that affect students' achievement in school mathematics" (Tate 2005, p. 19). Those characteristics include the quality of the instruction and the curriculum that students experience. *Time* and *quality* are considered malleable features of the learning environment that can be addressed to focus primarily on improved learning outcomes and that can be altered with intervention. Tate uses the term *design* to describe active engagement of educators in enhancing students' OTL. Design, according to Tate, consists of "innovative portfolios of strategies that will provide students appropriate content exposure, content coverage, content emphasis, quality instructional delivery" (p. 15). He asserts that a focus on time and quality is useful only if educators consider a coherent "design" that uses what is known about OTL to improve learning outcomes. This focus on design is essential if students' learning of mathematics is to improve.

Mathematics Instructional Programs

Foundational features of the instructional program directly influence how students learn. For example, the assumptions made about how one learns, the expectation for learning, and the tools (e.g., technology) to facilitate learning all influence how students will interact with the material to be learned. In fact, the NCTM (2000) Learning Principle states, "the kinds of experiences teachers provide clearly play a major role in determining the extent and quality of students' learning" (p. 21). In this section, the three essential features of a successful instructional program will be examined. Those features include (*a*) the learning community, (*b*) the development of mathematics literacy, and (*c*) an inclusive environment that addresses the needs of a diverse student population.

Learning Community

Learning is no longer considered an activity that an individual does alone nor the end result of engaging in a particular activity. Instead, some would argue that all learning takes place within, and is supported by, a "community of practice" (Lave and Wenger 1991; Wenger 1999). Those communities of practice share norms (common ways of doing and approaching things), values (common ideas about what is important), and expectations (assumptions about what is important to know). In fact, Lave and Wenger assert that all learning is *situated* within a community of practice. They suggest that for learning to take place, (*a*) knowledge needs to be presented in an authentic context (i.e., settings and applications that would normally involve that knowledge) and (*b*) learning requires social interaction and collaboration. Those notions about learning are consistent with ideals espoused in NCTM *Standards* documents (1991, 2000).

Mathematics educators and researchers have articulated a vision for learning communities in mathematics classrooms. For example, the NCTM (2000) goals for school mathematics support the development of learning communities as part of mathematics classrooms. The discussion on the community of learners in mathematics classrooms focuses on the norms or expectations that help students understand the means by which they are to interact with the mathematics that they are learning. Specifically, mathematics students are encouraged to participate in collaborative environments as a means of negotiating meaning and making sense of mathematics.

Normative practices that are associated with classrooms that promote mathematics understandings include four features (Hiebert et al. 1997). The first feature is that ideas and methods for approaching tasks are valued and are discussed as a means to contribute to the learning of all. The second feature is that students are permitted and encouraged to function autonomously in selecting solution methods or strategies. Students are supported in that effort and are encouraged to share their methods with their peers. That feature helps students understand that a variety of approaches may be appropriate for solving problems. A third feature is that mistakes are analyzed and viewed as opportunities for further learning. The final feature is that correctness is based on the logic and structure of the subject rather than on the authority of the teacher or the social or academic status of particular students. Collectively, those four features send a clear message regarding what is valued in the mathematics classroom. If those features are present, students will come to appreciate and value the fact that they can learn from their peers and can contribute to the learning of their peers.

The Development of Mathematics Literacy

Although initially associated with reading and writing, current notions about literacy include all the different ways people communicate, exchange information, and represent meaning within and about various subject areas (e.g., scientific literacy, computer literacy, or mathematics literacy). Literacy facilitates an individual's ability to understand, learn, and function in varied settings. Each literacy, or sign system (Chandler 2001), has its own particular elements, or ways of making meaning. In fact, literacy in a field is defined by the nuances, vocabulary, and symbolism of that field. Such content-specific literacy acknowledges that the means used for communicating, interpreting, and understanding subject matter within a field is specific to that field and requires specialized training to develop proficiency and fluency. Indeed, the abilities to listen, think, interpret, and speak mathematically and move across sign systems may affect students' access to the subject matter and have an effect on what they are able to learn. Accordingly, mathematics instruction should focus on the development of students' mathematical literacy. In its original *Standards* document, NCTM (1989, p. 5) emphasized the following about mathematics literacy:

> Educational goals for students must reflect the importance of mathematical literacy. Toward this end, the K–12 standards articulate five goals for all students: (1) that they learn to value mathematics, (2) that they become confident in their ability to do mathematics, (3) that they become mathematical problem solvers, (4) that they learn to communicate mathematically, and (5) that they learn to reason mathematically.

The document also adds that being mathematically literate "denotes an individual's abilities to explore, conjecture, and reason logically, as well as the ability to use a variety of mathematical methods effectively to solve nonroutine problems" (p. 5). How might the mathematics learning environments support the development of mathematics literacy? Building on recommendations made in *Principles and Standards for School Mathematics* (NCTM 2000), mathematics instructional environments should encourage the development of students' mathematics literacy by encouraging them to do the following:

- *Think and reason* mathematically by solving routine and nonroutine problems. The focus here is on building students' critical and higher-order thinking skills as they engage in the exploration and investigation of mathematics that requires them to make and test conjectures, form conclusions, and support claims that are made.
- *Communicate* mathematically by developing an understanding of the language of mathematics. This skill includes the ability to listen, read, speak, and write mathematically by generating solutions to problems using appropriate mathematics symbols and words.
- *Use* various *representations* to indicate and *connect* their understanding of mathematics. Such representations include the various methods and approaches that can be used to illustrate what students know and are able to do, such as diagrams, graphical displays, gestures, and symbolic expressions.

Generally speaking, mathematics literacy can be characterized as the use of oral or written language to make sense of mathematics and to communicate, solve problems, and engage in discussions and decision making.

Researchers have identified differences in the learning environments of high-performing and low-performing mathematics programs (Cooney 1998; North Central Regional Educational Laboratory 2000). A collection of those differences is reported in table 5.1. Clearly, learning environments that support quality mathematics programs are vested in developing and enhancing students' opportunities to learn and in developing students' mathematical literacy. Developing mathematics literacy requires establishing classroom norms and expectations that support and extend students' mathematics understandings as mentioned in the previous section.

Meeting the Needs of a Diverse Student Population

Today's schools are a microcosm of a society that is increasingly diverse. The diverse nature of a school or a classroom can be reflected by race, ethnicity, language proficiency, special needs, socioeconomic status, and gender. Clearly, the diversity of the student population results in dynamic school and classroom environments that are likely to produce some challenges. How do teachers address those challenges? Specifically, how do teachers address the requirements of a challenging curriculum while addressing the needs of *all* students? Each teacher's background is limited to the experiences he or she has had with different cultures, backgrounds, and populations. How does a teacher begin to understand, address, and use the cultural experiences of students with different backgrounds as springboards to enhance students' learning opportunities? How do teachers teach students who may not speak the same language as they do?

Issues related to the diversity of the student population are complex. Despite those complexities, NCTM's Equity Principle articulates a need to establish mathematics learning environments that meet the needs of *all* students. Those environments are expected to have high expectations for learning for *all* students, provide accommodations to promote access and attainment for *all* students, and furnish resources and support for *all* students. In short, the emphasis on *mathematics for all* requires learning environments that acknowledge and respect students' cultural[2] strengths and uses those strengths as part of instruction.

A recommended approach for addressing the needs of *all* students is the

---

2. Culture as used here is inclusive of various characteristics that describe or relate an individual's membership in a particular group. Culture is determined by a set of common customs, traditions, beliefs, behaviors, and practices. It is a lens through which a group interprets and relates to the world.

Table 5.1
Differences between Low-Performing and High-Performing Mathematics Programs

Low-Performing Mathematics Programs	High-Performing Mathematics Programs
• Tend to be teacher-centered (e.g., teacher telling, teacher-directed tasks) • Do not encourage the use of collaborative learning • Do not assign project work • Focus less on the problem-solving process • Suggest that mathematics is about memorizing facts • Provide students fewer opportunities to apply what they are learning • Require students to take more tests and take tests more often • Engage in fewer discussions of the underlying mathematics • Do not support the needs of low-performing students • Do not require students to explain their thinking or reasoning	• Use a program that has an easily followed format with consistent review of previously learned skills • Encourage students to make sense of mathematics • Engage students in mathematical activity (e.g., the use of manipulatives) to build meaning • Encourage the use of students' strategies • Link mathematics language with symbols, notations, and discourse • Encourage students to think deeply about the content they are studying • Work to make mathematics meaningful by establishing connections within and between mathematics domains • Embed mathematics in the lives of students • Encourage student-to-student interactions in addition to teacher-to-student interactions • Encourage students to explain and justify their reasoning

use of *culturally responsive teaching.* According to Gay (2000, p. 29), culturally responsive teaching of any subject matter has the following characteristics:

- It acknowledges the legitimacy of the cultural heritages of different ethnic groups, both as legacies that affect students' dispositions, attitudes, and approaches to learning and as worthy content to be taught in the formal curriculum.
- It builds bridges of meaningfulness between home and school experiences as well as between academic abstractions and lived sociocultural realities.
- It uses a wide variety of instructional strategies that are connected to different learning styles.

- It teaches students to know and praise their own and each other's cultural heritages.
- It incorporates multicultural information, resources, and materials in all the subjects and skills routinely taught in schools.

Gay (2000) asserts that culturally responsive teaching is a viable approach for addressing the educational needs of various student groups. Klump and McNeir (2005) provide a summary of research findings related to culturally responsive practices that respond to the needs of a diverse student population. They report the following common characteristics of the use of culturally relevant practices that have resulted in student success:

- A climate of caring, respect, and valuing of students' culture is fostered in the school and the classroom.
- Bridges are built between academic learning and students' existing understanding, knowledge, native language, and values.
- Educators learn from and about their students' cultures, language, and learning styles to make instruction more meaningful and relevant to their students' lives.
- Local knowledge, language, and culture are fully integrated into the curriculum and not added onto it.
- Staff members hold students to high standards and have high expectations for all.
- Effective classroom practices are challenging, cooperative, and hands-on, with less emphasis on rote memorization and lecture formats.
- Staff members build trust and partnerships with families, especially with families marginalized by schools in the past.

In sum, learning environments that are intended to meet the needs of all students must use approaches (e.g., culturally responsive teaching) that acknowledge, respect, and embrace students' backgrounds and experiences while ensuring that all students have equitable access to mathematics classes and instruction that are designed to engage them and offer them opportunities to learn.

Conclusion

If mathematics is to be seen as a pump that provides access to academic and economic opportunity rather than a filter that limits them, then many issues related to student learning need to be addressed. This article identifies the learning environment as one possible variable in the equation. To enhance students' learning opportunities in general, and in mathematics classrooms in particular, teachers need to take an active role in advocating for and developing the types of learning environments that have been shown to enhance students' learning outcomes. This recommendation is made with the recognition that the school administrator, the professional school culture, and the policy climate all influence the learning environment. However, in this section, I focus on how teachers can contribute to the formation of appropriate learning environments for students. All educators must realize that they can be change agents that play a role in shaping students' outcomes and experiences. To that end, teachers are encouraged to exert the influence they have to support the development of appropriate learning environments.

Below I list several actions that teachers can take to encourage the establishment of equitable mathematics learning environments for all students:

- Augment student learning by enhancing the classroom learning environment

Given the current political climate that emphasizes testing and meeting markers on particular assessments, we can easily lose sight of the overall mission of education—the development of a productive citizenry. As mathematics educators, we must consider the broader impact of the types of instruction and experiences that we provide students. Because of this influence, each mathematics classroom should be considered a venue for helping students acquire and use those skills that will enhance not only their mathematical understandings but also their ability to function in a vastly changing society that has access to more information and that can access information more readily than ever before. Specifically, students should be prepared to function in a society that demands that they be able to access, use, and make decisions based on a vast amount of information. Teachers of mathematics can help students develop those skills by establishing learning environments that support students' engagement in discourse, debate, and decision making.

When addressing mathematics topics, students should have opportunities to experience mathematics in context, interpret results of their mathematical thinking, and reflect on its use. Such experiences should be designed to help students realize that they are competent thinkers who can contribute to problem solutions, articulate their reasoning, and build new knowledge. Students who engage in such environments will realize that gaining knowledge is not a function of applying what an instructor has told them; instead, knowledge is generated by engaging in conversations about ideas and reconciling old information and ideas with new ones.

- Be an advocate for the development of appropriate learning environments

In the previous section, the call to action was related to any activity over which teachers have direct control—offerings in their own classroom. Here I call on teachers to move beyond their individual classroom to be an advocate for students and students' learning in the broader educational community. To influence the learning environment in schools and beyond, teachers must help others (other teachers, parents, school administrators, policymakers, and the like) understand what is to be gained by enhancing the learning environment and students' learning opportunities. One way to do so is to take every opportunity to showcase students' mathematical knowledge. To be successful in creating appropriate learning environments as discussed in *Principles and Standards for School Mathematics* (NCTM 2000), producing positive student outcomes, and developing students' mathematics literacy, teachers must help others realize what is possible when students are given appropriate opportunities. In my experience as a teacher educator, helping prospective and practicing teachers understand what is possible is often difficult if they are not given opportunities to "see it" themselves and to "see it working with students" (see, e.g., Kellogg and Kersaint 2004). I therefore encourage teachers to be advocates by providing evidence that others can use to begin to consider alternatives if the current environment is not effective. That evidence should illustrate students experiencing, interpreting, and reflecting on mathematics in context. It should also highlight student products, particularly those from students who are not typically expected to engage in such activities.

Several approaches can be used to foster such environments and their corresponding outcomes. As a first step, teachers can begin to have conversations about positive learning outcomes, learning environments, and student experiences. Through conversations with others, teachers reveal strategies used to establish appropriate learning environments, discuss early challenges that have resulted in positive changes, talk about students' development over time, and highlight the development of students' mathematical literacy. Such conversations serve as a means to introduce this topic to many and to help them begin to consider the influence of the established learning environment.

Another approach would be to engage in activities that formally highlight and display students' mathematics knowledge. For example, while modeling instruction in a twenty-nine-year veteran teacher's classroom who taught traditionally (talk and chalk), I demonstrated ways to engage students in problem solving. That teacher was concerned that students would not be able to solve problems unless they were first told the appropriate method to use. To elicit evidence of what was possible, in a demonstration lesson I invited students to use any method of their own design to solve a problem that I posed. Students were allowed to work in groups and attempt different approaches. During the lesson, I pointed out evidence of students' thinking and showed how students' reasoning can be accessed by using appropriate questions. As an outcome of that process, the teacher realized that the students in her class were seldom given an opportunity to really "show" what they know. As a result of participating in this and other demonstration lessons, the teacher learned that some students identified as poor achievers were able to thrive in the new environment. Although my initial conversations with this teacher were met with reluctance, once she saw evidence with her own students, she began to consider the possible benefits of a different approach.

Teachers can also invite others (other teachers, parents, and administrators) to visit their classrooms to observe how their students are working together to solve problems, the type of knowledge that they display, and the innovative approaches that students experience. An important component of such visits is helping the visitor interpret the interactions that are taking place. For example, if an administrator is visiting a class that appears to be louder than usual, the teacher may explain that as part of the lesson, students are working together to address a particular mathematics challenge. The teacher can then showcase students' knowledge or reasoning by asking students questions that reveal that they are in fact learning robust mathematics. This added information will help observers understand that they must consider the broader outcomes of such engagement (or "loudness") rather than look at superficial features of the classroom environment.

If educators want others to have different perceptions of the work that they do, then they must take actions to improve or enhance the current learning environment and must help others understand what is possible when those environments exist. By demonstrating mathematics knowledge, teachers and their students stand the best chance of influencing the beliefs, attitudes, and values of people who are setting the agenda for mathematics education. The actions recommended here reflect only a fraction of what teachers can do to influence the learning environment and experiences that students are provided. Only by engaging in such actions will we be able to make a difference.

REFERENCES

Bell, Jennifer A. "High-Performing, High-Poverty Schools." *Leadership* 31 (September–October, 2001): 8–11.

Chandler, Daniel. *Semiotics for Beginners.* London: Routledge, 2001. Retrieved on December 23, 2005, from www.aber.ac.uk/media/Documents/S4B/sem08.html.

Cooney, Sandra. *Education's Weak Link: Student Performance in the Middle Grades.* Atlanta, Ga.: Southern Regional Education Board, 1998.

Craig, Jim, Aaron Butler, Leslie Cairo III, Chandra Wood, Christy Gilchrist, Joe Holloway, Sheneka Williams, and Steve Moats. *A Case Study of Six High-Performing Schools in Tennessee.* Charleston, W.Va.: Appalachia Educational Lab, 2005. (ERIC Document Reproduction Service No. ED481122)

Fan, Xitao, and Michael J. Chen. "Academic Achievement of Rural School Students: A Multi-Year Comparison with Their Peers in Suburban and Urban Schools." *Journal of Research in Rural Education* 15 (Spring 1999): 31–46.

Finn, Jeremy, and Charles M. Achilles. "Tennessee's Class Size Study: Findings, Implications, Misconceptions." *Educational Evaluation and Policy Analysis* 21 (Summer 1999): 97–109.

Gay, Geneva. *Culturally Responsive Teaching: Theory, Research, and Practice.* New York: Teachers College Press, 2000.

Guiton, Gretchen, and Jeannie Oakes. "Opportunity to Learn and Conceptions of Educational Equality." *Educational Evaluation and Policy Analysis* 17 (Autumn 1995): 323–36.

Henchy, Norman. *Schools That Make a Difference: Final Report.* Kelowna, British Columbia: Society for the Advancement of Excellence in Education, 2001. (ERIC Document Reproduction Service No. ED467058)

Hiebert, James, Thomas P. Carpenter, Elizabeth Fennema, Karen C. Fuson, Diane Wearne, Hanlie Murray, Alwyn Olivier, and Piet Human. *Making Sense: Teaching and Learning Mathematics with Understanding.* Portsmouth, N.H.: Heinemann, 1997.

Kellogg, Matthew, and Gladis Kersaint. "Creating a Vision for the Standards Using Online Videos in an Elementary Mathematics Methods Course." *Contemporary Issues in Technology and Teacher Education* 4 (2004) [Online Serial]: 23–34. www.citejournal.org/vol4/iss1/mathematics/article1.cfm.

Klump, Jennifer, and Gwen McNeir. *Culturally Responsive Practices for Student Success: A Regional Sampler.* Portland, Ore.: Northwest Regional Educational Laboratory, June 2005.

Kozol, Jonathan. *Savage Inequalities: Children in America's Schools.* New York: Crown Publishers, 1991.

Lave, Jean, and Etienne Wenger. *Situated Learning: Legitimate and Peripheral Participation.* Cambridge: University of Cambridge Press, 1991.

Lee, Jaekyung, and Walter G. McIntire. "Interstate Variation in the Mathematics Achievement of Rural and Nonrural Students." *Journal of Research in Rural Education* 16 (Winter 2000): 168–81.

McDonnel, Lorraine M. "Opportunity to Learn as a Research Concept and Policy Instrument." *Educational Evaluation and Policy Analysis* 17 (Autumn 1995): 305–22.

National Council of Teachers of Mathematics (NCTM). *Curriculum and Evaluation Standards for School Mathematics.* Reston, Va.: NCTM, 1989.

———. *Professional Standards for Teaching Mathematics.* Reston, Va.: NCTM, 1991.

———. *Principles and Standards for School Mathematics.* Reston, Va.: NCTM, 2000.

North Central Regional Educational Laboratory. "A Study of the Differences between Higher- and Lower-Performing Indiana Schools in Reading and Mathematics." Naperville, Ill.: North Central Regional Educational Laboratory, February 2000.

Nye, Barbara, Larry V. Hedges, and Spyros Konstantopoulous. "The Effects of Small Classes on Academic Achievement: The Results of the Tennessee Class Size Experiment." *American Educational Research Journal* 37 (Spring 2000): 123–53.

O'Day, Jennifer A., and Marshall Smith. "Systemic Reform and Educational Opportunity." In *Designing Coherent Education Policy,* edited by Susan Furhman, pp. 250–312. San Francisco: Jossey-Bass, 1992.

Schmidt, William D., and Curtis C. McKnight. "Surveying Educational Opportunity in Mathematics and Science: An International Perspective." *Educational Evaluation and Policy Analysis* 17 (Autumn 1995): 337–54.

Shannon, G. Sue, and Pete Bylsma. "Nine Characteristics of High Performing Schools." Olympia, Wash.: Office of Superintendent of Public Instruction, 2005. Retrieved on December 23, 2005, from www.k12.wa.us/research.

Tajalli, Hassan, and Cynthia Opheim. "Strategies for Closing the Gap: Predicting Student Performance in Economically Disadvantaged Schools." *Educational Research Quarterly* 28 (2005): 44–54.

Tate, William. "Access and Opportunities to Learn Are Not Accidents: Engineering Mathematical Progress in Your School." Southeast Eisenhower Regional Consortium for Mathematics and Science at SERVE (University of North Carolina at Greensboro), 2005. (Available at www.serve.org)

Wenger, Etienne. *Communities of Practice: Learning, Meaning and Identity.* Cambridge: Cambridge University Press, 1999.

6

Examining School Mathematics through the Lenses of Learning and Equity

Celia Rousseau Anderson

> Imagine a classroom, a school, or a district where all students have access to high quality, engaging mathematics instruction. There are ambitious expectations for all, with accommodation for those who need it. Knowledgeable teachers have adequate resources to support their work and are continually growing as professionals. The curriculum is mathematically rich, offering students opportunities to learn important mathematical concepts and procedures with understanding.
>
> —*Principles and Standards for School Mathematics*

THE AUTHORS of *Principles and Standards for School Mathematics* (National Council of Teachers of Mathematics [NCTM] 2000) open the document with a vision—a mental picture of the end goal of mathematics education, as they see it. It is a powerful image in which "*all students* ... have the opportunity and support necessary to learn significant mathematics with depth and understanding" (p. 5; emphasis added). It is an exciting image in which all students are able to experience success in school mathematics. It is an image that involves the interrelated goals of equity and learning.

Yet, as the authors of *Principles and Standards* acknowledge, that vision is highly ambitious (NCTM 2000). Several indicators demonstrate how far we currently are from achieving those goals, particularly with respect to equity. Despite some gains, gaps in opportunity to learn and achievement persist (Oakes et al. 2000; Tate 2005; Tate and Rousseau 2002). According to the NCTM Research Committee (2005), equity in mathematics education has proved to be an elusive goal. "No simple solutions exist [to the problems of inequity]. The difficulties in improving the situation make it apparent that the issues are complex and resistant to easy solutions" (p. 95). Thus, in some ways, we have a vision of the ends that we seek without a clear picture of how to get there.

As the NCTM Research Committee (2005) has argued, achieving equity in mathematics education will require attention and effort on the part of all involved. They assert that researchers, for example, can play a crucial role in the search for equity through applying a "critical equity lens" to research in mathematics education. Such a lens offers a means to gain more understanding of the implications of research for equity. In this article, I build on the idea of a "critical equity lens" to highlight various considerations that are important for all those who seek to achieve the vision of mathematics learning described by the authors of *Principles and Standards for School Mathematics.*

The View through an Equity Lens

What is an equity lens? As the members of the NCTM Research Committee (2005) describe it, an equity lens, when applied to a research study or other educational situation, would highlight equity-related concerns or issues. The lens would delineate those conditions and bring them to the fore. Rather than fade into the background, where they have a chance (if not a likelihood) of remaining unnoticed, those concerns would become the focal point.

An important consideration, however, is to make clear exactly what aspects of school mathematics that lens would highlight. No single generally accepted definition for equity exists. Thus, I state at the outset the definition from which I am working, because it will determine the conditions on which I train my lens. The different conceptions of equity (Gutierrez 2002b; NCTM Research Committee 2005; Secada 1989; Tate and Rousseau 2002) are beyond the scope of this article to examine. For that reason, I look to the Equity Principle of *Principles and Standards* (NCTM 2000) for the framework to be used for the purposes of this article. Although the Equity Principle does not provide an explicit definition of the meaning of equity, it focuses specific attention on access and opportunities to learn. For example, the authors of *Principles and Standards* (NCTM 2000) assert that "all students, regardless of their personal characteristics, backgrounds, or physical challenges, must have opportunities to study—and support to learn—mathematics. Equity ... demands that reasonable and appropriate accommodations be made to promote access and attainment for all students" (p. 12). Following this introduction to the concept of equity, the authors cite the importance of access and opportunity to learn several more times in the text of the Principle (NCTM 2000). Thus, in an effort to be consistent with the view of equity reflected in *Principles and Standards,* I ground my discussion of equity in the issues of access and opportunity.

Equity and Learning: The View through Both Lenses

I teach the mathematics methods course for preservice elementary school teachers at my university. Each semester I begin the course with an activity related to the six Principles from *Principles and Standards* (NCTM 2000). I assign the students to groups, with each group responsible for a specific Principle.[1] After they discuss their assigned Principle and summarize what it means to them, they watch a video of a middle school classroom. As they view the video, one of their tasks is to look for evidence related to their Principle. They are to process what is happening in the lesson as a whole, but with a specific focus on the aspects of classroom interaction related to their Principle. Essentially, they watch the events of the classroom through a particular lens. Some are assigned to watch with an "equity" lens and others with a "learning" lens. During the discussion that follows, the groups have the opportunity to share the view through their assigned lens and to note the interactions or connections between those perspectives. For example, both the equity group and the learning group might highlight the same classroom practice or action of the teacher. They are then able to see how the same practice can serve dual purposes. Although the students begin the activity focused only on the view through a single lens, to see the whole picture with clarity, they must overlay a second lens.

The need to combine lenses applies beyond that introductory activity to the examination of all aspects of mathematics education. The application of only one lens is problematic because the potential exists for the picture to become distorted. For example, we can view classrooms through only a learning lens. With that lens, we ask a particular set of questions: Are students learning mathematics with understanding? Are they connecting new knowledge with existing knowledge? Can they apply what they are learning in a variety of situations? If we view the same class through an equity lens, we ask a potentially different set of questions: Who has access to the learning that is occurring? Are all students able to participate in the learning process? Who has access to the resources that support learning? Our understanding of the situation is more complete when we begin to combine those questions to examine learning and equity in conjunction.

We have several examples in which the learning lens has been applied to situations in the absence of the equity lens. However, when the equity lens was superimposed, the picture became clearer. Members of the NCTM Research Committee (2005) cite the example of the research involving Cognitively Guided Instruction (CGI). The CGI studies provided evidence that teachers can use knowledge extracted from studies of learners' thinking to strategically influence students' learning (Carpenter et al. 1988; Carpenter et al. 1989). Thus, the picture, as seen through the learning lens, appeared in a very positive light. However, when the CGI researchers overlaid an equity lens and examined more closely the strategies of boys and girls, they found gender-related differences in the abstractness of the strategies used (Fennema et al. 1998). The application of the equity lens shifted the image to reveal a slightly different picture, at

1. I first experienced this activity in a professional development session led by Linda McQuillan of the Madison Metropolitan School District..

least with respect to the learning of the girls in those classes. Similarly, Secada (1996) applied an equity lens to a classroom that is often viewed as an exemplar in the use of student discourse to promote learning. Viewed through a learning lens, the picture of that classroom appeared one way. When the equity lens was applied, a different image emerged. Secada's application of the equity lens revealed that limited English proficiency was a potentially marginalizing factor in the classroom, excluding students from activities intended to support learning. Thus, situations in school mathematics should not be viewed exclusively through a learning lens. An equity lens must also be superimposed to clarify the picture with respect to the learning of *all* students.

Conversely, the importance of the learning lens is highlighted in an examination of achievement trends. Reviews by Secada (1992) and by Tate (1997) have noted increased achievement by all student groups and a reduction in the achievement gaps on national assessments over the past two decades. Through an equity lens, that trend would appear to be positive. However, an overlay of the learning lens qualifies that progress. As both Secada and Tate note, the growth and reduction of the achievement gaps have occurred largely on basic skills assessment items. Viewed through a learning lens, success on basic skills is an insufficient indicator of understanding; thus the picture is not as rosy as it might initially have appeared. As we imagine those lenses, then, we must keep in mind that they cannot operate in isolation. The consistent application of both lenses is necessary not only for researchers but for all those involved in mathematics education.

Seeing the Whole Picture

Another characteristic of a critical equity lens, as I imagine it, is the necessity to apply that lens at different levels. In other words, the lens must have the capacity not only to zoom in but also to zoom out to consider broader policy-level issues and conditions. That capability is particularly important as we conceptualize equity with respect to learning.

Research related to students' learning in mathematics is often framed within a psychological paradigm and focused on individual students' cognition (Tate and Rousseau 2002). However, according to Secada (1991), a focus on the individual is problematic from an equity standpoint. "As the external and social worlds get transformed into the inner and personal worlds, external and social issues are transformed into internal and personal states" (p. 22). Secada (1993) argued, in fact, that the psychological focus on the individual fails to adequately explain the nature of student learning, in part because external factors that shape student learning fall outside the view of a lens focused strictly on the individual. Secada asserted that a full picture would require attention to larger social and cultural forces. To get a clear picture, we must shift the focus beyond the individual student to the larger spheres in which the student's learning is situated.

Classroom Practices

Although mathematics education has traditionally maintained cognitive focus on the individual, more recent scholarship has begun to attend to the social dimension of learning (e.g., Carpenter et al. 1999; Cobb and Yackel 1996). Work attending to the social dimension of learning has generally broadened the focus from the individual to include the classroom practices that influence students' learning. Attention to social processes is also embedded in the recommendations included in *Principles and Standards* (NCTM 2000). For example, the Learning Principle explicitly names classroom discourse and social interaction as means of promoting students' understanding. Thus, a learning lens, applied at the classroom level, would likely take into consideration the role of discourse in students' learning.

Yet we must also examine interactions at the classroom level through an equity lens. Secada and Berman (1999) suggest, for example, that the classroom norm of having students articulate their thinking is of potential significance from an equity standpoint. They note that the expectations for communication can put some students at a disadvantage. Similarly, Cobb (2001) argues that "students' home communities can involve differing norms of participation, language, and communication, some of which might actually be in conflict with those that the teacher seeks to establish in the classroom" (p. 471).

Such a scenario prevailed in a study by Murrell (1994) of middle school mathematics classrooms. In his study, he focused on the mathematics learning experiences of African American males. He characterized the classes as implementing the kind of discourse, or "math talk," advocated in the NCTM (1989, 1991) *Standards* documents. Murrell found that those communication expectations were not benefiting African American males in the ways intended. He notes that the students did not interpret math talk in a manner consistent with the goals of the teacher. As teachers engaged in more math talk as a means of exploring and elaborating mathematics principles and concepts, the students did not regard the discussion as an increased focus on mathematical learning. Rather, they tended to regard the emphasis on math talk simply as a new regimen to be mastered to meet their teachers' requirements. Murrell goes on to suggest that this mismatch between the students' views and the teacher's intentions would likely have occurred not only for African American males but for any student who "has not been socialized into the mathematics discourse forms assumed by mainstream teachers" (p. 565). Lubienski (1996) found similar results in her study of the impact of classroom discussion on her students' learning experiences. In particular, Lubienski noted that the higher-SES (socioeconomic status) students in her class shared more of her expectations for discussion, seeming to share her beliefs about the purpose of discussion as an arena to "create, share, analyze, and validate mathematical ideas" (p. 185). The lower-SES students were more apt to focus on right and wrong answers and to limit their participation to those instances in which they could share answers rather than ideas.

Lubienski (1996) suggested that such differences between the teacher's intentions and students' perceptions are a manifestation of "cultural confusion." She asserted that the changes in the curriculum and classroom practices advocated by NCTM are cultural in nature. "They involve central beliefs and norms regarding ways of knowing and communicating" (p. 246). However, that culture is more closely aligned with that of the middle class. Zevenbergen (2000) makes a similar assertion, arguing that success in the mathematics classroom requires cultural knowledge that can serve to privilege the middle class or those who are in the linguistic majority. Thus, an equity lens reveals that classroom practices intended to promote understanding have the potential to maintain, rather than eliminate, learning differences.

An important point to note, however, is that the potentially negative impact of "cultural confusion" can be mitigated when teachers make explicit the cultural knowledge and expectations of the classroom and attend to any potential factors that might restrict student participation. For example, Boaler (2002) examined the classrooms of two different reform programs in which access to mathematics learning appeared to be equitably distributed. She found that teachers in those classrooms focused attention on teaching students *how* to explain and justify their thinking. Her findings suggest that such explicit attention to the norms of participation can ensure that all students have access to the mathematical learning promoted by those practices. Similarly, Gutierrez (2002a) presented evidence that language differences need not be a constraint in the learning of mathematics. She notes that teachers can discover how language influences the specific participation patterns and learning of individual students. Such attention to the impact of language can allow the teacher to develop strategies that better support students' learning of mathematics.

Secada (1991) has argued that a cognitive focus on student learning, in the absence of considerations of culture or language, can cast students' nonparticipation in the classroom in a negative light, characterizing the student as deficient. In other words, a failure to apply an equity lens can lead to a distorted picture of classroom learning experiences. However, with an equity lens, students' participation can be understood within the broader context of the cultural and linguistic norms that students bring to the classroom. Moreover, working with those larger influences in mind, teachers can create classrooms that promote access for all students (Boaler 2002).

Knowledge Policies

However, student learning is shaped by more than the interactions in the classroom. Thus, the equity lens must be able to take in a wide-angle picture that includes broader impacts on students' opportunity to learn with understanding. For example, even as we examine mathematics learning in classrooms, we must acknowledge that students' opportunities to learn with understanding are not equitably distributed across those classrooms. Some of those inequities relate to what Tate and Johnson (1999) refer to as "knowledge policies," which include "the folkways of schooling, regulations of policymakers, and practices of teachers that influence student opportunity to learn" (p. 222).

Tracking is one example of a knowledge policy with the potential to affect students' opportunity to learn. In fact, multiple studies have indicated that students' learning opportunities are constrained by tracking practices. According to Darling-Hammond (1997), "students in lower tracks receive a much less challenging curriculum that accounts for more of the disparities between what they learn and other students learn than does their entering ability level" (p. 127). In mathematics, research indicates that students in high-track classes have access to "high-status" knowledge (ideas and concepts), whereas students in low-track classes repeat the same basic computational skills year after year (Oakes 1985, 1995). Students in low-track classes are significantly more likely than those in high-track classes to spend time each week doing worksheet problems, and students in low-track classes are less likely to be asked to write their reasoning about solving a mathematics problem (Weiss 1994). Thus, through a learning lens, students in low-track classes do not have the opportunities to learn "significant mathematics with depth and understanding" (NCTM 2000).

Viewed through an equity lens, the picture becomes even bleaker, as we observe that minority and low-income students are more likely to experience such inequities in opportunities to learn. For example, Oakes (1995) examined the tracking policies of two school districts and demonstrated that the track placements in both districts were racially skewed. African American and Latino students were much less likely than white and Asian students with the *same* test scores to be placed in high-track classes. She noted two mechanisms that appear to support those disparities: (1) differences in parents' knowledge of their power to intervene in the placement process and (2) teacher and counselor recommendations that included both "objective" data and more subjective judgments about behavior, personality, and attitudes. She concluded from those findings that "grouping practices have created a cycle of restricted opportunities and diminished outcomes, and exacerbated the differences between African American and Latino and White students" (p. 689). As a result of those practices, minority students within racially mixed schools continue to be disproportionately overrepresented in low-track mathematics courses (Oakes et al. 2000). Thus, although tracking is a knowledge policy with the potential to constrain the learning opportunities of all students placed into low-track classes, the impact is felt more profoundly by certain student groups.

Another knowledge policy described by Tate and Johnson (1999) is teacher quality. According to Darling-Hammond (2000), "substantial evidence from prior reform efforts indicates that changes in course taking, curriculum content, testing or textbooks make little difference if teachers do not know how to use these tools well and how to diagnose their students' learning needs" (p. 37). Without teacher quality, curriculum and instructional strategies have little chance of being effectively implemented to support student learning with understanding. Several studies of mathematics reform have demonstrated the importance of teacher knowledge of both content and pedagogy for the effective use of strategies and materials associated with student learning with understanding (Borko et al. 1992; Clarke 1995; Clarke 1993; Eisenhart et al. 1993; Johnson 1995; Manouchehri 1998; Manouchehri and Goodman 1998; Stein, Baxter, and Leinhardt 1990). Moreover, the importance of teacher quality for students' learning is further demonstrated through studies of student achievement. For example, in an analysis of data from the National Assessment of Educational Progress, Darling-Hammond (2000) found that teacher certification status and holding a degree in the field being taught were significantly and positively correlated with students' achievement. In fact, "the most consistent highly significant predictor of student achievement in reading and mathematics in each year tested is the proportion of well-qualified teachers in a state; those with full certification and a major in the field they teach" (p. 27). Similarly, Fetler (1999) analyzed the mathematics scores of high school students in California. He found that after controlling for student poverty, teacher experience and preparation were significantly related to student achievement. In particular, schools with larger numbers of teachers on emergency permits had lower average achievement scores. Thus, when viewed through a learning lens, the qualifications and knowledge of the teacher are clearly important.

Moreover, the impact of the teacher on students' learning must also be considered in conjunction with the inequitable distribution of teacher quality. In fact, Darling-Hammond (1997) argued that "perhaps the single greatest source of inequity in education is [the] disparity in the availability and distribution of well-qualified teachers" (p. 273). For example, Sanders and Rivers (1996) found that although African American and white students made comparable progress when assigned to teachers of comparable effectiveness, African American students were nearly twice as likely to be assigned to the most ineffective teachers. Similarly, although more than one-fourth of high school mathematics students are taught by out-of-field teachers, those proportions are highest in high-poverty schools and high-minority classes (Darling-Hammond 1997). In a national survey, Weiss (1994) found that more than half of high school mathematics classes with populations that are at least 40 percent minority are taught by teachers without a degree in the field. Ingersoll (1999) found similar results in high-poverty schools (schools in which 50 percent or more of the students receive free or reduced-price lunch). Whereas 27 percent of secondary mathematics teachers in low-poverty schools (schools in which 10 percent or fewer of the students receive free or reduced-price lunch) have neither a major nor a minor in the field, approximately 43 percent of teachers in high-poverty schools do not have at least the equivalent of a minor in the field. Certainly, such measures as certification status and degree in the field are but proxies for the knowledge and depth of understanding necessary to help students learn. However, the predictive power of those indicators with respect to student achievement and the disparities in their distribution highlight the significance of teacher quality to any discussion of equity and opportunity to learn in mathematics education. In fact, Darling-Hammond (1997) asserted that "much of the difference in school achievement among students is due to the effects of substantially different school opportunities and, in particular, greatly disparate access to high-quality teachers and teaching" (p. 270). Those differences in opportunity to learn are highlighted when an equity lens is applied to teacher quality.

Students' opportunities to learn mathematics are also shaped by the assessment policies of the school district, since assessment policies often influence the nature of content and pedagogy in the classroom. In particular, the influence of high-stakes standardized tests can lead teachers to curtail the amount of time given to the types of curriculum and instruction that are associated with learning with understanding (NCTM 2000). For example, Gutstein (1998) found that pressure to cover the material to be included on district-mandated standardized tests prevented middle school mathematics teachers from implementing the reform curriculum Mathematics in Context (MiC). "The district emphasis on these tests has a chilling effect on teachers' willingness to let go of the old (even if it did not work) and embrace the new (e.g., MiC)" (p. 22). Similar examples of the impact of standardized tests can be found throughout the literature on mathematics education (Ball 1990; Clarke 1994; Eisenhart et al. 1993; Knapp and Peterson 1995; Manouchehri 1998; Manouchehri and Goodman 1998; Putnam et al. 1992; Romberg 1997; Schifter and Fosnot 1993; Stein and Brown 1997; Webb and Romberg 1994; Wilson 1990). In general, the foregoing examples demonstrate that the emphasis of many standardized tests on computational skills makes teachers less likely to attempt to change their practices to focus more on conceptual understanding. Thus, a learning lens applied to assessment policies highlights the negative impact of high-stakes tests on what happens in the classroom.

An equity lens, however, reveals that the negative impact is felt more acutely in certain classrooms. The influence of high-stakes standardized tests is often greater in high-minority classrooms (Darling-Hammond 1994; Madaus 1994; Shepard 1991; Weiss 1994). For example, in a nationwide survey, teachers in high-minority classrooms reported using test-specific instructional practices more often than teachers in low-minority classrooms (Madaus et al. 1992). In high-minority classrooms, about 60 percent of the teachers reported teaching test-taking skills, teaching topics known to be on the test, increasing emphasis on tested topics, and starting test preparation more than a month before the examination. Those practices were reported significantly less often by teachers in low-minority classrooms. Moreover, mathematics teachers in high-minority classes indicated more pressure from school district officials to improve test scores than teachers in low-minority classes. The example of the impact of the Texas Assessment of Academic Skills (TAAS) also illustrates the effects that high-stakes tests can have on students' opportunity to learn. The results of surveys of Texas teachers indicate that they devote a substantial amount of instructional time to preparing students specifically for the TAAS, with most teachers reporting that they begin test preparation more than a month before the test (Haney 2000). According to McNeil and Valenzuela (2000), teachers report spending several hours a week drilling students on practice examinations. In that effort, commercial test-preparation materials become the de facto curriculum in many schools, reducing mathematics to sets of isolated skills. Teachers report that the time devoted to instructional activities that engage students in higher-order problem solving is severely reduced (or disappears completely) in the press to prepare students for the TAAS. Instructional focus on test preparation is unevenly distributed and more likely to exist in schools attended by low-income and minority students. Thus, an equity lens reveals the disproportionate impact of testing policies on the learning of certain student groups.

Conclusion

In this article, I have attempted to build on the argument made by the NCTM Research Committee (2005) of the need to apply a "critical equity lens" to mathematics education. I have used their metaphor as a means to highlight some of the important considerations with respect to learning and equity. I argue that realizing the vision outlined in *Principles and Standards* will require ongoing examination of mathematics education through the dual lenses of equity and learning. Moreover, I suggest that this examination must occur at the level of the classroom and at broader policy levels.

The overlay of an equity lens on the classroom reveals that classroom processes intended to promote students' learning with understanding do not necessarily contribute to the deeper learning of all students. However, other examples demonstrate that such differences in participation and learning can be overcome with greater attention to the factors that contribute to those differences (e.g., unfamiliarity with the expectations for communication). As Secada (1996) noted, unless we attend to the impact of those practices on all students, we risk the possibility that our efforts intended to promote student learning will, in fact, work against equity. We must be aware of the potential for differential impact. With such awareness and sensitivity, we are better positioned to create classrooms that promote the learning of all students. This goal requires that we simultaneously attend to issues of learning and equity in the classroom.

In addition to a focus on the classroom, the equity lens must also have the capacity to "zoom out" from the classroom to consider other factors that shape students' opportunities to learn mathematics with understanding. For mathematics education researchers, that perspective requires a blurring of the lines that we have traditionally imposed between classroom research and policy. To have a full understanding of equity concerns, we must be able to take in the wide-angle picture that includes the knowledge policies that affect students' learning. The same is true of classroom teachers and others involved in the day-to-day operation of schools. We must be careful not to fall into the trap outlined by Secada (1991) of attributing students' learning strictly to conditions within the individual or the individual's family. Rather, we must consider the big picture to include the student's prior mathematics learning experiences and the influences of the different knowledge policies on those experiences.

Finally, equity is not a stand-alone consideration that can be examined in isolation. If we seek to fulfill the vision outlined in *Principles and Standards for School Mathematics* (NCTM 2000), we must examine equity in conjunction with the other facets of school mathematics. We must consider learning, teaching, curriculum, technology, and assessment with the equity lens as an overlay. The members of the NCTM Research Committee (2005) note that "equity concerns exist in any educational endeavor, whether or not they are made explicit" (p. 95). If we treat the equity lens as something to be used only in isolation, only by a few people, or only with respect to certain learning situations, the vision of school mathematics outlined in *Principles and Standards* will remain nothing more than a mirage.

REFERENCES

Ball, Deborah. "Reflections and Deflections of Policy: The Case of Carol Turner." *Educational Evaluation and Policy Analysis* 12, no. 3 (1990): 263–75.

Boaler, Jo. "Learning from Teaching: Exploring the Relationship between Reform Curriculum and Equity." *Journal for Research in Mathematics Education* 33 (July 2002): 239–58.

Borko, Hilda, Margaret Eisenhart, Catherine A. Brown, Robert G. Underhill, Doug Jones, and Patricia C. Agard. "Learning to Teach Hard Mathematics: Do Novice Teachers and Their Instructors Give Up Too Easily?" *Journal for Research in Mathematics Education* 23 (May 1992): 194–222.

Carpenter, Thomas P., Elizabeth Fennema, Penelope L. Peterson, and Deborah Carey. "Teachers' Pedagogical Content Knowledge of Students' Problem Solving in Elementary Arithmetic." *Journal for Research in Mathematics Education* 19 (November 1988): 385–401.

Carpenter, Thomas, Elizabeth Fennema, Penelope L. Peterson, Chi-Pang Chiang, and Megan Loef. "Using Knowledge of Children's Mathematics Thinking in Classroom Teaching: An Experimental Study." *American Educational Research Journal* 26 (November 1989): 499–531.

Carpenter, Thomas, Cristina Gomez, Celia Rousseau, Olof Steinthorsdottir, Carrie Valentine, Lesley Wagner, and Peter Wiles. "An Analysis of Student Construction of Ratio and Proportion Understanding." Paper presented at the Annual Meeting of the American Educational Research Association, Montreal, Que., 1999.

Clarke, Barbara. "Expecting the Unexpected: Critical Incidents in the Mathematics Classroom." Unpublished doctoral dissertation, University of Wisconsin—Madison, 1995.

Clarke, David. "Influences on the Changing Role of the Mathematics Teacher." Unpublished doctoral dissertation, University of Wisconsin—Madison, 1993.

————. "Ten Key Principles from Research for the Professional Development of Mathematics Teachers." In *Professional Development for Teachers of Mathematics,* 1994 Yearbook, edited by Douglas B. Aichele, pp. 37–48. Reston, Va.: National Council of Teachers of Mathematics, 1994.

Cobb, Paul. "Supporting the Improvement of Learning and Teaching in Social and Institutional Context." In *Cognition and Instruction: Twenty-five Years of Progress,* edited by Sharon M. Carver and David Klahr. Mahwah, N.J.: Lawrence Erlbaum Associates, 2001.

Cobb, Paul, and Erna Yackel. "Constructivist, Emergent, and Sociocultural Perspectives in the Context of Developmental Research." *Educational Psychologist* 31 (1996): 175–90.

Darling-Hammond, Linda. "Performance-Based Assessment and Educational Equity." *Harvard Educational Review* 64, no. 1 (1994): 5–30.

————. *The Right to Learn: A Blueprint for Creating Schools That Work.* San Francisco: Jossey-Bass, 1997.

————. "Teacher Quality and Student Achievement: A Review of State Policy Evidence." *Education Policy Analysis Archives* 8, no. 1 (2000): 1–48.

Eisenhart, Margaret, Hilda Borko, Robert Underhill, Catherine Brown, Doug Jones, and Patricia Agard. "Conceptual Knowledge Falls through the Cracks: Complexities of Learning to Teach Mathematics for Understanding." *Journal for Research in Mathematics Education* 24 (January 1993): 8–40.

Fennema, Elizabeth, Thomas Carpenter, Victoria Jacobs, Megan Franke, and Linda Levi. "New Perspectives on Gender Differences in Mathematics: A Reprise." *Educational Researcher* 27, no. 5 (1998): 19–21.

Fetler, Mark. "High School Staff Characteristics and Mathematics Test Results." *Education Policy Analysis Archives* 79, no. 9 (1999): 1–22.

Gutierrez, Rochelle. "Beyond Essentialism: The Complexity of Language in Teaching Mathematics to Latina/o Students." *American Educational Research Journal* 39 (2002a): 1047–88.

———. "Enabling the Practice of Mathematics Teachers in Context: Toward a New Equity Research Agenda." *Mathematical Thinking and Learning* 4, no. 2–3 (2002b): 145–87.

Gutstein, Eric. "Lessons from Adopting and Adapting Mathematics in Context, a Standards-Based Mathematics Curriculum, in an Urban, Latino, Bilingual Middle School." Paper presented at the Annual Meeting of the American Educational Research Association, San Diego, Calif., 1998.

Haney, Walt. "The Myth of the Texas Miracle in Education." *Education Policy Analysis Archives* 8, no. 41 (2000).

Ingersoll, Richard. "The Problem of Underqualified Teachers in American Secondary Schools." *Educational Researcher* 28, no. 2 (1999): 26–37.

Johnson, Loren. "Extending the National Council of Teachers of Mathematics' Recognizing and Recording Reform in Mathematics Education: Documentation Project through Cross-Case Analysis." Unpublished doctoral dissertation, University of New Hampshire, 1995.

Knapp, Nancy, and Penelope Peterson. "Teachers' Interpretations of 'CGI' after Four Years: Meanings and Practices." *Journal for Research in Mathematics Education* 26 (January 1995): 40–65.

Lubienski, Sarah. "Mathematics for All? Examining Issues of Class in Mathematics Teaching and Learning." Unpublished doctoral dissertation, Michigan State University, 1996.

Madaus, George. "A Technological and Historical Consideration of Equity Issues Associated with Proposals to Change the Nation's Testing Policy." *Harvard Educational Review* 64, no. 1 (1994): 76–95.

Madaus, George, Mary West, Maryellen Harmon, Richard Lomax, and Katherine Viator. *The Influence of Testing on Teaching Math and Science in Grades 4–12.* Boston: Boston College, Center for the Study of Teaching, Evaluation, and Educational Policy, 1992.

Manouchehri, Azita. "Mathematics Curriculum Reform and Teachers: What Are the Dilemmas?" *Journal of Teacher Education* 49 (1998): 276–86.

Manouchehri, Azita, and Terry Goodman. "Mathematics Curriculum Reform and Teachers: Understanding the Connections." *Journal of Educational Research* 92, no. 1 (1998): 27–41.

McNeil, Linda, and Angela Valenzuela. "The Harmful Impact of the TAAS System of Testing in Texas: Beneath the Accountability Rhetoric." Cambridge, Mass.: Harvard Civil Rights Project, 2000.

Murrell, Peter. "In Search of Responsive Teaching for African American Males: An Investigation of Students' Experiences of Middle School Mathematics Curriculum." *Journal of Negro Education* 63, no. 4 (1994): 556–69.

National Council of Teachers of Mathematics (NCTM). *Curriculum and Evaluation Standards for School Mathematics.* Reston, Va.: NCTM, 1989.

———. *Professional Standards for Teaching Mathematics.* Reston, Va.: NCTM, 1991.

———. *Principles and Standards for School Mathematics.* Reston, Va.: NCTM, 2000.

NCTM Research Committee. "Equity in School Mathematics Education: How Can Research Contribute?" *Journal for Research in Mathematics Education* 36 (March 2005): 92–100.

Oakes, Jeannie. *Keeping Track: How Schools Structure Inequality.* New Haven, Conn.: Yale University Press, 1985.

———. "Two Cities' Tracking and Within-School Segregation." *Teachers College Record* 96 (1995): 681–90.

Oakes, Jeannie, Kate Muir, and Rebecca Joseph. "Coursetaking and Achievement in Mathematics and Science: Inequalities That Endure and Change." Madison, Wis.: National Institute of Science Education, 2000.

Putnam, Ralph, Ruth Heaton, Richard Prawat, and Janine Remillard. "Teaching Mathematics for Understanding: Discussing Case Studies of Four Fifth-Grade Teachers." *Elementary School Journal* 93 (1992): 213–28.

Romberg, Thomas. "Mathematics in Context: Impact on Teachers." In *Mathematics Teachers in Transition,* edited by Elizabeth Fennema and Barbara Scott Nelson, pp. 357–80. Mahwah, N.J.: Lawrence Erlbaum Associates, 1997.

Sanders, William, and June Rivers. *Cumulative and Residual Effects of Teachers on Future Student Academic Achievement.* Knoxville, Tenn.: University of Tennessee Value-Added Research and Assessment Center, 1996.

Schifter, Deborah, and Catherine Fosnot. *Reconstructing Mathematics Education: Stories of Teachers Meeting the Challenge of Reform.* New York: Teachers College Press, 1993.

Secada, Walter. "Educational Equity versus Equality of Education: An Alternative Conception." In *Equity and Education,* edited by Walter Secada, pp. 68–88. New York: Falmer Press, 1989.

———. "Diversity, Equity, and Cognitivist Research." In *Integrating Research on Teaching and Learning Mathematics,* edited by Elizabeth Fennema, Thomas Carpenter, and Susan Lamon, pp. 17–53. Albany, N.Y.: State University of New York Press, 1991.

———. "Race, Ethnicity, Social Class, Language, and Achievement in Mathematics." In *Handbook for Research in Mathematics Teaching and Learning,* edited by Douglas Grouws, pp. 623–60. New York: Macmillan Publishing Co., 1992.

———. "Equity and a Social Psychology of Mathematics Education." Paper presented at the Annual Meeting of the International Group for the Psychology of Mathematics Education, North American Chapter, San Jose, Calif., 1993.

———. "Urban Students Acquiring English and Learning Mathematics in the Context of Reform." *Urban Education* 30 (1996): 422–48.

Secada, Walter, and Patricia Berman. "Equity as a Value-Added Dimension in Teaching for Understanding in School Mathematics." In *Mathematics Classrooms That Promote Understanding,* edited by Elizabeth Fennema and Thomas Romberg, pp. 33–42. Mahwah, N.J.: Lawrence Erlbaum Associates, 1999.

Shepard, Lorrie. "Will National Tests Improve Student Learning?" *Phi Delta Kappan* 73, no. 3 (1991): 232–38.

Stein, Margaret, Juliet Baxter, and Gaea Leinhardt. "Subject-Matter Knowledge and Elementary Instruction: A Case from Functions and Graphing." *American Educational Research Journal* 27 (1990): 639–63.

Stein, Margaret, and Catherine Brown. "Teacher Learning in a Social Context: Integrating Collaborative and Institutional Processes with the Study of Teacher Change." In *Mathematics Teachers in Transition,* edited by Elizabeth Fennema and Barbara Scott Nelson, pp. 155–92. Mahwah, N.J.: Lawrence Erlbaum Associates, 1997.

Tate, William F., "Race-Ethnicity, SES, Gender, and Language Proficiency Trends in Mathematics: An Update." *Journal for Research in Mathematics Education* 28 (December 1997): 652–79.

———. "Access and Opportunities to Learn Are Not Accidents: Engineering Mathematics Progress in Your School." Southeast Eisenhower Regional Consortium for Mathematics and Science at SERVE (University of North Carolina at Greensboro), 2005. (Available at www.serve.org)

Tate, William, and Howard Johnson. "Mathematics Reasoning and Educational Policy: Moving beyond the Politics of Dead Language." In *Developing Mathematical Reasoning in Grades K–12,* 1999 Yearbook of the National Council of Teachers of Mathematics (NCTM), edited by Lee V. Stiff, pp. 221–33. Reston, Va.: NCTM, 1999.

Tate, William, and Celia Rousseau. "Access and Opportunity: The Political and Social Context of Mathematics Education." In *Handbook of International Research in Mathematics Education,* edited by Lyn English, pp. 271–99. Mahwah, N.J.: Lawrence Erlbaum Associates, 2002.

Webb, Norman, and Thomas Romberg. *Reforming Mathematics Education in America's Cities: The Urban Mathematics Collaborative Project.* New York: Teachers College Press, 1994.

Weiss, Iris. "A Profile of Science and Mathematics Education in the United States: 1993." Chapel Hill, N.C.: Horizon Research, 1994.

Wilson, Suzanne. "A Conflict of Interests: The Case of Mark Black." *Evaluation and Policy Analysis* 12 (1990): 309–26.

Zevenbergen, Robyn. "'Cracking the Code' of Mathematics Classrooms: School Success as a Function of Linguistic, Social, and Cultural Background." In *Multiple Perspectives on Mathematics Teaching and Learning,* edited by Jo Boaler, pp. 201–24. Westport, Conn.: Ablex Publishing Corp., 2000.

7

Academically Productive Talk: Supporting Students' Learning in Mathematics

Suzanne H. Chapin
Catherine O'Connor

THE National Council of Teachers of Mathematics (NCTM) has emphasized communication as an essential component of learning mathematics for well over a decade (NCTM 1989, 2000). More recently, a number of studies have suggested that discourse-intensive approaches to mathematics instruction may have tremendous potential for students' learning (Lampert 2001; O'Connor and Michaels 1993, 1996; Michaels and Sohmer 2001; Cobb et al. 1997; O'Connor 1999, 2001; Wells 1999; Rosebery, Warren, and Conant 1992; Lampert and Ball 1998). The skillful use of classroom talk is potentially a very powerful tool in creating a respectful and orderly classroom and in helping students to learn complex material. In this article we will present a brief sketch of work we have conducted in middle school mathematics classes, pointing to implications for mathematics teaching and learning more generally.

At the most fundamental level, we know that teachers rely on spoken and written language to facilitate learning. But not all are maximally effective in this usage. We have had the good fortune to work with teachers who are able to leverage the potential of classroom talk. We have seen their students develop habits and behaviors characteristic of inclusive, respectful communities of learners. Most important for this volume, these teachers have told us they believe that by participating in academically productive forms of talk, their students have gained a relatively greater grasp of mathematical concepts and procedures. Further, we have evidence that after using these academically productive forms of talk for one, two, or three years, these students' scores on standardized tests of mathematics improved significantly (Chapin and O'Connor 2004).

How could a teacher's skillful use of classroom talk help his or her students learn mathematics? In these teachers' view (and in ours), there are several ways that classroom talk may support learning. Students learn by clarifying and organizing their thoughts in order to present their own ideas. Having to talk about their own thinking often compels students to strive to make sense of what others think. Students learn by striving to make sense of what their classmates conjecture. They cannot know what their classmates conjecture unless they talk about it together. And they learn by actively reflecting on public talk about mathematical concepts, procedures, strategies, and vocabulary (Bransford, Brown, and Cocking 2000).

In our work with Project Challenge[1] (Chapin and O'Connor 2004; Chapin, O'Connor, and Anderson 2003), we have put a great deal of emphasis on having students talk with one another and with the teacher in particular ways, ways that we and others (Michaels et al. 2003; O'Connor and Michaels 1993, 1996) have found to be academically productive. In this article, we'll briefly discuss the results of that intervention and then describe the ways that talk was used in the Project Challenge classes.

From 1998 to 2002, in one low-income urban school district in the Northeast, Project Challenge enrolled about 100 students a year, starting in grade 4. By the end of our funded research, we had worked with eighteen teachers in grades 4–7 and almost 400 students. These classrooms emphasized communication by supporting discussions, both lengthy and brief, and by maintaining a constant focus on explanations for students' reasoning. During each class period, Project Challenge teachers orchestrated discussions in which mathematical ideas were the explicit focus of attention. The majority of these students (65%) were English language learners; students were predominantly of Latino heritage. More than 78 percent of our students qualified for free or reduced-price lunches.

Every year we administered a nationally normed standardized test—the California Achievement Test (CAT)—in order to gauge students' progress. Students in every cohort, years 1 through 4, made striking achievement gains on both the computation and the concepts and applications subtests of the CAT. Here we report only the percentile performance. This shows the position of Project Challenge students in relation to the national norming sample. After about three years in the project, our students were scoring better than 90 percent of the students in the national sample. Summary data from the four cohorts of students are presented in table 7.1.

Although Project Challenge was not designed as a randomized controlled field trial, we nevertheless were able to conduct a post hoc quasicontrolled comparison, which showed that students who were matched for test scores in third grade and who were not included in Project Challenge but who received the regular mathematics instruction provided by the school district performed significantly less well on the CAT and the state assessment given each year (see Chapin and O'Connor 2004 for details).

---

1. Project Challenge was funded by the U.S. Department of Education, Jacob K. Javits Act, 1998–2002. It was an intervention to develop mathematical talent in at-risk, urban elementary and middle school students.

Table 7.1
California Achievement Test of Mathematics (Total) Mean Percentile Rank

	Fourth-Grade Cohorts I–IV after 3 Months in Project Challenge	Fifth-Grade Cohorts I–IV after 15 Months in Project Challenge	Sixth-Grade Cohorts I–III after 27 Months in Project Challenge	Seventh-Grade Cohort I after 39 Months in Project Challenge
Percentile rank	75	78	87	90
*n**	304	261	176	*70*

*Only students who started Project Challenge in grade 4 and remained in the project are included. New students joined Project Challenge throughout each year but were not included in these data.

What contributed to these achievement gains? There were many components of the Project Challenge intervention, such as the use of a reform-based mathematics curriculum and professional development opportunities for teachers, which may have contributed to the success of the project. However, we and the teachers with whom we worked have speculated that the ways of communicating in the classroom that we instituted had a powerful impact on students' comprehension of concepts, on their understanding of computational procedures, and on their ability to reason about mathematical problems. Although these speculations must be tested in further research, we will present the contents of this aspect of our intervention for consideration by mathematics educators who might want to think further about the potential impact of classroom communication on students' learning.

First, many educators have observed that talk, in and of itself, may or may not support students' learning. We aimed to put in place a particular type of talk by both teachers and students—talk that is academically productive in that it supports the development of students' reasoning and students' abilities to express their thoughts clearly. We'll illustrate this type of talk with a brief discussion in Mrs. Holden's Project Challenge sixth grade, stopping along the way to point to ways we think this type of talk might help students learn.[2]

The students in Mrs. Holden's class are discussing how they used a set of clues to deduce a three-digit mystery number. One student, Kimberly, uses a clue that tells the class that the mystery number is odd, but as she talks, it becomes clear that she does not know how to differentiate even from odd numbers. The discussion then focuses on how to determine whether a number is even or odd. Through Mrs. Holden's expert use of classroom discourse, she is able to turn Kimberly's misconception into a powerful site for learning. As this segment of talk begins, Mrs. Holden has written the number 215 on the board.

2. All names used here are pseudonyms..

0. *Mrs. Holden:* So, two hundred fifteen, Kimberly, is what type of number, even or odd?

1. *Kimberly:* Um, even.

2. *Mrs. Holden:* Why?

3. *Kimberly:* Because there's a two.

4. *Mrs. Holden:* Okay, so two hundred fifteen is an even number. Dan?

5. *Dan:* Um, two hundred fifteen divided by two is, if you divided it, it would be an answer and it will have a remainder of one.

Notice that Dan's contribution could potentially help the class to deconstruct Kimberly's misconception. However, what he says is less than completely clear; Mrs. Holden's responses help him restate and refine his idea.

6. *Mrs. Holden:* So are you saying that this number would have a remainder of one if we divide it by two?

7. *Dan:* Yes. It is odd because it has a remainder.

8. *Mrs. Holden:* So is two hundred fifteen even or odd, Dan?

9. *Dan:* Odd.

10. *Mrs. Holden:* Brian, do you agree or disagree with what Dan said, and why?

11. *Brian:* I agree, but it's also because the last number is odd.

As the teacher asks Brian to weigh in on what Dan has suggested, he is able to offer another piece of the puzzle. This gives all students listening more to work with as they try to remember and clarify their own thoughts about what is an even or odd number.

12. *Mrs. Holden:* Rita, what do you think of these ideas?

13. *Rita:* If the ones, the ones digit, if it's an odd, it, the number is going to be an odd number because it's the first number that is important.

Even though Rita has been quite precise in using the term *ones digit,* her final statement about "the first number" is potentially confusing. The first number could be understood as being the first number reading from left to right. Here, that would be a 2, the number that Kimberly had focused on in her decision that 215 was even.

14. *Mrs. Holden:* I'm a little confused. First number? Last number? [*A five-second pause*]

15. *Rita:* The ones digit has to be odd for the number to be odd.

At this point, Kimberly is aware that she is confused. The preceding talk does not fit in with what she had assumed. She feels comfortable breaking in:

16. *Kimberly:* But how can you tell if a number is odd?

17. *Rita:* All you have to do is look at the ones digit and if the ones digit is odd, it's going to have to be odd. The other numbers don't matter.

At this point, Mrs. Holden could simply move on. But she asks a question to determine whether Kimberly really heard and understood what Rita said.

18. *Mrs. Holden:* Kimberly, would you repeat what Rita said?

19. *Kimberly:* Um, you know that if you look at the last number and it's odd, then the whole number is odd.

20. *Mrs. Holden:* Which number is the last number?

21. *Kimberly:* Five.

22. *Mrs. Holden:* Which place is it in?

23: *Kimberly:* The ones.

24. *Mrs. Holden:* OK, so why is the ones digit the important one? Talk to your partner for a minute.

Mrs. Holden has done far more with this little stretch of talk than simply correct Kimberly. She has brought the discussion to the deeper question of why the ones digit determines whether the number is even or odd. By asking students to talk with their partner, each student is required to try and make sense of how the number in the ones digit is related to evenness or oddness. Mrs. Holden knows that it will be easier for her students to share their thinking with the whole class after they have had the opportunity to talk with another student.

Mrs. Holden walks around the room listening while pairs of students sort this out. After about a minute, she calls on one pair.

25. *Mrs. Holden:* Carlos. Tell us what you and your partner discussed.

26. *Carlos:* Well, if the last place, hm... the ones is a one, three, five, seven, or nine, the number is odd. We aren't sure why.

27. *Mrs. Holden:* Carlos and Amy have noticed that only odd digits can be in the ones place, but they aren't sure why. Who has an idea about why we only have to look at the ones place?

28. *Kayla:* I think the ones digit is important because it means—when you divide, that's where the leftovers come from. The remainders are always from the ones.

Examples like these suggest that talk can support and promote students' learning in mathematics in a number of ways. First, as we can see in Mrs. Holden's classroom discussion, students do not always remember or fully understand concepts, procedures, and rules. Yet in a discussion like this one, ideas and facts are repeated and revisited, giving students a second and third chance to catch important details or clarify their interpretation of what they thought had been said. Second, as students respond to questions about why they believe something, their explanations become clearer and more highly elaborated, supporting both their own and other students' understanding.

Finally, having students engage in an explicit discussion of ideas supports their learning in another way: discussions often bring to the surface students' gaps in understanding or their misconceptions. It is only when teachers know what students do and do not understand that they can structure instruction to support students' learning. So talk by students can help teachers refine their instruction to meet students' learning needs.

One powerful reason to use talk-intensive instruction is that discussions offer a mechanism to engage students in reasoning about mathematics. The ability to communicate clearly and precisely is a hallmark of mature mathematical reasoning. Classroom talk provides a context for the socialization of students into this practice. In the dialog above, Mrs. Holden asks students to discuss the reason behind the procedure for identifying an odd or even number—why can we examine the ones digit and determine whether the number is even or odd? Students must connect the knowledge that odd numbers have a remainder when divided by 2 with the fact that the remainder results from an extra unmatched unit in the ones place. Notice that Mrs. Holden does not supply the "correct answer." This does not mean that she doesn't value the correct answer. Instead, she makes sure that the correct answer emerges out of the students' efforts. She keeps pressing for students to explain their ideas, methods, and solutions. She lets the correctness of the students' statements be determined by the logic of the mathematics, drawn out of what some of them know or speculate about.

The discussion above continued, with many students questioning why the digits in the other places do not matter.

29. *Rachel:* Look, if it is two hundred forty-five (245), think about it. Two hundred can—is divided evenly into two groups, four tens can be divided into twenty and twenty, but five will have a remainder.

30. *Arjun:* But what about one hundred ninety-five?

31. *Kayla:* Take each place. One hundred can be divided by 2 evenly. So can 90. But five ones gives two remainder one.

Instruction that features consistent use of academically productive discourse also assists students in developing their language abilities. Students must clarify their thinking and articulate their ideas as clearly as possible if others are to understand them. This is hard work, and often students' comments are unclear. In lines 13 and 15 of the discussion, we see how when asked to restate an idea, Rita is more focused and direct. Another advantage of using academically productive talk is that students' vocabularies improve. Not only do students acquire and use mathematical terms and phrases, but also by working to express themselves precisely and without ambiguity, their overall ability to describe mathematical situations and ideas improves. What is said also provides input for students learning English as a second language. As students repeat the words of their peers, language learners get a chance to receive input that is increasingly comprehensible to them.

Discourse-Based Instructional Tools

What tools are available to help facilitate academically productive talk? In Project Challenge classrooms, teachers and students repeatedly used five "talk moves" to focus discussion on important academic topics (Chapin, O'Connor, and Anderson 2003; Michaels et al. 2003). Talk moves are simple conversational actions that have the potential to make discussions productive. The five talk moves we used (discussed further below) were as follows:

1. "Revoicing" by both teacher and students—restating a previous speaker's utterance and asking whether the restatement is correct (e.g., "So you're saying.... Is that right?") (O'Connor and Michaels 1993, 1996)
2. Teacher-initiated requests that a student repeat a previous contribution by another student (e.g., "Sandra, can you repeat what Giselle just said?")
3. Teacher's elicitation of a student's reasoning (e.g., "Do you agree with Sandra's suggestion? Why do you think that?")
4. Teacher's request for students to add on (e.g., "Does anyone have more to add to that?")
5. Teacher wait time

To illustrate these moves, we revisit Mrs. Holden and her students as they discuss the following probability problem: Consider two spinners, A and B (see fig. 7.1). If we spin both A and B, we can record the number from each spin. Spinner A has landed on a 5 and Spinner B has landed on a 3. Let's form a fraction from these numbers such that the number from spinner A is the numerator and the number we get from spinner B is the denominator: $\frac{5}{3}$. We can then look at the fraction $\frac{A}{B}$ and assess whether when simplified, it is a mixed number or a whole number. In our example, $\frac{5}{3}$ simplifies to a mixed number: $1\frac{2}{3}$. By considering the numbers on each spinner, students can assess the probability that a fraction generated by spinning A and B will be a mixed number or a whole number. Students worked with a partner to answer the questions, and then Mrs. Holden conducted a discussion of the problem.

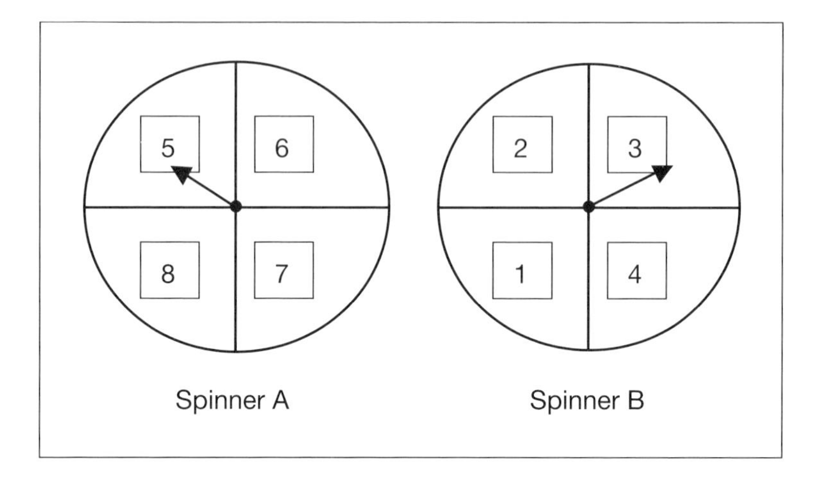

Fig. 7.1

Revoicing—"So You're Saying.... Is That Right?"

Revoicing is a talk move that can be used when a student's contribution is unclear. Sometimes when students' words or phrases are ambiguous, teachers interpret the utterance and restate it in their own words. However, there is no guarantee that this reinterpretation is actually what the student originally meant to say. The goal is *not* simply to render a student's contribution as meaningful at any cost. The goal is to help students clarify their thinking and improve their understanding. Therefore, we recommend that teachers usually try to repeat exactly at least some of the student's statements. The second part of the revoicing move asks the teacher to explicitly check with the student that what they revoiced was in fact what the student intended to say. This helps the student clarify his or her own contribution both in meaning and in language and tends to make the ideas more accessible to the rest of the class. A revoicing move is in boldface below.

1. *Mrs. Holden:* How did anyone solve this problem?
2. *Catherine:* We made a tree diagram.
3. *Mrs. Holden:* Why did you make a tree diagram?
4. *Catherine:* Um, so we could get all of the answers. So that we can ... um ... find all the combinations.
5. *Mrs. Holden:* Okay, so, I'm curious about this tree diagram. **You said you made a tree diagram so you could find all the combinations?**
6. *Catherine:* Yes, we, we wanted, we wanted to see, um, what are all

the ways of getting a mixed and whole number so we decided to make a tree diagram, which is easy to see. We got sixteen combinations of numbers from Spinners A and B.

Although the students who created the tree diagram may have understood what they were doing, other students might be confused. In order to learn, one must first understand. By getting these students to repeat and clarify, the teacher has helped other students learn from what this group did.

Repeating—"Can You Repeat What Roxanna Just Said?"

Another talk move that we have found especially useful is getting students to repeat what someone else has said. There are a number of reasons why we asked students to repeat their classmates' contributions during a discussion. First, this move makes everyone aware that the discussion is truly a discussion among the whole class, not simply a conversation between the teacher and one student. Students must pay attention if they are going to be able to repeat another's contribution. They must listen to understand what everyone has said. Second, repeating enables teachers to highlight a contribution, not because it was confusing, but because a student made an important point. This slows down the pace of the instruction to focus on the important ideas. Students are able to reconsider a statement they didn't quite understand the first time it was mentioned or ask a question they didn't think of at first. Third, by repeating another student's ideas, individuals must reflect on those ideas; they are pushed to make sense of the contribution.

7. *Mrs. Holden:* Jose, **would you repeat what Catherine said about what a tree diagram helps us do?** [*Mrs. Holden waits eight seconds.*]

8. *Jose:* Well, it helps you find out all the probabilities there because you write out all the combinations that you can get, so that helps you find out what's the probability.

9. *Mrs. Holden:* Does that make sense to you, Roxanna? **What did Catherine say a tree diagram can help you to do?**

10. *Roxanna:* It's like a systematic way to, um, see what's the probability. It's a way to get all the answers. And we found out that, for, um, the probability for getting a whole number was eight over sixteen, which equals one-half, and the probability for a mixed number was eight over sixteen, which equals one-half.

11. *Mrs. Holden:* Okay, so a tree diagram helps us to organize information in a systematic way. How many people tried the tree dia-

gram? Okay. Anybody try another method? Tarrell, what did you try?

Eliciting Students' Reasoning—"Do You Agree or Disagree and Why?"

Once students become comfortable listening to one another and repeating each other's ideas, the focus shifts to helping them reason about the content under discussion. Students are asked if they agree or disagree with results or ideas and why. The importance of this talk move is that it causes students to explicitly use their reasoning to analyze other students' statements. Students also learned that in order to support their statements and convince other students, they needed to use evidence. We found that after students had been in classrooms that regularly used these talk moves, they internalized them and explained the reasons behind their thinking without being prompted. This supports robust learning because students are not simply accepting or memorizing without thinking but are making a habit of reasoning about why their mathematical claims or suggestions are valid.

12. *Tarrell:* We made a list, and we put the whole numbers on one side and the mixed numbers on the other side. We got one, two, three, four, five, six, seven, eight wholes, and eight mixed numbers.

13. *Mrs. Holden:* So you got eight wholes and eight mixed numbers? **Do you agree or disagree with Roxanna's answer [that the probability is one-half]? Why?**

14. *Tarrell:* We kind of agree. But we got eight.

15. *Mrs. Holden:* Eight what?

16. *Tarrell:* Eight mixed numbers.

17. *Alex:* I agree with Roxanna but disagree with Tarrell's answer. The probability is one-half, not eight. Tarrell has to give the answer as a probability, he can't just give the number of mixed numbers.

18. *Tarrell:* Yeah, I meant eight-sixteenths. The probability is eight-sixteenths. I agree with Roxanna's probability. I have to compare it [the eight] to all of the possibilities. There are sixteen possibilities.

Adding On—"Can Anyone Add to This Conversation?"

The fourth talk move that effectively contributes to supporting academically productive talk can be called "adding on." This is a simple method of eliciting contributions from other students. We have found that there is great value in involving many students in the conversation. This move enables students to carefully consider the ideas, to think about what they understand, and to put it into their own words. Sometimes the contributions are repetitious, but this is not problematic; we have found that students need time to individually grapple with the ideas. And often, as in the example below, a new point or nuance will emerge.

19. *Mrs. Holden:*	Okay... how about Natasha and Mario? **What can you add to this conversation?**
20. *Mario:*	We, what we did, we did something like Tarrell. We first found the whole numbers and then we found the mixed numbers, and we got, we got eight-sixteenths, but I put it in a percent of 50 percent. Fifty percent were mixed numbers.
21. *Mrs. Holden:*	So can we express the probability as a percent?
22. *Mario:*	Yes. Fifty percent of the time the spinners will land on a mixed number.

Teacher Wait Time

The final talk move isn't a move at all but a reminder to provide plenty of wait time. Teachers need to wait after they have posed a question before calling on students, so that everyone has time to compose her or his responses. Teachers also need to wait after asking students to respond (e.g., after asking them to repeat or add on). Too often teachers become uncomfortable with more than three seconds of silence. However, three seconds is not enough time in a discourse-intensive situation to figure out what to say. We think that when learning new and complex material, students have the right to sufficient wait time. Thus, teachers have an obligation to offer it.

Mrs. Holden uses different techniques to remind herself to wait for students, such as silently tapping her leg or watching the clock. In line 7 above, she waits for Jose for eight seconds. Usually students just need extra time to figure out what they want to say. When we began using these discourse-intensive moves in Project Challenge, many students thought that if they didn't answer, the teacher would move on to another student. However, if we expect all students to have an obligation to participate, then we have to modify the structure so that they do participate. Students learned that they must contribute, and wait time gave them the time to articulate their knowledge. Many educators have discussed the importance of wait time in supporting students' learning, and we would add that it is particularly important in mathematics, where abstract ideas, technical language, and complex concepts are always present.

As described, we found that using the five "talk moves" helped Project Challenge teachers and students engage in academically productive talk about mathematics. But the talk moves alone were not sufficient. We discovered that for some teachers, the outcome was not as positive as it was in other classrooms. Students would refuse to talk, or some students would dominate the conversation. In these classrooms, sometimes students openly ridiculed one another. We learned that in addition to mathematical content and skillful use of the five strategies, teachers had to address another area, at least as important as the first two: classroom conditions had to support respectful and equitable participation.

Creating the Learning Environment

In order for academically productive talk to be used effectively, certain conditions must be in place within the learning environment—conditions that will support both language development and mathematics development. Foremost is the requirement that teachers establish conditions for *respectful* discourse. Discourse is respectful when each person's ideas are taken seriously and no one is ridiculed or insulted. Discourse is respectful when no one is ignored.

Most people are unwilling to venture an idea or opinion in a situation where they feel they may be teased or made to feel inadequate or even stupid. Teachers in Project Challenge classrooms were encouraged to support safe, positive learning environments and to immediately put a stop to overt displays of disrespect, such as "That's stupid." Every teacher created explicit policies about this matter and discussed these policies throughout the year. However, we also discussed the more subtle behaviors that can occur in classrooms—behaviors that work against positive interactions but that often go uncorrected by teachers. Students in a variety of ways pass judgment on their classmates' contributions. Whether by using body language to suggest boredom, making noises to indicate impatience, violently waving a hand to suggest superiority ("how can everyone else not know the answer?"), or surreptitiously making a derogatory comment, students directly and indirectly send signals that stop other students from participating.

A number of conditions must be established in order for an environment of respectful discourse to develop: (1) explicit expectations for students in regard to their role in classroom discussions; (2) clear rules about what constitutes respect and disrespect so that even subtle negative messages are not permitted; and (3) clear sanctions against disrespectful behavior with zero tolerance during classroom talk sessions. In our project, expectations for students regarding their role took the form of a list of students' rights and students' obligations: students have the right to ask questions, the right to make a contribution to an interested audience, and the right to expect support rather than embarrassment when confused. Below is the list of basic rights that all students have in Project Challenge mathematics classes.

Rights

- You have the right to ask questions.
- You have the right to make a contribution to an attentive, responsive audience.
- You have the right to be treated civilly.
- You have the right to have people discuss your ideas, not you.

The classroom teachers in our project also explicitly agreed on a range of behaviors that were not allowed, since they violated students' important right of civility. These were described to the students and discussed at length at the beginning of the year. Finally, when disrespect was observed, these teachers stopped the discussion and immediately dealt with the infraction.

As mentioned earlier, one of the advantages of using discourse-based instructional strategies is that students have an opportunity to extend their understanding of ideas or solution methods by discussing them in depth. The goal of a discussion might be to assist students in clarifying an idea or in generalizing a solution method to a set of problems. Sometimes during these discussions, mistakes are presented, faulty reasoning is revealed, or misconceptions arise. Students can feel extremely vulnerable and may see another student's disagreement with their ideas as a value judgment about themselves. For some students, "I don't think Jasmine's method will work" is equated with "I don't think Jasmine ever has good ideas." This occurs even in situations where students are very respectful of one another. Without the teacher's intervention, the social norm that was so carefully established can be damaged. Teachers must explicitly ask students to reflect on the ideas, methods, and facts that are presented, not on the people who presented them: "What do you think about the *idea* Peter has just put forth?" and "What do you think about what Peter just *said*?" This changes the focus—contributions are evaluated according to their mathematical validity or their accuracy—and the purpose of the interaction shifts to sense making and justification.

Just as important as establishing the conditions of respectful discourse is the need to establish the conditions of *equitable participation*. Participation is fair and equitable when each person has a fair chance to ask questions, make statements, and express his or her ideas. Academically productive talk is not just for those students with strong verbal skills or who are confident about what they know and don't know. We believe that every student should participate in discussions.

Why don't students talk in math class? First, many students don't have the opportunity to talk—teacher-sanctioned student discourse is not present in their classrooms. But even in classrooms where there are opportunities to ask questions, discuss solution strategies, and make conjectures, there are always students who choose not to participate. Some students adopt the role of passive recipient of knowledge; their job is to listen respectfully to the teacher, not to debate or argue positions. We believe that these students have an obligation to participate in the learning environment—students participate both by listening for understanding and by sharing their own ideas and reasoning. In Project Challenge classrooms, students were called on to respond to the ideas that were being discussed, regardless of whether or not they volunteered. By regularly calling on all her students, Mrs. Holden sends a message to the whole class: you all are expected to make sense of this content; you all are expected to participate. In line 13 of the first transcript, Mrs. Holden expected that Rita had been paying attention and would have something important to say.

The requirements of equitable participation must be explained to students. Students must understand that everyone will be called on because the instructional goal is for everyone to learn the content, not because the teacher is picking on any one student. We presented these expectations to students in the form of students' obligations—obligations to try to understand all material that is presented as well as obligations to the learning community to participate, to speak distinctly, and to respond to others' contributions.

Obligations

- You are obligated to listen for understanding when others speak.
- You are obligated to speak loudly enough for others to hear.
- You are obligated to agree or disagree with the speaker's comments and explain why.
- You are obligated to be civil and to challenge ideas without criticizing the speaker.

Summary

We have briefly described a complex system: to really get the full benefits of this kind of classroom talk and discussion, teachers must first establish the conditions of respectful and equitable participation. After these behavioral norms are in place, teachers can gradually begin using the five talk moves and strategies described above. Finally, as the students grow familiar and comfortable with this new way of interacting, the mathematics learning they are capable of will begin to emerge, coming into full view as they discuss their ideas and confusions and give audible evidence that their understandings are developing.

Many teachers who participated in our project started out with grave doubts about whether this approach would work in their classrooms. Like many teachers in low-income urban schools, these teachers were accustomed to teacher-centered classrooms where lectures and quiet were the norm. However, as they put into place the conditions for respectful discourse and as their students began to discuss and talk in the ways described here, the teachers saw evidence of more robust and extensive student learning. They also told us that they found their own teaching taking on a new depth and interest as they saw what their students could do. We have described some of their struggles in a book for teachers (Chapin, O'Connor, and Anderson 2003).

But how did the students feel about taking part in this daily talk? We'll close here with two comments from students in Project Challenge talking about how this way of using classroom talk helped them learn mathematics:

> I like the talking to your neighbor or partner thing because if somebody doesn't really understand something and they talk to their partner, their partner can help them understand it so you get to go on without being confused or having trouble. (Chapin, O'Connor, and Anderson 2003, p. 161)

> It was helpful to listen and talk to each other because I got to listen to how people thought about problems and why they agreed or not and I got to talk to people and tell them what I thought about certain things. (P. 163)

During the year, students did not always enjoy the process of repeating what others had said, nor did they always like it when classmates disagreed with their ideas, however respectfully. Nevertheless, we believe the vast majority of the Project Challenge students thought as these two students did. They appreciated the opportunities they got every day to talk with one another about their mathematical thoughts, and they—like their teachers—had the sense that engaging in this talk actually supported their learning.

REFERENCES

Bransford, John, Ann Brown, and Rodney Cocking, eds. *How People Learn: Brain, Mind, Experience, and School.* Washington, D.C.: National Academy Press, 2000.

Chapin, Suzanne H., and Catherine O'Connor. "Project Challenge: Identifying and Developing Talent in Mathematics within Low-Income Urban Schools." Research Report No. 1. Boston: Boston University School of Education, Autumn 2004.

Chapin, Suzanne H., Catherine O'Connor, and Nancy Anderson. *Classroom Discussions: Using Math Talk to Help Students Learn, Grades 1–6.* Sausalito, Calif.: Math Solutions Publications, 2003.

Cobb, Paul, Ada Boufi, Kay McClain, and Joy Whitenack. "Reflective Discourse and Collective Reflection." *Journal for Research in Mathematics Education* 28 (May 1997): 258–77.

Lampert, Magdalene. *Teaching Problems and the Problems of Teaching.* New Haven, Conn.: Yale University Press, 2001.

Lampert, Magdalene, and Deborah Loewenberg Ball. *Teaching, Multimedia and Mathematics: Investigations of Real Practice.* New York: Teachers College Press, 1998.

Michaels, Sarah, Catherine O'Connor, Megan Hall, and Lauren Resnick. *Accountable Talk: Classroom Conversation That Works.* 3 CD-ROM set. Pittsburgh, Pa.: Institute for Learning, University of Pittsburgh, 2003.

Michaels, Sarah, and Richard Sohmer. "Discourses That Promote New Academic Identities." In *Discourses in Search of Members,* edited by David Li, pp. 171–219. New York: University Press of America, 2001.

National Council of Teachers of Mathematics (NCTM). *Curriculum and Evaluation Standards for School Mathematics.* Reston, Va.: NCTM, 1989.

———. *Principles and Standards for School Mathematics.* Reston, Va.: NCTM, 2000.

O'Connor, M. Catherine. "Language Socialization in the Mathematics Classroom: Discourse Practices and Mathematical Thinking." In *Talking Mathematics,* edited by Magdalene Lampert and Merrie L. Blunk, pp. 17–55. Cambridge: Cambridge University Press, 1999.

———. "Can Any Fraction Be Turned into a Decimal? A Case Study of a Mathematical Group Discussion." In *Learning Discourse: Discursive Approaches to Research in Mathematics Education,* edited by Carolyn Kieran, Ellise A. Forman, and Anna Sfard, pp. 143–85. Dordrecht, Netherlands: Kluwer Academic Publishers, 2002. (Reprinted from *Educational Studies in Mathematics* 46 (2001): 143–85.)

O'Connor, M. Catherine, and Sarah Michaels. "Aligning Academic Task and Participation Status through Revoicing: Analysis of a Classroom Discourse Strategy." *Anthropology and Education Quarterly* 24 (4) (1993): 318–35.

———. "Shifting Participant Frameworks: Orchestrating Thinking Practices in Group Discussion." In *Child Discourse and Social Learning,* edited by Deborah Hicks, pp. 63–102. Cambridge: Cambridge University Press, 1996.

Rosebery, Ann, Beth Warren, and Faith Conant. *Appropriating Scientific Discourse: Findings from Language Minority Classrooms.* Cambridge, Mass.: TERC, 1992.

Wells, Gordon. *Dialogic Inquiry: Towards a Sociocultural Practice and Theory of Education.* Cambridge: Cambridge University Press, 1999.

8

Engaging Students in Collaborative Discussions: Developing Teachers' Expertise

Jeffrey Choppin

Mathematics reformers aligned with the National Council of Teachers of Mathematics (NCTM) *Standards* (1989, 2000) express a vision for learning that includes meaningful student participation in the classroom discourse. For example, Forman, McCormick, and Donato (1998) state that "new forms of instruction include more active participation of students in providing explanations, conducting arguments, and reflecting on and clarifying their thinking" (pp. 313–14). The *Standards* (2000) state that students should be able to "communicate their mathematical thinking coherently and clearly to peers, teachers, and others" and "analyze and evaluate the mathematical thinking and strategies of others" (p. 60). Realizing this vision of classroom discourse has proved to be challenging. Educators have recognized both the complexity of involving students meaningfully in classroom discourse and the new demands placed on teachers concerning knowledge and skills. In this article I discuss opportunities and challenges for teachers and students in classrooms attempting discourse reform. I use examples from two classrooms to illustrate and suggest teachers' specific actions that facilitate the development of students' active roles.

Reform Discourse as Collaborative or Dialogic

Sociocultural theorists suggest that learning is facilitated by participating in activities that have a collaborative or conversational dynamic (Nystrand 1997; Tharp and Gallimore 1988) in which students' explanations become "thinking devices" for the classroom community (Wertsch and Toma 1995). The conversational or collaborative characteristic of discourse is enhanced by teachers' efforts to build from students' explanations to develop mathematical ideas (Sherin 2002). This serves several purposes: building from students' contributions positions students as competent mathematical thinkers, students are offered the opportunity to influence the thinking in the classroom community and to receive feedback on their ideas, and teachers learn more about how their students think about mathematics. Teachers can help students participate more meaningfully not only by eliciting explanations but also by supporting students' efforts to reflect on their peers' mathematical claims. I suggest that the practice of students explaining solutions is evident in many reform classrooms; however, the collaborative practices of reflecting on and building from students' explanations are both largely absent and crucially important to the active learning of mathematics.

Transforming discourse toward a truly collaborative set of interactions is a difficult and long-term endeavor. Although the literature includes instances in which teachers have managed to increase students' participation in the classroom discourse, this has not necessarily led to the collaborative type of discussion in which students' ideas serve as the catalyst for developing mathematical concepts. On occasion, teachers have either focused the discussions narrowly, thus minimizing the students' voice (Forman, McCormick, and Donato 1998), or they have emphasized students' participation at the expense of the development of mathematical ideas (Williams and Baxter 1996; Nathan and Knuth 2003).

Establishing Collaborative or Dialogic Discourse Patterns

Collaborative or dialogic discourse requires that the teacher and students understand and react to one another's thinking. Consequently, classroom discourse becomes a process of negotiating understanding, which is a complex and difficult process. Orchestrating or directing such a process requires a great deal of knowledge and experience on the part of the teacher. The challenge for the teacher is not necessarily to help students explain their thinking—which students readily do—but rather to support the consequent development of collaborative discussions that lead to the emergence of mathematical ideas.

Below, I present examples that help illustrate the possibilities for involving students in collaborative discourse. In presenting these examples, I outline the opportunities as well as the challenges for helping students elaborate and reflect on mathematical explanations. The two examples include discussions from two seventh-grade classrooms, which I have modified to suit the purposes of this article.

Example 1: "Slowing Down" versus "Stopping"

In this example, the students were presented with four graphs and asked to match them to descriptions of verbal, time-versus-distance scenarios. Below, two students presented competing graphical interpretations for a given scenario. See figure 8.1 for the two graphs that are referenced in the discussion.

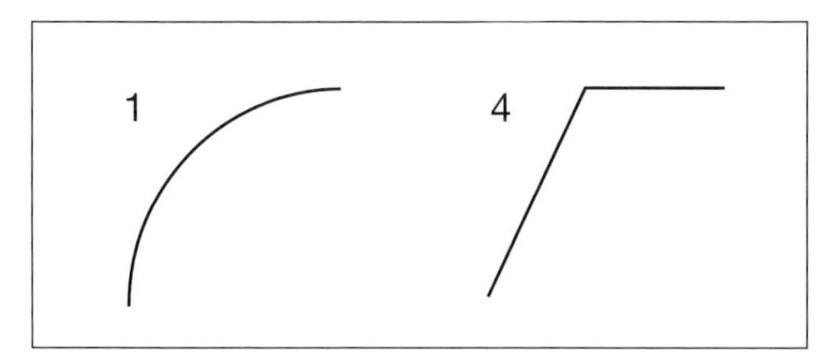

Fig. 8.1. Two distance-versus-time graphs of bicyclists

Teacher:	How about the second one? Tony and Liz rode real quickly and reached the campsite early. Which one shows that, Doug?
Doug:	Number 1.
Teacher:	Number 1? Going very quickly and reached it early. Okay. How many people agree with that? Okay, we've got a few people who agree. How many people disagree with that? What would you say? Neal?
Neal:	Well, 'cause it said, like, they rode really quickly, and I picked 4 'cause they rode really quickly and everyone else had to catch up. That's why there's a straight [horizontal] line at the end.
Doug:	Yeah, but they'd have to slow down.

Doug's interpretation of the scenario is that Tony and Liz would need to slow down, which would indicate graph number 1, whereas Neal's interpretation implicitly suggests that Tony and Liz would need to have stopped once they reached the campsite, which would indicate graph number 4. These explanations provide an opportunity for the class to connect "slowing down" with a curved graph and "stopping" with a horizontal graph. However, the explanations are not fully elaborated and do not adequately make the important connections; consequently, an ensuing discussion would provide important opportunities to help students collaboratively build those connections. The teacher, however, chooses to resolve the disagreement by presenting his own explanation for the correct response, as is shown below.

Teacher:	They rode real fast and it said they got there early so if this was ... Let's say this was half an hour increments ... this was at zero and this was at point 5 so we have zero and .5....

The teacher then divides the graph into the two segments and discusses each piece separately. This explanation ends the discussion. See figure 8.2 for the new graph displayed by the teacher.

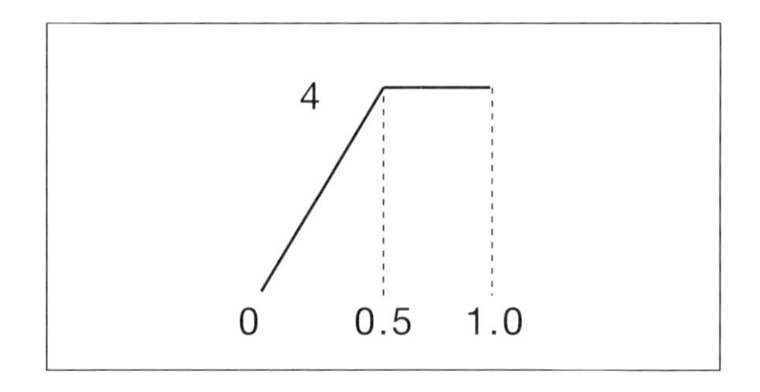

Fig. 8.2. Graph as augmented by the teacher

This example illustrates the tension teachers often feel about completing tasks quickly and ensuring that a correct and clear solution has been presented. However, there is a cost for such efficiency, and that is a lost opportunity for students to establish meanings collaboratively. This exchange is a manifestation of a bigger question related to the learning of mathematics, which is: What is the value or purpose of student engagement in mathematical activity? Schleppegrell (2004) states that an important function of education is to help students understand and appropriate academic forms of discourse. Engaging students in collaborative discussions is a crucial form of mathematical activity.

This example also demonstrates the often cryptic nature of classroom discourse. The discussion consisted of brief interactions and lasted less than a minute. Nevertheless, the students' explanations offered plausible alternative interpretations that underscored mathematics central to the themes in the unit. In order to develop collaborative discussions from such brief interactions, the teacher will need to slow down the discussion by, for example, asking for further elaboration or by publicly recording the two explanations.

In the next example, the exchanges are lengthier and less cryptic. Although this next discussion is more interactive and student-centered, it illustrates additional challenges and issues related to engaging students in collaborative discussions.

Example 2: "A Straight Line Means You're Not Going Anywhere"

In this example, similar to the one above, the students were provided a series of speed-versus-time graphs and asked to match those graphs with scenarios for a group that is biking from one location to another. The graph for number 4 is given in figure 8.3. The discussion begins with Jaakko interpreting the graph.

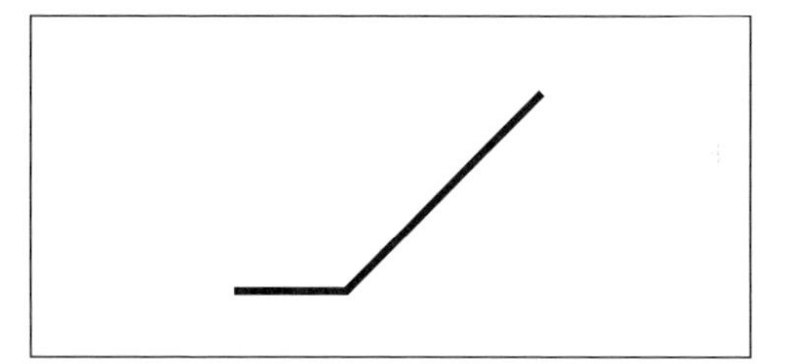

Fig. 8.3. Graph of cyclist's speed over time

Jaakko:	And then number 4, the bottom, it starts steady and then as it goes up it starts to go up faster gradually ...
Adrian:	Ah, for number 4, I think they started late. That's why there's a straight [horizontal] line, 'cause they never went anywhere.
Jackie:	When you have a straight line it means you're not going anywhere.
Adrian:	Right.

The teacher has a choice at this point to correct Jackie's and Adrian's misinterpretation of the graph as a distance-versus-time graph. This would serve to eliminate the confusion over the nature of the dependent variable and would allow the discussion to focus more quickly on Jaakko's interpretation. However, it would also diminish the opportunity for students to discuss collaboratively what it means for a line segment to be horizontal. The teacher opts to let the students attempt to resolve the disagreement.

Teacher:	What do you think, Jaakko?
Jaakko:	Well if you're going through a straight line you could be going steady, too.
Melinda:	No.
Jaakko:	Yeah, you could be going at the same speed just straight.
Annie:	He's right.
Teacher:	Annie, you agree with Jaakko? Okay. Jackie?
Jackie:	If you're going steady you'll have a line that's between the y-axis and the x-axis. You'd have a line that would go like right up the middle, which would mean some movement, but a straight [horizontal] line means you're not going anywhere. Staying at the same pace or speed or whatever ...
Jackie:	You're not going anywhere if you're going in a straight [horizontal] line ...
Adrian:	Just think back to that one where they stopped for lunch. It was a straight line for like an hour and a half.
Raymond:	Yeah, and they're not going anywhere.

Jackie: If you made a table of what their speeds were in an hour or whatever, then it would show at point 5 hours it would be like 20 miles and at one hour it would be 20 miles. That means they're not going anywhere ...

Jackie and Adrian offer warrants for their explanations, which provide an opportunity for the class to consider their interpretations. Jackie, for example, has given an explanation about the meaning of horizontal line segments in a distance-versus-time graph, which she backs up by offering a tabular representation. Adrian has warranted his explanation by referring to a prior task in which a horizontal segment indicated no change in distance traveled. These warrants offer an opportunity for other students to consider the explanations and highlight mathematical themes of the instructional unit—namely, the connections between forms of representation and interpretations of graphs.

Another notable feature of this exchange is that both Jaakko's and Jackie's explanations contain ambiguities. Jaakko's "going at the same speed, just straight" comment possibly confounds the direction of the graph with the direction of the cyclists, whereas Jackie's "staying at the same pace or speed or whatever" comment contradicts her own claim that a horizontal line indicates no movement.

The teacher facilitates the discussion by maintaining a nonevaluative stance and by continuing to seek students' explanations. Although this results in an active discussion, it fails to clear up the confusion about the nature of the dependent quantity, which is speed and not distance. The value in letting students attempt to resolve the disagreement is that they gain experience participating in an advanced academic discourse, that of mathematical argumentation. If properly reflected on, this experience could shed light on how to conduct arguments such as what constitutes acceptable justification and how to identify and evaluate warrants and assumptions. The danger for the teacher in not stepping in is that mathematical ideas central to the instructional unit may not clearly emerge.

The teacher begins to record Jackie's table, but instead uses Jackie's measurements to augment the graphical representation. The teacher proceeds to draw a grid over the graph (see fig. 8.4), showing the horizontal part to have a value of 20.

The discussion orients to the augmented graphical representation, which helps change students' interpretation of the quantities from distance versus time to speed versus time.

Annie: I think he's right, because it shows 20, and that's how much miles per hour he's going, and it just stayed the same miles per hour. It doesn't show that it stopped....

Raymond: It stayed the same.

Jaakko: If they were stuck it'd be down by the zero wouldn't it?

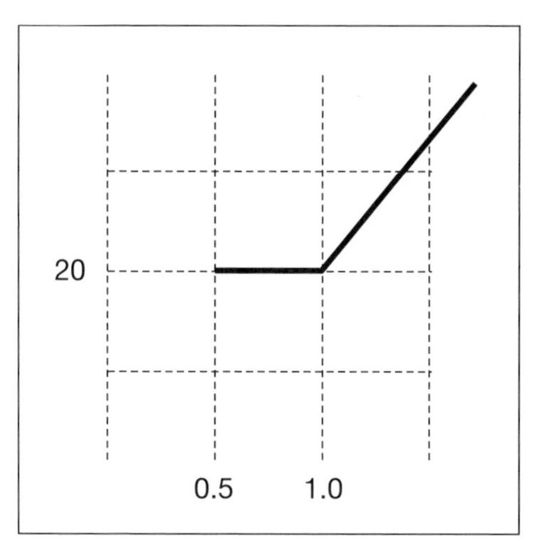

Fig. 8.4. Diagram with grid imposed by the teacher

Teacher:	Okay, so they ... Jaakko is saying if they had stopped ...
Jaakko:	The line would go straight down to a zero.
Teacher:	The line would go down to a zero as opposed to staying at 20.
Annie:	Yeah.

The teacher's actions provided an opportunity for students to evaluate their peers' explanations; however, the interactions were rapid and occasionally ambiguous. Although students offered several forms of evidence for their claims, this increased the complexity of the discussion and made it difficult for classmates to follow. In everyday conversation participants often attempt to clarify ambiguous remarks in order to establish common understandings; in whole-class discussions, social considerations often preclude such attempts and confusing statements are left unchallenged. Consequently, the role of the teacher is very important in helping students clarify and reflect on one another's thinking.

For example, the teacher could have publicly recorded the various explanations, with accompanying representations and further elaboration by the students. Jackie claimed that a horizontal graph segment represents no change in distance traveled and offers as her warrant a table showing how a constant *y*-value indicates no change in distance. The warrant offers a productive point of discussion for students to evaluate Jackie's claim. By asking students to elaborate claims and by highlighting the nature of these claims, the teacher acts to potentially slow down discussions and allow a broader range of students to access the explanations, much like "wait time" after questions increases participation.

The teacher's special and influential role in the classroom discourse is further highlighted in both examples by the teacher's ultimate attempts to resolve the disagreements. In both examples, the teacher narrowed the possible range of interpretations by adding additional information to the graph. This served to focus the discussion on a correct interpretation but also shifted the control of the discussion to the teacher.

A Synthesis of Examples

These two examples describe challenges as well as opportunities for engaging students in collaborative discussions. In both instances, the rapid, ambiguous, and cryptic nature of interactions created barriers for most participants to fully understand the explanations, especially the underlying mathematics. The teacher can ameliorate the possible confusion by more explicitly modeling effective ways to engage in an academic discourse. For example, the teacher can help students identify different claims and warrants in order to ascertain agreements or disagreements. The two examples also suggest that establishing collaborative discussions will require a lengthy learning process and the development of new expertise on the part of students and the teacher.

There are no simple suggestions or set of techniques for becoming proficient at establishing collaborative discussions. Much like learning mathematics, learning how to understand and discuss how others think about mathematics is a recursive process that at times will be frustrating and seemingly unproductive. Both classrooms above show some promising characteristics, such as students presenting warranted claims and teachers offering opportunities for students to do so, but the attempts are formative. In order to develop further competency, students and teachers will need to accumulate additional experiences attempting collaborative discussions, experiences that will need to be supported. Below, I suggest ways to support such efforts.

Supporting the Development of Collaborative Discussions

I divide the recommendations in this section into two sections: "Initiating Collaborative Discussions" and "Sustaining the Development of Collaborative Discussions." Much of the literature on classroom discourse concerns how to initiate collaborative discussions; consequently, this section represents a synthesis of best practices. The second section more directly relates to professional development efforts.

Initiating Collaborative Discussions

The teacher's most obvious action to initiate collaborative discussions is to seek students' explanations. Without an initial effort to have students explain their solutions, there will be no opportunities for discussion. However, in order for other students to reflect on an explanation, the teacher will need to seek warrants for students' explanations. The emphasis on warrants provides a basis for establishing what constitutes appropriate evidence and for developing norms for acceptable explanations.

The second recommendation is for the teacher to maintain a nonevaluative stance. Instead of indicating whether or not a solution is correct, the teacher may seek comments from other students. Once again, the teacher should seek not just different claims but warrants to those claims. This reduces the possibility of negative social implications by focusing on mathematical qualities of an explanation.

The third recommendation is for the teacher to slow down and clarify the discussion. By highlighting and recording different explanations, the teacher allows a broader range of students to participate. This is also an opportunity for the teacher to clarify ambiguities and to seek greater elaboration of claims, so that students can fully reflect on the mathematical thinking of their classmates. As teachers become more experienced at this practice, they will gain greater insight into students' thinking and can begin to anticipate the various claims that may prove productive.

The fourth, and most difficult, recommendation is to attempt a synthesis or summary of a discussion, both in the mathematics and in the process of collaborative discussions. This reflective piece is crucial in modeling learning practices, especially if there were events in the discussion that the teacher identified as especially productive for future attempts at collaborative discussions or in establishing mathematical concepts.

Sustaining the Development of Collaborative Discussions

Learning to develop and direct collaborative discussions effectively is a form of expertise that is new to most teachers. It is complex and requires a well-connected sense of mathematics (what Ma [1999] calls "knowledge packages"), an understanding of mathematical argumentation, great patience, and an accumulated set of experiences attempting collaborative discussions. This will not happen in isolation.

The same learning principles that support students' engagement in collaborative discussions apply to developing teachers' expertise in conducting collaborative discussions. It also involves participants attempting to negotiate a mutual understanding of a complex topic. In this instance, it would involve a sustained community of mathematics educators who are interested in developing expertise in collaborative discussions. This community of mathematics educators would regularly discuss effective practices, share resources, and support one another's efforts. Such activities are characteristic of any group of professionals who wish to develop expertise in a complex field (Wenger 1998).

Discussion and Implications

The theme of this yearbook is how students learn mathematics. In this article, I have suggested that mathematics classrooms should have a collaborative or dialogic quality. In an era of globalization, it will be even more important for students to go beyond the mastery of standardized skills to understanding and appropriating academic forms of discourse. Participating in collaborative discussions gives students opportunities to develop proficiency in advanced forms of mathematical literacy.

The article lays out some of the challenges for students and teachers who attempt to engage in collaborative discourse. This is a complex and difficult process that will involve frustrating and seemingly unproductive moments. In order to develop expertise at establishing collaborative discourse, teachers will need to form sustained communities of practice.

It is evident from these examples that further development in these teachers' and students' practices will require yet more difficult transformations, but the potential is there to build from students' ways of thinking to develop mathematical ideas. Developing collaborative discourse communities is a difficult process; yet, if we recognize the qualities inherent in learning contexts in which people develop high levels of competency, we must continue to advocate for students' meaningful participation in classroom discourse.

REFERENCES

Forman, Ellice Ann, Dawn E. McCormick, and Richard Donato. "Learning What Counts as a Mathematical Explanation." *Linguistics and Education* 9, no. 4 (1998): 313–39.

Ma, Liping. *Knowing and Teaching Elementary Mathematics: Teachers' Understanding of Fundamental Mathematics in China and the United States.* Mahwah, N.J.: Lawrence Erlbaum Associates, 1999.

Nathan, Mitchell J., and Eric J. Knuth. "A Study of Whole Classroom Mathematical Discourse and Teacher Change." *Cognition and Instruction* 21, no. 2 (2003): 175–207.

National Council of Teachers of Mathematics (NCTM). *Curriculum and Evaluation Standards for School Mathematics.* Reston, Va.: NCTM, 1989.

———. *Principles and Standards for School Mathematics.* Reston, Va.: NCTM, 2000.

Nystrand, Martin. *Opening Dialogue: Understanding the Dynamics of Language and Learning in the English Classroom.* New York: Teachers College Press, 1997.

Schleppegrell, Mary J. *The Language of Schooling: A Functional Linguistics Perspective.* Mahwah, N.J.: Lawrence Erlbaum Associates, 2004.

Sherin, Miriam Gamoran. "A Balancing Act: Developing a Discourse Community in a Mathematics Classroom." *Journal of Mathematics Teacher Education* 5 (2002): 205–33.

Tharp, Roland G., and Ronald Gallimore. *Rousing Minds to Life: Teaching, Learning and Schooling in a Social Context.* Cambridge: Cambridge University Press, 1988.

Vygotsky, Lev. *The Collected Works of L. S. Vygotsky: Volume 1. Problems of General Psychology.* New York: Plenum Press, 1987.

Wenger, Etienne. *Communities of Practice: Learning, Meaning, and Identity.* Cambridge and New York: Cambridge University Press, 1998.

Wertsch, James V., and Chikako Toma. "Discourse and Learning in the Classroom: A Sociocultural Approach." In *Constructivism in Education*, edited by Leslie Steffe and Jerry Gale, pp. 159–74. Hillsdale, N.J.: Lawrence Erlbaum Associates, 1995.

Williams, Steven R., and Juliet A. Baxter. "Dilemmas of Discourse-Oriented Teaching in One Middle School Mathematics Classroom." *Elementary School Journal* 97, no. 1 (1996): 21–38.

9

Putting It All into Context: Students' and Teachers' Learning in One Mathematics Classroom

Jon D. Davis

PLACING a mathematical problem within the confines of a context other than a strictly mathematical or abstract one is a recurring phenomenon in mathematics textbooks in the United States. For example, during the nineteenth century students routinely worked with money. Hence, this was a face of mathematics that students came to know, as seen in the following example from Pike (1821) as cited in Michalowicz and Howard (2003, p. 93).

> A grocer bartered 5 cwt. [hundred weight, or 112 pounds] of sugar at 6d. [pence] per pound for cinnamon at 10s. [shillings] 8d. per lb; how much cinnamon did he receive?

A similar problem from the twentieth century is shown below (McConnell et al. 1996, p. 172).

> When Val works overtime at the zoo on Saturday, she earns $9.80 per hour. She is also paid $8.00 for meals and $3.00 for transportation. Last Saturday she earned $77.15. How many hours did she work?

These two examples contrast significantly from those used in *Contemporary Mathematics in Context* (Core-Plus) (Coxford et al. 2003).[1] At the heart of Core-Plus lies the view that mathematics is the study of patterns. Accordingly, Core-Plus students experiment, collect, and analyze data set in real-world contexts as they search for underlying mathematical patterns. This is similar to how mathematicians engage in mathematics on a daily basis (Steen 1990). Research suggests that solving problems set in these contexts can promote connections between the real world and mathematics and help students develop understanding (Hiebert et al. 1996). In addition, since real-world contexts often involve several mathematical areas, these problems have the potential to

1. Core-Plus was one of five secondary school mathematics curriculum projects funded by the National Science Foundation in 1992.

develop connections among these areas, which is the essence of a deep understanding of mathematics (Donovan and Bransford 2005).

Context, as represented by the two problems in this section, formed a container into which the mathematics was placed. These contexts could have easily been replaced with another without affecting how students would solve the problem. Indeed, students often ignore the context when solving such problems (Silver 1986). However, in modeling, the context is necessary in order to solve the problem, the solution to which depends on students' understanding of the context and the mathematics that constitutes the model. Since students have had different life experiences, the role of context in helping them to create mathematical models and invent mathematics in the process can be unpredictable in many ways. The first section of this article describes these surprising advantages and nonobvious barriers that secondary school students encounter when creating mathematical models in Core-Plus. The second section describes how the teacher drew on these contexts while teaching to promote students' understanding of mathematical concepts, modeling, and the real world itself. The data in this article came from a high school algebra classroom whose teacher had just begun implementing the first course of Core-Plus. The vignettes described in this article furnish a revealing glimpse into how students interact with context when creating models in the confines of a secondary school mathematics classroom.

Opening Possibilities for Students

Context to Shed Light on Unexpected Mathematical Topics

In one investigation students are introduced to systems of linear equations through the context of the percentages of male and female doctors from 1960 to 2000. At one point in the investigation students entered the equations $y = 98 - 0.54x$ and $y = 2 + 0.54x$ into their graphing calculators with x representing the year since 1960 and y representing the percent of male and female doctors, respectively. As the students began investigating the different representations of these two equations on their calculators, the following conversation occurred.

Steve:	Oh, this is weird, because if you look at the table, because it's not going to start at like 1960, it's going to start at like 1.
Teacher:	Right.
Steve:	So how are we going to get it to 1960?
Teacher:	Just assume that 0 is 1960, 1 is 1961, 2 is 1962.
Steve:	Yeah.

Teacher: And so the number, the x number is how many years after 1960.
Steve: I know, but how can we just put it to be 1960, 1961, 1962?
Jim: Can't you just add 1960 to the number?

In this example, students felt a need to transform the equation so that the table representation more closely matched the context and their understanding of the problem. The context and a desire to make sense of the models that represented it set the stage for a meaningful lesson about horizontal translations. Although a proper translation would have involved subtracting 1960 from the equation, the teacher could have asked students to make guesses about how "adding 1960" would have changed the different function representations. Next, these guesses could have been followed up by having the students alter the equation by adding 1960 to the independent variable and noting its effects. The procedure for transforming an equation to perform a horizontal translation is very counterintuitive for students as seen in Jim's suggestion to add instead of subtract 1960 to the number (Zazkis, Liljedahl, and Gadowsky 2003). These translations are traditionally taught in an abstract mathematical context that can hold little meaning for students. Here the context promotes sense making, which can support learning with understanding. Moreover, as students realize a need to use this procedure later, the context may act as a seed from which the algorithm can grow.

Kilpatrick and Silver (2001) compare and contrast a contingent teaching model with an anticipant model. In the latter, teachers carefully map out a lesson and know in advance how students are likely to interact with the material. In the former, the path that students take through the material is not completely known before class begins. In the excerpt above, the necessary conditions were in place for contingent teaching. Although the teacher only intended for students to learn how to solve a system of linear equations, the context itself along with students' sense making opened up a pathway to learn about horizontal shifts. Thus problems in which aspects of the context are needed to make sense of the solutions provide opportunities for students to travel through mathematical content in a sequence different from that laid out in most textbooks.

Developing Sensitivity to the Nuances of Mathematical Modeling

In unit 2, students were asked to create a table from the graph shown in figure 9.1 below. Initially, students created a table with negative y-values; however, they soon began interpreting this answer in light of the context as seen in the dialogue below.

Steve: Wait, you can't ever have negative fuel, can you?
Teacher: Class, we have a question. Can you have negative fuel?
Christine: On the chart, should we put zero then?

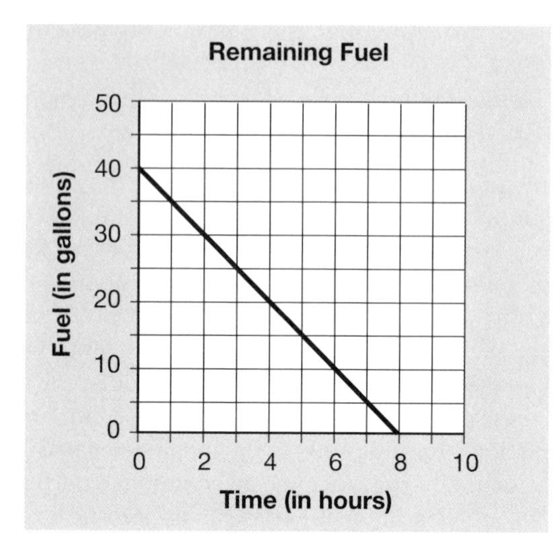

Fig. 9.1. Fuel in gallons as a function of time for an airplane in flight (Coxford et al. 2003, p. 155). (Reprinted with permission of Glencoe McGraw-Hill Publishing, © 1998)

Steve: Yeah, I don't think we can put negative five.

In mathematical modeling, students must carefully move back and forth between a real-world context and the mathematics that is used to represent it. It is easy to become fixated on either one or the other. For instance, in the vignette above, students initially applied the pattern of subtracting 5 each time to represent the amount of fuel in the tank. However, Steve kept his eye on both realms as he worked on this problem. This comment, voiced to the entire class by the teacher, helped other students learn that the context must be kept in mind when working with a mathematical model. Indeed, the connections between the real world and the mathematical one were illustrated by a student when explaining his formula $F = 40 - 5H$, relating the amount of fuel left in the tank, F, to the number of hours of the flight, H.

Student: Uhmm, well, for every hour five gallons of fuel that are used and we started at 40, so you know that's just the five gallons for every hour, so H times 5, and then we started at 40, so 40 minus that number.

Although this was the answer in the textbook that other students in class validated, Steve found it problematic.

Steve: It doesn't say like how much gallons the tank can have if it is full. It only assumes that it is 40.

Teacher:	Yeah that's true. Does it say ... you're right, it doesn't say.
Steve:	It could be 80 gallons that it can take or it could be 12.
Teacher:	Right, well it can't be 12 because in this one ...
Steve:	Oh, yeah. So what should I, what should I think for how much the tank is full or whatever?

Once again, Steve never lost sight of the context in this problem. In the previous problem, he connected the tabular representation to the real-world context. Here he was connecting an equation to the real-world context. Although he learned about the subtleties of modeling with the aid of a context that he could relate to, it also gave him a surprising opportunity to understand the role of the y-intercept in a linear function better.

Context as Barrier to Understanding Mathematics

Silver (1986) described students' difficulties in bringing aspects of the real world to bear in determining the number of buses needed to transport a group of soldiers. Students learning from Core-Plus for the first time experienced the opposite problem. They brought a rich understanding of the real world based on their experiences when attempting to model certain problems. These real-world entailments interfered with their ability to find and interpret the y-intercept even though they had previously solved similar problems with little difficulty. The first instance of this behavior occurred in the ice-cream problem, where the teacher asked students to describe the relationship between the number of scoops and the price, given that five scoops of ice cream in a cone costs $3.00 and nine scoops of ice cream costs $3.80.

Finding the slope of 20 cents per scoop was straightforward for the students, but using the context to think about and find the y-intercept relating number of scoops and price proved to be much trickier.

Christine:	One scoop would cost 20 cents and five times 20, because, like, the difference in the number of scoops—you know, the difference between five and nine—is, like, four and so between that many scoops there is 80 cents difference.
Jim:	Because now, zero has to be a one because you never ... you can't buy an ice cream that doesn't have a scoop, right?
Teacher:	Well, think ...
Christine:	They are charging you for like a cone, maybe.
Jim:	The cone costs $1.00.
Christine:	Which makes a lot of sense, then.
Teacher:	Is 1 our initial cost[2] then?
Christine:	No, 1.2 is our initial value.

2. The students later discovered their error and changed the initial price of the cone to $2.00.

Teacher: Why?
Jim: Because you can't buy an ice cream that doesn't have a scoop in it.
Teacher: Can you buy a cone?
Jim: Yeah, but why would you want to?

These two students had successfully used the initial value to find equations modeling real-world contexts in the past. In this instance, however, they were unwilling to do so, since one could not purchase a cone without a scoop of ice cream. In fact, it had such a strong effect on their mathematical reasoning that Christine placed the price of a cone and one scoop ($1.20) as the initial value.

A similar barrier for students appeared when discussing the equation $H = 1.4 + 1.25C$ relating miles per gallon in the city, C, to the miles per gallon on the highway, H. Although students had successfully made sense of the y-intercept on earlier occasions such as in the Palace Theater problem, they had difficulty thinking of the y-intercept in this problem. The teacher's decision to ask about the meaning of the y-intercept led to the following conversation.

Teacher: If I have a car that gets zero miles per gallon in the city ...
Steve: That is horrible.
Teacher: How many would I get on the highway?
Steve: That is horrible—usually, you usually have like 15.
Carl: You get zero miles. You can't go anywhere if you don't have any gas.
Teacher: Right, but according to this equation I'm not saying it will actually work ....
Steve: Well, that equation won't work.
Teacher: But if it did, just go with me on that. If it did if I got zero miles per gallon in the city, how much would I get on the highway?
Steve: Zero.
Carl: Zero.
Steve: You'd get zero because you couldn't get out of the city to get on the highway.

Here the teacher wanted students to interpret the meaning of the y-intercept in context as they had done in other problems. In order to do this, he set the independent variable to zero by asking students to consider a car getting zero miles per gallon in the city. However, this ran counter to students' experiences with cars. They equated a car getting zero miles per gallon in the city with a car that was not working.

From one point of view context was a nonobvious barrier for students to enrich their understanding of the y-intercept. From another, it was an unexpected opportunity for the students to learn the important idea that many mathematical models do not correspond completely to a real-world context and in creating a model, one must discard some aspects of the situation. The conditions were set for students to learn this idea, but it was not brought to the surface for students to examine at this point in the unit. This was, in part, due to the choices of the textbook authors and the teacher's limited experience with modeling. At the end of the semester, however, the students made more progress in understanding this idea. The teacher noted, "If my Core-Plus students encounter an outlier [a minor aspect of the real-world situation that doesn't quite fit the model] in fitting a mathematical concept to a context, they view it as an anomaly, and they ignore it. On the other hand, if my traditional students encounter an outlier in the modeling process, they will discard the model." The students in Core-Plus began to see the first step in modeling as selecting the essential aspects of the real-world situation that were absolutely necessary in order to model it.

Teachers' Moves

The focus of this paper now shifts to the teacher, who was using the Core-Plus materials for the first time. Although he had been teaching mathematics for three years before this study began, he had used very few problems drawing data from the world outside of the classroom. In the paragraphs that follow, I will describe how he used context to facilitate students' understanding of mathematics, modeling, and the real-world contexts themselves.

Using Context to Facilitate the Construction of an Equation

The teacher became more aware of the context as a tool to develop mathematical concepts and achieve specific goals as his experience with Core-Plus grew. In one activity, students were trying to find an equation relating the length of a projected image on a screen to the overhead projector's distance from the screen. The data for this experiment are shown in figure 9.2.

Distance from Screen (in cm)	Length of Projected Image (in cm)
210	35
300	48.5

Fig. 9.2. Data gathered from the overhead-projector experiment

Some students chose to look at the difference between successive values, as had been done on previous problems, finding that when the distance from the screen changed by 90 cm, the length of the projected image changed by 13.5

cm, resulting in a rate of change of 6.67cm in the length of the projected image for each centimeter the projector is moved away from the screen. Another student, Christine, however, noticed that 210/35 and 300/48.5 were both approximately 6. Instead of simply telling students that they should assume that there was a y-intercept, the teacher used the context to get students thinking about whether a y-intercept existed in this instance.

Teacher: I want you to imagine this in your head right now. I take the projector and put a 5-cm line on the surface of the projector and bring that right up to the screen so the light part is touching the screen part. Is it going to project something?

Christine: No.

Teacher: It's not going to project anything?

Joe: Well, it will, but we can't see it.

Christine: Well, it might project something. It won't be exactly zero.

Teacher: What's the size of the little bulb at the top here, the lens at the top, right?

Christine: Three inches by three inches.

Teacher: I want you to think about that. The light comes from the surface and is collected in this lens, and a mirror reflects it out of this lens. So does the entire image come off of this lens, this surface?

Ann: No.

Teacher: We've got the surface of the projector, right, so we've got this square between my fingers. This square between my fingers, okay, has to be collected in this lens, right? The lens is curved so it picks up light from here and shrinks it to this size, and the same image that is brought here is just reflected out of this lens, so it means that our square is now this big. So if I put this right up against the screen, what's going to come out will be however big this thing is, right?

Ann: Yeah.

Teacher: Is that going to be zero?

Ann: No.

Teacher: No, it can't be zero.

Christine: So we have to do something with the five [the initial size of the segment that was drawn on the overhead screen].

In the conversation above, the teacher engaged in the modeling process with students. Through his questioning he had students focus on an important piece of the context that needed to have a mathematical counterpart in the model (equation) that was being created. The teacher's learning occurred in two areas. First, he became more adept at the modeling process itself. Second, he developed a tool for teaching students how to create an equation by tapping their sense-making abilities in relation to a specific context.

Moving beyond the Textbook

In the past, the teacher had predominately used fantastical contexts such as pirates to motivate students' study of linear functions and modeling. As the teacher began using Core-Plus, he followed the textbook closely and his goals focused more on the mathematics the contexts were designed to teach. However, as his experience with the contexts in Core-Plus grew, his goals expanded over time as evidenced by his desire to move beyond the explicit questions of the textbook. The conversation below took place while students were using linear equations to describe the relationship between the ranking of a show and the viewership in millions of people, as shown by the graph in figure 9.3.

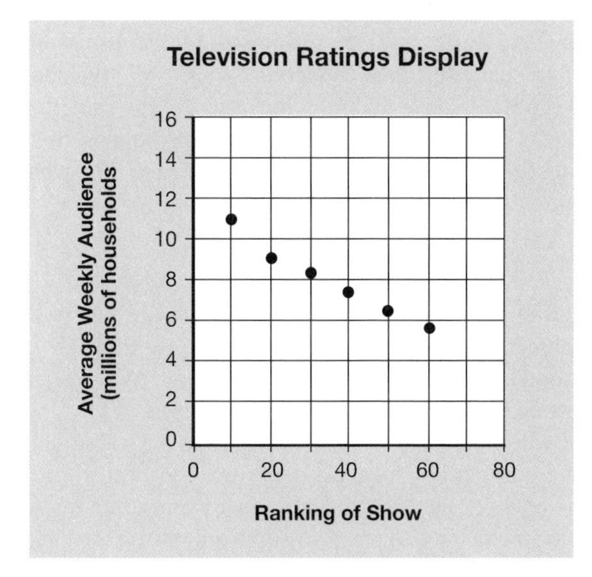

Fig. 9.3. Graph depicting the relationship between show ranking and average weekly audience (Coxford et al. 2003, p. 162). (Reprinted with permission of Glencoe McGraw-Hill Publishing, © 1998.)

Teacher:	If this is linear, what's going to happen the further we get out? What's going to happen?
Steve:	Keep shrinking.
Teacher:	Keep shrinking.
Ann:	It's going to roughly continue to be linear.
Teacher:	It's going to roughly continue to be linear, ... uhmm, so what's going to happen as it goes out?
Ann:	Eventually negative people are going to be watching.

Teacher:	If we were doing this as real as possible, would this graph ever hit zero?
Ann:	No.
Teacher:	What's it going to do?
Christine:	It would hit a point and go like sooo [*Christine makes a gently downward sloping movement with her hand like an exponential decay graph.*].

In this activity, the textbook's focus was on determining the advantages and disadvantages of different representations of the data, fitting a line to the data, finding an equation for the set of data, and making predictions using the equation. The teacher addressed these goals earlier, then he went beyond the textbook to ask questions concerned with mathematical modeling. Specifically, he pushed students to think about the viability of a linear function to make predictions beyond the data that were collected. Not only did the contexts encourage students to tap their sense making, they also did the same for the teacher. After several weeks of instruction, he began moving beyond the purely mathematical goals of the curriculum to think about where and when they made sense in the context they were designed to represent.

Understanding the Real World

In addition to the modeling goals above, the teacher believed that contexts were worthy subjects of study in their own right. He justified his actions by stating, "If they don't hear it here, when will they get it? When we work with the real world so much, I want them to have a good picture of it." A prime example of this occurred while students were discussing the rates charged by two competing telephone companies. After students worked on the problems in this investigation, he pushed them to think about why one company would provide a cheaper rate when you have to pay more up front. Using the data in figure 9.3, he asked students to think about why companies would still purchase air time for shows that had very low rankings. The students' discussion touched on the possibility that the show may have been ranked lower in the past but recently had increased in popularity. Other students brought up the loyalty of certain fans that had common interests such as technology. This may lead specific advertisers to buy air time in order to market a specific product to this group.

Conclusion

Not all problems use context in similar ways. In some problems, it represents the transparent wrapping on a brightly colored package. Students can spy the package's mathematical contents through the wrapping and, consequently, they may disregard the contextual veneer. These experiences may cause difficulties for students when working with problems where the context is an important part of finding and interpreting an answer (Silver 1986). Context plays just such a pivotal role in the mathematical modeling problems described here.

Context can be a powerful bridge between students' informal understandings and formal mathematical ideas, but their use in the classroom can also lead classroom lessons into unexpected mathematical domains like horizontal translations that the teacher may not be prepared to enter. At the same time, the secondary school students in this study occasionally struggled with applying and understanding mathematical ideas within different contexts. This was seen in the context of finding a linear relationship relating the number of scoops to the price of an ice-cream cone. Because of the disconnection between the concept of y-intercept in this problem and students' experiences of this context, they were unwilling to denote the y-intercept as the price of the cone. For these students, the utility of a context in promoting an understanding of mathematical ideas was directly related to its agreement with their out-of-school experiences of these situations. However, this led to another unintended learning opportunity for students about mathematical modeling. These situations gave students a chance to learn about the disconnections between mathematics and real-world situations that often happen when modeling. The fickle nature of lessons involving context is unavoidable because of the many different experiences students bring to the classroom, which are not always possible to determine before lessons are constructed.

Not only did the students grow in their understanding of mathematics and modeling, so did the teacher. He used contexts as a pedagogical device to help students learn about y-intercepts and the process of modeling. The contexts in the materials enabled him to tap his own sense making, which led him beyond the curriculum to focus on the limits of a mathematical model. He also believed that if students were working within a medium rich in context, they should also come away with a better understanding of the world around them.

This paper provides specific examples of the role that context plays in how students learn mathematics and modeling. In addition, it initiates a focus on how the use of context in mathematics curricula affects teachers. As more students and teachers experience curricula where real-world situations are an indispensable aspect of solving mathematical problems, we will become better prepared to seize the unexpected opportunities that wait in classrooms where it's all put into context.

REFERENCES

Coxford, Arthur F., James T. Fey, Christian R. Hirsch, Harold L. Schoen, Gail Burrill, Eric W. Hart, and Ann E. Watkins. *Contemporary Mathematics in Context: A Unified Approach.* Course 1. New York: Glencoe/McGraw-Hill, 2003.

Donovan, M. Suzanne, and John D. Bransford, eds. *How Students Learn: History,*

Mathematics, and Science in the Classroom. Washington, D.C.: National Academies Press, 2005.

Hiebert, James, Thomas P. Carpenter, Elizabeth Fennema, Karen Fuson, Piet Human, Hanlie Murray, Alwyn Olivier, and Diana Wearne. "Problem Solving as a Basis for Reform in Curriculum and Instruction: The Case of Mathematics." *Educational Researcher* 25 (May 1996): 12–21.

Kilpatrick, Jeremy, and Edward A. Silver. "Unfinished Business: Challenges for Mathematics Educators in the Next Decades." In *Learning Mathematics for a New Century*, 2000 Yearbook of the National Council of Teachers of Mathematics (NCTM), edited by Maurice J. Burke, pp. 223–35. Reston, Va.: NCTM, 2000.

McConnell, John W., Susan Brown, Zalman Usiskin, Sharon L. Senk, Ted Widerski, Scott Anderson, Susan Eddins, Cathy Hynes Feldman, James Flanders, Margaret Hackworth, Daniel Hirschhorn, Lydia Polonsky, Leroy Sachs, and Ernest Woodward. *Algebra.* Glenview, Ill.: Scott Foresman, 1996.

Michalowicz, Karen D., and Arthur C. Howard. "Pedagogy in Text: An Analysis of Mathematics Texts from the Nineteenth Century." In *A History of School Mathematics*, Vol. 1, edited by George Stanic and Jeremy Kilpatrick, pp. 77–109. Reston, Va.: National Council of Teachers of Mathematics, 2003.

Pike, Stephen. *The Teacher's Assistant, or a System of Practical Arithmetic.* Philadelphia: Johnson & Warner, 1821.

Silver, Edward A. "Using Conceptual and Procedural Knowledge: A Focus on Relationships." In *Conceptual and Procedural Knowledge: The Case of Mathematics*, edited by James Hiebert, pp. 181–98. Hillsdale, N.J.: Lawrence Erlbaum Associates, 1986.

Steen, Lynn A., ed. *On the Shoulders of Giants: New Approaches to Numeracy.* Washington, D.C.: National Academies Press, 1990.

Zazkis, Rina, Peter Liljedahl, and Karen Gadowsky. "Conceptions of Function Translation: Obstacles, Intuition, and Rerouting." *Journal of Mathematical Behavior* 22 (December 2003): 435–48.

10

Shared Reflection in an Online Environment: Exposing and Promoting Students' Understanding

Debra M. Dosemagen

MY PERSONAL interest in the topic of conceptual understanding grew out of my interaction with my own students. One such interaction serves to illustrate its importance. A student and I were working through a problem involving a rational expression, absolute value, and an inequality. In the process she had to evaluate an inequality when the variable took on certain values. At that point she asked, "Isn't there some formula I can use? Do I have to think about it?" The student's question highlighted an important principle: Students need to really understand the concepts of mathematics in order to apply them flexibly. Without the understanding of important concepts, mathematics is merely a set of magical manipulations that produce a correct but meaningless result.

To explore this principle in my own classroom and with my own students, I designed an action research project to investigate my students' own perceptions of their mathematical understanding. My objective as the teacher-researcher was to use what I learned about the students' perceptions and the instructional strategies they thought helped them actually understand mathematical concepts to make adjustments to my practice to facilitate deeper understanding.

I formulated my action research question using a combination of models (McNiff, Lomax, and Whitehead 1996; Altrichter, Posch, and Somekh 1993; Schwalbach 2003; Stringer 2004), and as suggested in several of these models, my research question focused on changes in my own practice: How can I use the content of students' reflections (shared in a Web-based environment) about their learning experiences to improve my instructional strategies as they relate to students' mathematical understanding? The intent of the question and, therefore, of the action research study was to explore the students' experiences and the nature of the instruction, as well as the use of an online environment as a forum for shared reflection.

An Advanced Placement (AP) Calculus course provided the context for this study. The College Board (2003) philosophy of AP Calculus emphasizes students' understanding of concepts rather than mere "manipulation [or] memorization of an extensive taxonomy of functions, curves, theorems, or problem types" (p. 5). The specific content of the calculus course explored during the project included the application of the derivative to related rates, optimization, economics, techniques of integration, and the application of the techniques of integration to area and volume problems.

The school in which I conducted the study was a Catholic high school with an enrollment of slightly fewer than fifteen hundred students, and it was located in a large urban area. The school attracted students from all parts of the city and suburbs and from Catholic, private, and public junior high schools. The flexible nature of the school's schedule insured that all students had Internet access during the course of the school day.

In the context of action research, students were given the opportunity to reflect on their learning, to articulate their observations about their learning, and to share these observations online with their classmates and teacher. The ultimate goal of these metacognitive exercises was twofold. First, I hoped that reflection on, and communication about, mathematical concepts would reinforce learning by prompting students to examine their own understanding of the concepts. A Web-based format furnished a public forum for students' reflections, thus making my students' thinking visible to other students and to me, the teacher. Second, as a result of this insight into my students' thinking, I intended to adapt my instruction to address specific needs.

Theoretical Context

Many conversations in the professional arena formed a backdrop for my personal observations and the action research study that grew out of them. Investigations of topics such as understanding, metacognition and reflection, communication and writing, and Web-based learning environments not only influenced my view of my students' learning but also directed my choice of instructional strategies.

Hiebert and Carpenter (1992) articulate what many other educators and authors (Carpenter and Lehrer 1999, 2001; Hiebert et al. 1997; Moyer 2001; Skemp 1971, 1987) promote. They state, "One of the most widely accepted ideas within the mathematics education community is the idea that students should understand mathematics" (p. 65). However, the National Council of Teachers of Mathematics (NCTM) in its 2000 document *Principles and Standards for School Mathematics* acknowledges, "learning mathematics *without* understanding has long been a common outcome of school mathematics instruction" (p. 20).

Because of the importance given to teaching and learning with understanding, a variety of educational models have emerged that attempt to define understanding and to provide teachers with some direction regarding ways to help students develop it. Models such as the Understanding by Design (UbD) framework developed by Wiggins and McTighe (1998) and the Teaching for

Understanding (TfU) framework created by Gardner (1999) and his colleagues at Harvard's Project Zero emphasize external, performance-based views of understanding critically shaped by goals and objectives. In contrast to understanding-as-doing (the performance view), the representational view (Perkins 1998; Skemp 1971; Hiebert and Carpenter 1992; Hiebert et al. 1997) supports an understanding-as-seeing premise rooted in cognitive psychology. When ideas "make sense" to learners or when they "get it," understanding is being described from this representational view (Gardner 1999).

From this perspective, students' learning is primarily a function of the organization of concepts, which are essentially invisible structures. Since existing knowledge embedded in these structures is so crucial to the development of new learning, students and teachers must have some way of accessing and assessing it. One strategy involves the use of metacognition, literally defined as "cognition about cognition." When individuals become aware of and express their own thought processes, they are engaged in the process of metacognition (Flavell 1985; Anderson and Krathwohl 2001; Marzano 2001).

Authors like Carpenter and Lehrer (1999) describe reflection in much the same way that other authors describe metacognition. They state that "to be reflective in their learning means that students consciously examine the knowledge they are acquiring and, in particular, the way it is related both to what they already know and to whatever other knowledge they are acquiring" (p. 22). If students are not aware of gaps in their knowledge base, "it is unlikely they will make any effort to learn the new material" (p. 60). Similarly, if a teacher is unaware of gaps in students' knowledge, it may be unlikely that she will design instruction to target those gaps.

Reflection alone is yet another internal mental act. Combined with communication, however, reflection becomes an important indicator of what is and is not understood. Carpenter and Lehrer (1999) state, "The ability to communicate or articulate one's ideas is an important goal of education, and it also is a benchmark of understanding. . . . Articulation requires reflection, and, in fact, articulation can be thought of as a public form of reflection" (p. 22). In *Principles and Standards for School Mathematics*, the NCTM (2000) notes that reflection and communication are, in fact, integrated processes.

Writing is one form of communication. Zinsser (1988) and others (see also Connolly 1989) discuss the trend toward writing across the curriculum that Zinsser asserts is "based on two principles: learning to write and writing to learn" (p. 16), which Connolly suggests was "fundamentally about using words to acquire concepts" (p. 5). Zinsser believes that students' writing presents a teacher with a "window into the brain" (p. 46), because he viewed "a piece of writing [as] a piece of thinking" (p. 50). If this written record of mathematical thinking is shared among students and their teacher, the advantages are multiplied.

What becomes potentially challenging for a teacher is the construction of a system to facilitate sharing among a group as large as a high school class. The use of online discussions can serve as a solution to logistical problems posed when twenty-five people or more attempt to share a number of their written reflections. According to Donaldson and Conrad (2002), "The power of online learning lies in its ability to enhance learning through interaction with the instructor and peers" (p. 113). Swan (2002) notes that just as face-to-face student-teacher interaction in a classroom has been shown to support students' learning, student-teacher interaction in an online environment has the same effect.

Theory into Practice

The research cited in the previous section directly and indirectly suggests a variety of things related to students' learning and instructional practice. To begin with, deep understanding is a crucial component of mathematics learning if the concepts and skills learned are to be applied flexibly to real problems. Understanding, however, is an invisible construct that can be made visible to an extent if students are prompted to reflect on the clarity and accuracy of what they know and can do and to communicate in writing the results of this metacognition. In fact, the acts of reflection and writing may, in and of themselves, strengthen the students' understanding. When students share their written reflections with peers and the teacher, students' thinking becomes an object for discussion and analysis. If a teacher uses this thinking-made-visible as a means to assess her own practice, she can adjust her instruction to better support students' learning.

Prompting Reflection

A significant element in the scenario above is student reflection. I developed reflection prompts with a variety of authors' work (Marzano 2001; Wiggins and McTighe 1998) in mind, and I intentionally asked students to formulate *written* responses to the prompts based on other authors' findings. For example, Brandenburg (2002), in her own classroom-based study, concludes that when students were forced "to demonstrate their comprehension in writing, they learned to pinpoint any confusion, compare and contrast mathematical methods, and ultimately deepen their understanding and retention" (p. 68). Reason (2003) asks students to tell her "a time when they felt they understood something mathematical" (p. 6). Marzano (2001) suggests reflection questions to monitor for clarity, such as "Can you identify those things about the concept about which you are confused?" and "What do you think is causing your confusion?" He also suggests questions to monitor for accuracy: "Can you identify those things about which you are sure you are accurate?" "How do you know?" and "What is the evidence of your judgment?" Gopen and Smith (1989) report efforts at Tulane and Duke universities to focus on reflection and writing in calculus classes specifically. The purpose of this focus was to emphasize conceptualization, and "this new hope assumes that thought and expression of thought are so closely interrelated that to require the latter will engender the former" (p. 210).

In the action research project, my students chose from a selection of prompts. This choice was an instructional decision on my part. I was concerned about keeping students engaged in the activity. I wanted to offer enough flexibility to give students something worthwhile to think about regardless of their unique experience that week. Several prompts specifically challenged them to monitor for clarity. For example, in one option, they were asked to describe an "aha!" moment when they made the transition from nonunderstanding to understanding. Another option directed them to reflect on something they had difficulty understanding. Other prompts asked them to consider the decisions they made about which problems to do and what they did when they were uncertain about the best way to solve a problem. Students posted a message at least twice a week. One posting was a student's original reflection. The second message was a response to a classmate's message. I read students' reflections and responded to them.

Analyzing Responses

It is important to note that each of the messages students posted could be viewed through two different lenses. Through the lens of a teacher, I viewed students' reflections for cues related to classroom instruction; simultaneously, I noted, through the lens of a researcher, my insights into the students' use of metacognitive strategies. As a result, I analyzed their reflections and responses in two stages. During the first stage, I read their electronic messages on an almost daily basis, checking them in the morning before I went to school. This cycle of data analysis was relatively informal. If a question or issue emerged in the forum, I addressed it that same day in class if appropriate. I was looking on the one hand for specific questions or confusions and on the other hand any hints that could influence and guide my instruction and support my students' understanding.

The second stage of my analysis was done from the perspective of a researcher interested not only in mathematical understanding but also in students' use of metacognition. This phase occurred after students completed the course. I revisited students' postings with an eye toward dominant patterns. A theme that emerged in this analysis illustrates the importance of the connections students made as they thought through their responses. To label this connection, I borrowed language from current literature on instruction in reading (Atwell 1987; Wilhelm 2002; Buehl 2001; Harvey and Goudvis 2000; Tovani 2000). This literature refers to text-to-text connections, text-to-world connections, and text-to-self connections. I adapted this terminology to describe (1) math-to-math connections, which are discussed in the NCTM (2000) *Prin-*

ciples and Standards document, (2) math-to-world connections, or the procedural knowledge students were developing, and the contexts in which they would apply this knowledge, and (3) math-to-self connections, which I have defined according to metacognitive awareness regarding mathematical understanding.

Shaping Practice

The math-to-math and math-to-world connections—or lack thereof—implied in students' responses posted in the Web-based forum had a direct impact on the instructional strategies I employed in the classroom. In some situations, students' reflections affirmed what I had been doing and encouraged me to look for new ways to use an instructional strategy. In other instances, comments challenged me to rethink how I was presenting material and what learning experiences I was developing for my students.

Affirming Practice

Many of the math-to-math connections students discussed involved the connections between a visual image and a mathematical concept. They were, for instance, seeing equations represented as graphs, and graphs as defining area. In other words, they were using multiple representations as promoted by the NCTM (2000).

A visual strategy I had been applying at this time involved the use of a graphing calculator and a projection device that could display the contents of my calculator screen, so it could be seen by the whole class. Several early references to visualization addressed the use of the calculator. Doug wrote the following:

> the calculator is a really helpful visual tool. none of that was making much sense until I saw what we were actually doing on my calculator. It's also reassuring to know I can check some of my work.[1]

Given comments like these, I tailored my instruction to emphasize, to an even greater degree, multiple representations—particularly graphing ones—related to problems with using the graphing calculator display or with manually drawing graphs for problems explored in class.

In an early message, Christie confirmed the value of this approach and noted,

---

1. Throughout this article, when students' online messages were quoted, the message was quoted exactly as the student posted it. In an attempt to encourage the students to participate in the forum, I told them the format would be informal, so they could respond in a way that was comfortable for them. This often meant ignoring formal conventions like capitalization, punctuation, and checking for correct spelling.

> When Ms. D drew that curve on the board and showed how to find the area between two curves, and everything turned out the same, it made everything really click for me. When I see something applied visually it really helps me to understand what I am doing.

Christie's observation prompted a number of her classmates to agree. For example, Maureen wrote,

> I couldn't agree with you more! Visual tools help me too!! I actually understand what Ms. D. is talking about when I can see it. It clicked for me too when she drew it out.

Nora posted a message that illustrated the value of using multirepresentations as well as the multidimensional nature of a posting. In the message, she described a hypothetical problem as a context for her question.[2] Nora wrote,

> ok. so say there's a math problem . . . it was something like a parabola concave upward and a linear equation with a positive slope that never crossed through the parabola but there was one point of tangency [*as an aside, it is important to note that Nora is accurately using mathematical language to describe her problem scenario*] (I wish I could draw a picture. . .) [*Nora recognized the value of a graphic representation of her problem and the limitation of using only words to describe the problem*] and there aren't any boundary points and I was supposed to find the area between the two curves . . . but it seemed like the area was almost infinity because the graphs never touched anywhere else and there wasn't this "area between two curves" because the "between" seem to go on forever . . . [*Nora gives evidence that she understands the importance of boundaries in a problem of this type.*] so maybe my general question is can you have an infinite area? Would such a problem like this exist? [*Nora is able to move from a specific example—the problem she describes—to a general question about infinite area.*]

In the context of a discussion of visualization, Nora explicitly articulated the value of a picture to convey a message. But beyond visualization, she demonstrated her understanding of the concept by (1) posing a question related to a key element in such problems—boundaries—and (2) creating a scenario in which the key element is absent, and then speculating about the possibility of a solution. Nora's message not only highlights the importance of connections between concrete representations and verbal descriptions but also shows the complex nature of an individual message.

As the course progressed from two-dimensional graphs, three-dimensional forms would emerge. I demonstrated ideas using a variety of manipulatives to represent concepts and provided students with manipulatives as well so that they could explore relationships. For example, when considering the volume

---

2. To help the reader digest Nora's example, my italicized comments are inserted in brackets.

of a three-dimensional object created by rotating part of the graph of a function around both vertical and horizontal axes, I supplied students with plastic eggs (roughly the shape of the solid) and modeling clay so they could create the objects themselves and explore how they related to each other and the original graph.

This activity seemed to spark even more frequent references to visualization in the forum. For example, Linda initiated this discussion when she wrote,

> Today in class I really liked using the play Dough . . . [because] it helped me visually. When we were doing things on the board, I got most of it, but making the vortex triggered my brain. What made everything come together was adding the noodles for the axis. Hopefully this will make things easier when solving volume problems because I will have a visual image in my head.

Donna continued this discussion in the forum:

> I think that the thing that has been most helpful through this entire class, so far, has been Ms. D's visualization help. From the tangent line light saber to the recent play Dough area, all the visualizations have helped me to really understand a concept much better.

A group of messages posted in the forum referred to visual strategies that students used on their own when working independently. As already mentioned, students would make graphs of equations in order to represent them visually. As Doug noted,

> I don't know what I'd do without the visual graphs. That's really the only way I know at all what is going on in class. sure you can memorize methods, but as soon as it gets more complicated you don't know where to go.

Don concurred, "The graph can show things that are not easily seen with just an equation." Maureen also agreed, "I always make a graph of the problem before I try to solve it. It makes me clear on what is being asked of me." In another message Maureen suggested that a classmate "zoom in [reference to a calculator function] on the graph, [so] you can see that they really do cross each other."

The last series of students' messages considered here explicitly illustrate several things. First, students are demonstrating a math-to-math connection because they are relating a visual representation to an algebraic representation in the form of an equation. Second, they are demonstrating what I have defined as math-to-self connections because they are able to metacognitively reflect on strategies they employ and the effectiveness of their use.

Challenging Practice

In contrast to those messages that affirmed an instructional approach I already used, other messages highlighted areas that clearly needed further development. A prominent area related to the relationships between the procedural knowledge students were developing and the contexts in which they would apply this knowledge. I labeled these connections math-to-world connections. I first became aware of this important theme when I began reading messages related to finding the area under a curve. As Kelly wrote,

> I really don't understand the need to know how to get area under a curve, etc. I guess a few limited people in the world need to know this kind of thing for headlights, etc. but is there any other reason why we should be learning this? Does it connect to something else that would be useful later? . . . I don't understand the use of this.

Maureen, too, expressed her confusion about this particular area concept. She wrote,

> Today in class we did a lot more of the "area under a curve" problems. This really didn't make sense to me and it still is a little confusing for me. I don't understand exactly why we need to find the area under a curve.

Mary explicitly expressed her need for a practical context. She also articulated her process when one wasn't provided. She explained,

> When I'm able to see the big picture I'm able to better understand a concept. I'm still stuck with this area between two curves. i looked in the book and I saw that there was a connection to profit. I really understood the profit and cost problems that we did in the economics problems. Does anyone know how these two things connect or what we are finding (in the big picture) when we are finding the area between two curves.

After reading messages of this type, it was apparent to me that I had not been deliberate enough about contextualizing the procedures we were studying. I made a concerted effort to discuss applications of the topics. For example, I was reminded of an application of area under a curve that related to biology. The application involved finding the surface area of a leaf—an irregular shape much like the graphs we were studying—in order to make a link to photosynthesis.

I also reviewed connections to physics that linked acceleration, velocity, and object position to area under a curve. As part of this discussion, I showed the class a documentary on the space shuttle program, which led into work with escape velocity as an application of the process of integration.

As the concepts in the course shifted from the consideration of two-dimensional applications to three-dimensional applications, the comments in the forum shifted as well. Students began to comment on the applications of calculus they saw in the three-dimensional world around them. For example, Mary observed,

> Now that we're doing all this area and volume stuff ... when I was putting away the dishes, I was looking at the different odd shapes of the containers that we have. I think that I can determine the method I should use to find the area and volume ... now I just need to finish doing the take home rather than analyzing my Tupperware.

Similarly, Caryn reported,

> as soon as we started working with volume and area and such i realized the real life applications for calculus. we talked about animation in class and even though the animators themselves dont use calculus someone else does! i thought that doing the volume of the duck helped me see the "real phenomena" of calculus. someone out there is using calculus to make all that stuff we have like dishes and vases and even duck figurines.

Caryn's message referenced finding the "volume of a duck," which was an activity done late in the class. According to the plan for the course, one of the final assessments for students involved the completion of a set of problems typical of problems posed on the Advanced Placement exam in calculus. Since the exploration of real objects and situations had become so important in the context of our discussions, I offered students an option in the final assessment. I went to a local department store and purchased a variety of objects of different shapes, sizes, and uses. These objects included vases, bottles, candles, toys, even lawn ornaments—including Caryn's duck.

Students who chose this option were to select an object, identify a function or functions that would closely approximate the two-dimensional shape or shadow of the object, and then use calculus algorithms to approximate the volume of the three-dimensional object. Approximately half of the students chose this assessment option. Several of these students commented on their experience and in general saw it as a positive one.

Doug described the process he and his partner used to find volume and demonstrated his understanding of the procedures involved. He wrote,

> Almost all symmetrical type objects can use calculus to find volume. For example Ms. D went to [a local department store] . . . and grabbed a bunch of miscellaneous objects that one can find the volume of the object. One of the objects included a yo-yo which is much like a square root function in the first quadrant. Take the function and integrate to find the area under the curve for however wide half the yo-yo is. Now rotate it around the x-axis causing the circular shape. then simply multiply times two to get the other half of the yo-yo. there you have it, calculus in a yo-yo.

In stark contrast to forum discussions earlier in the semester, messages posted during and after the unit on volume and rotation no longer asked the question, "When does anyone use this?" Instead, students commented on the ubiquitous nature of calculus. Christie observed,

> I totally understand. Everything can be applied to calculus. Everything can have its volume found in some way or another.

Doug concurred,

> where isn't calculus used???? thats the more difficult question because calculus is the mathematics of change, and what is not changing ... calculus can be applied to probably everything in one way or another, even if it hasn't yet.

Illuminating Metacognition

In light of the literature on metacognition, I reviewed postings I had categorized as examples of math-to-self connections when there was evidence that students displayed a degree of self-knowledge about their own understanding, or a self-assessment of their ability to perform or apply a particular technical skill. In some messages, metacognitive knowledge was explicitly stated. In the vast majority, however, including those already cited, the self-assessment process was implied in the description of a concept. Examples discussed here are meant to be illustrative, not exhaustive, since by the very fact that students could capture their thinking in words, they engaged, to one degree or another, in a self reflective, metacognitive act.

From my students' responses, I identified three categories within the theme of math-to-self connections. The first category includes those responses in which students raise a question. In order to pose a question, students must be aware of a gap in their understanding. The second category includes those responses that explicitly identify a concept that a student does or does not understand, like those noted earlier. Often responses in this category indicate a movement from nonunderstanding to understanding, but that movement is attributed to something unknown, a mysterious process.[3] Students report, "It just clicked." Finally, in the third category, much like the first one, students identify a transition from nonunderstanding to understanding, but they also suggest some type of strategy—either instructional or an independent learning strategy—that helps them make that transition.

Questioning

A number of students posted messages that raised general or specific questions about mathematical concepts or processes. When students pose an authentic question, they demonstrate two things. First, they demonstrate their ability to assess their own understanding and recognize that a gap existed. Second, when they formulate a specific question, they pinpoint a particular area that is causing them confusion. They also demonstrate a degree of metacogni-

3. Skemp (1987) and Reason (2003) both discuss the role of intuition, which can be either an ungrounded leap of faith or a true insight into the mathematical process.

tive self-knowledge as described by Marzano (2001) and others (Anderson and Krathwohl 2001; Wiggins and McTighe 1998) because they identify a concept that was unclear to them and also recognize what helps them to clarify their confusion. In Marzano's terms, they monitor the clarity of their thinking.

For example, Stephanie posed a question but implied she had a sense of what the answer might actually be. By suggesting a viable answer, she demonstrates an understanding of the processes, in this example using the area of rectangles to estimate area under a curve. Stephanie asked,

> My question is whether or not there is a way to estimate the area between curves the same way you can estimate the area under a curve. Does the rectangle thing work? I think I get the integration way of finding the area, but what happens when there are no equations, or only one?

In contrast, Christie posed a general question, but she gave no indication that she had a conjecture to confirm. Her question suggests that she is having difficulty with some of the algebraic processes demanded in the problems. She asked,

> I am really having a hard time trying to do some of those optimization problems ... I guess the real problem I have is that I cannot tell which formula or how to go about solving certain types of problems. Are there any quick and easy hints to tell what kind of optimization problem it is? I also don't [know] how to get two variables in a problem. I can't get two variables to work with. There is always three in there. It really bothers me. Are there any great hints for me? It seems like there should just be an easier way but I can't find it! Help!

Some of the questions students posed were speculative rather than directed at some specific process or problem. For example, Nancy's question implies that she may have discovered an alternative approach to a type of problem, and she is checking to see if she might be right. She wrote,

> I was just wondering if when you are trying to find the area between two curves and it is in an interval in which one equation is above the other for some time, but then under for the rest (I'm not too sure I described that well) if you can some how solve them together instead of having to set up and solve for two different areas and then add them together. I am pretty sure that you have to solve for the different areas, but I was just wondering because if so that would save some time.

Nancy also recognizes the general value of asking questions. She commented as follows:

> Isn't it always fun to all of a sudden have a realization about how to solve things. I think this just shows the importance of asking questions, because now not only do you understand how to do the problem but it is something that you will not forget.

Movement

In many of their reflections, students indicate that their understanding had changed, but they do not articulate what may have caused the shift. For example, Stephanie wrote,

> I finally understand this substitution thing. I was in [class] and realized that I had been dividing wrong. Well, not dividing wrong, but leaving du all by itself instead of getting the coefficient on that side and isolating dx and whatever other variable there was inside the integration notation. . . . Anyways, once I started to substitute the right way this stufff didn't seem that hard anymore.

In her message, Stephanie "realized" what she had been doing incorrectly, but she does not indicate how she came to that realization. However, she is able to accurately describe the mathematics involved and articulated the specific mistake she is making as well as how to fix it. Linda posted a similar message:

> Yes, when I was doing the homework, I forgot to find boundary points on the first one. I don't know what I was doing, I think I was making them up. But then I realized my mistake when it didn't work out, and I'm glad that we can integrate the two functions together, or else it would take a really long time.

Unlike the others, she makes an attempt to analyze her thinking more deeply when she stated, "I don't know what I was doing. I think I was making them up." On some level her message implies that "knowing what she was doing" incorrectly would help her be more accurate.

Several students described their "aha!" moments with language that had an almost magical quality to it. For example, Henry wrote,

> It wasn't so much of an "aha!" moment as it was a "d'uh" moment. I couldn't figure out why we were supposed to set dirivitives equal to zero in order to find the rectangle of greatest area, or sphere of greatest volume, or whatever of greatest such and such. Then it hit me, because the the slope of the graph equals zero when the graph hits a relative maximum, d'uh. I guess the concept just needed to incubate in my mind for a little while.

Henry wrote that the insight "hit" him and may imply, with the use of the term *incubate*, that an almost unconscious process had taken place. Maureen, too, described a moment of insight when a concept "clicked." She wrote,

> I love examples too!! Once I get a few good examples in, it usually clicks for me. Seeing which numbers go where and why is really clarifying. . . . When there is a formula to work with, it seems to flow so easily. Without getting a visual and practice with an equation, I don't think I would be able to do any of my math homework.

In contrast to Henry, however, Maureen is able to identify several specific things that may have affected her understanding—modeling, visuals, and practice.

Maureen's reflection highlights a significant issue. There was a range of insight evident in the reflections the students shared. Like many of her messages, in the example above Maureen identifies what she understood and to some degree what helps her understand it. By noting what helps her understand, she also suggests a general strategy to use when learning a concept. In contrast, some students, like those quoted at the beginning of this section, identified only a shift in their understanding. Others were so general in their remarks that it may even be difficult to discern the topic they were discussing. Students' responses illustrate the range of what Wiggins and McTighe (1998) called "innocent" to "wise" self-knowledge.

Strategies

In many of the online reflections, students identify an area in which they have difficulty and discuss what they did to adjust their process. In some instances students suggest a general learning strategy, which could be transferred to other subject areas, or a specific problem-solving strategy, or both. For example, Lisa describes a general strategy. She writes:[4]

> When I am faced with a problem that I have to solve I first **look at the direction** to see what they are asking for. Once I **figure out what type of problem** it is I **recall the formulas**. Next, I **plus [plug] in the information** I have and **see what I am missing**. From there I **solve the problem**.

More specifically, Linda wrote,

> At first I did not understand how the *du* was playing a double role. But what helps me is **taking each problem slow**. I look **back to the orginal problem** to see what I have and I try to **find things that will easily sub back in**. And **putting the coefficient outside** the integral signs helps a lot and avoids confusion.

In this message, Linda suggests several general strategies—going slowly, checking the original problem—and several problem-solving strategies—paying attention to substitution and relocating the coefficient—that are specific to problems of this type. Other students also posted messages that identify an area of confusion for them and then suggest a strategy to help them work more accurately. These strategies, culled from students' reflections, are listed below:

- Taking a step back from problems
- [Don't] look too deep
- Pay more attention to notation
- All I need to do is a little algebraic manipulation
- Paying careful attention to which axis we are rotating around
- Separating fractions

4. I've inserted boldface type to highlight Lisa's and Linda's specific references to the use of a strategy.

- Basic things like breaking a fraction apart
- Looking at the thing I would replace with u
- Make a graph or draw a picture
- Use examples and practice problems
- Look for simple substitution
- [Start with the] part with the most terms
- Write out all the steps
- Look at things in different ways . . . $dx/3$ is the same as $1/3\ dx$.
- Check my work—mentally find it's derivative
- Look for the easy math first
- Distinguish between derivative and integration
- Think it through in your head or on paper and see if the derivative of the part you are choosing is still in the problem so you can substitute for dx

The strategies students identified could, in general, be grouped into two categories: pay attention to detail and avoid overcomplication. A theme running throughout these strategies relates to the cumulative nature of mathematics. Even in AP Calculus, fundamental ideas about areas like fractions and algebra get revisited. In addition to highlighting strategies students used during problem solving, this list also offered me a set of directions for those things I needed to emphasize when modeling problems in class. I was particularly conscious of these approaches when verbalizing, in a think-aloud manner, the steps I would use to complete a problem.

Online Environment

As a teacher, the reflections students posted online were invaluable to me. By reading their messages, I became aware of the need to set a context for the procedures we were studying, to use visual representations to clarify abstract concepts, to demonstrate a variety of sample problems, and to connect explicitly mathematical concepts to the context, to the visuals, to the examples, and to one another. My hope also was not just to gather information about students' understanding but also to further it by asking students to reflect on it.

Logistically, the online environment presented an efficient way for the students and me to exchange information and share reflections. If I intended to respond to questions or misconceptions in a timely manner, I needed ongoing access to my students' reflections. The online connection allowed me that access.

However, when I interviewed a small group of students as a follow-up to the classroom-based action research project and asked general questions about their understanding, such as, "Could you describe something that helped you understand a concept?" none of them referenced the online reflection. In fact, students didn't refer to that activity until specifically asked to think about it.

Mark had the most positive reaction of those who were interviewed. He commented,

> I know every time I went on, I would just read everyone's. And, you know, whoever posted this question about this problem, I'm sure at least 60 percent of the people had that question. And it was nice that somebody took the time to write it out. Then usually you answered it. That was nice. . . . I went on. I couldn't give you a number, but just to read, just out of curiosity, you go online, you check the forum, check the discussion boards.

Kelly honestly spoke of the difficulty she had doing the online assignments. She stated,

> The honest truth for me is that . . . I felt like some of the time I had to make up what I was going to say because there was really nothing I felt I needed to say at the time on the subject we were talking about. . . . That's where it got difficult. But then when we didn't have the answers, and I got the answers, and it was awesome, and I'd learn even more and stuff. But sometimes we were just like "I don't have anything to say about the topic."

Even though the interviewees did not explicitly describe a value in their online reflections, an interesting development emerged in a final follow-up activity. I asked the interviewees to categorize classroom Indicators of Lesson Quality as identified by Horizon Research (see Weiss et al. 2003) in a study of mathematics and science instruction. I gave students a list of quality indicators and asked them to select and rank the eight items that they thought were most helpful in supporting their understanding in our calculus class. It was especially gratifying for me to note that the only item selected by all these students as important was the following: "Teacher is able to 'read' the students' level of understanding and adjust instruction accordingly." This was, after all, the whole point of the online reflections: to give me the opportunity literally to "read" what students had written about their understanding in order to tailor my instruction to support that understanding.

Final Observations

In this project, reflection for the purpose of improving my own practice was done in conjunction with the processing of the students' reflections, and it led me to several observations. First, what may seem like an obvious step, I had to ask the students to reflect. While reading their responses as they posted them, I was struck by how much more I knew about student thinking—individually and collectively—than I ever had in the past. Certainly, traditional assessments I used and even the performance assessments of students' understanding suggested by Gardner (1999) and his colleagues and by Wiggins and McTighe (1998) would not have given me this same kind of window into a student's thinking. As Skemp (1987, p. 160) notes,

> From the marks on the paper, it is very hard to make valid inference about the mental processes by which a pupil has been led to make them.... In a teaching situation, talking with the pupil is almost certainly the best way to find out; but in a class of over 30, it may be difficult to find the time.

Online communication like that used in this study may offer a viable alternative to the conversations Skemp suggests.

The second and third observations were almost inseparably linked. The efficient and effective processing of students' responses—in the words of the National Board for Professional Teaching Standards (NBPTS) (1999), "managing and monitoring student learning"—became crucial, as did the resulting adjustments in classroom practice. Students posted messages nearly every day during the eleven weeks of the project. Course topics sometimes changed quickly. In order to respond efficiently in class or online to students' ideas, questions, and misunderstandings, it was best for me to read their messages on a daily basis. This demanded flexibility on my part. Had I felt tied to a lesson plan schedule that focused on the coverage of material mapped out months in advance, I may have viewed alterations in planned lessons as disruptive rather than instructive. Although some adjustments to instruction involved simply reemphasizing a concept like the axis of rotation in the three-dimensional problems or more consistently using strategies like graphing, others involved almost starting over by reintroducing a topic.

Throughout this process, I recognized that just because students said they understood a concept, there was no guarantee that they did, in fact, understand it completely. It was important for me to keep in mind that the reflections were just that, reflections, not actual concepts. The students' written reflections were the filters through which I viewed their conceptions. In these reflections students were revealing their *thinking about their mental representations, not their representations.* Understanding could only be inferred, and students' metacognitive reflections furnished one source. However, the value of students' reflections as a means of formative assessment of their understanding in conjunction with more traditional forms of assessment cannot be overstated.

Throughout the study, I was conscious of the fact that I was working with a group of exceptionally talented students. I was particularly cautious about what my findings might suggest about students who had different characteristics. I am reminded of a first-grade classroom I observed. When the teacher asked a student to explain the way she arrived at her answer to a math problem, the student responded: "I did it in my head." Even though calculus students were more articulate in their explanations, translating mathematical representations into words sometimes posed a challenge. Scott noted this difficulty in one of his observations:

> If I get the concept, a lot of the time, I think the problem was just putting it into

> words. [It] just seemed incomplete in some way. There were a lot of ones where you said "explain how you understand integration or differentiation," and I think whatever I seemed to write just didn't seem to fully encompass it all. I think that's more a deficiency of language than anything else.

Embedded throughout the discussion of students' reflections and interviews are suggestions for teaching practice. A summary of these suggestions is offered here as a reminder to me as well as food for thought for other mathematics teachers and teachers in other subject areas and grade levels:

- There is a great deal of power in asking the simple question, What do you understand? In order to respond, students reflected on their own thinking, formulated their responses, and then communicated them. I wouldn't have known what they were thinking if I hadn't asked.
- When I asked the question, I had to be prepared for the responses. On a basic level, that meant being prepared to process the responses in a useful way. The reflections are only valuable as windows into students' thinking if the teacher has time to process them.
- Related to the item above, it also meant I needed to be prepared to adapt and to change my lesson plans in order to respond to what students did not understand. This flexibility would challenge a priority on coverage as opposed to an emphasis on students' learning.
- To really understand students' understanding, I needed more than one window from which to view it. Analyzing traditional problem-solving activities, reading students' reflections, and viewing nonverbal renderings of concept maps would create a more complete picture of students' learning.

Students' mathematical understanding is complex and multifaceted. It is rooted in past experience and can be supported in a comfortable learning environment carefully and intentionally orchestrated by a teacher. Traditional and performance assessments of problem-solving capability offered only limited views of what students potentially understood conceptually. Asking students to reflect on their own understanding, an act often equated with intelligent behavior (Marzano 2001), opens additional windows, and the power of that reflection is multiplied when it is articulated, shared, and processed. In a practical way, the Web-based environment offers a concrete arena in which that discourse occurs.

REFERENCES

Altrichter, Herbert, Peter Posch, and Bridget Somekh. *Teachers Investigate Their Work: An Introduction to Action Research.* London: Routledge, 1993.

Anderson, Lorin W., and David R. Krathwohl, eds. *A Taxonomy for Learning, Teaching, and Assessing: A Revision of Bloom's "Taxonomy of Educational Objectives."* New York: Longman, 2001.

Atwell, Nancie. *In the Middle: Reading, Writing and Learning with Adolescents.* Upper Montclair, N.J.: Boynton/Cook, 1987.

Bloom, Benjamin, ed. *Taxonomy of Educational Objectives: Handbook 1—Cognitive Domain.* New York: David McKay, 1974.

Brandenburg, M. Luka. "Advanced Math? Write!" *Educational Leadership* 60 (November 2002): 67–68.

Buehl, Douglas. *Classroom Strategies for Interactive Learning.* Newark, Del.: International Reading Association, 2001.

Carpenter, Thomas P., and Richard Lehrer. "Teaching and Learning Mathematics with Understanding." In *Mathematics Classrooms That Promote Understanding,* edited by Elizabeth Fennema and Thomas A. Romberg, pp.19–32. Mahwah, N.J.: Lawrence Erlbaum Associates, 1999.

———. "Teaching and Learning Mathematics with Understanding." In *Planning Curriculum in Mathematics,* edited by Wisconsin Department of Public Instruction (WDPI), pp. 21–33. Madison, Wis.: WDPI, 2001.

College Board. *AP Advanced Placement Program Course Description: Calculus AB, Calculus BC.* Electronic version. New York: College Board, 2003.

Connolly, Paul. "Writing and the Ecology of Learning." In *Writing to Learn Mathematics and Science,* edited by Paul Connolly and Teresa Vilardi, pp. 1–13. New York: Teachers College Press, 1989.

Donaldson, J. Ana, and Rita-Marie Conrad. "Moving F2F Activities Online." In *18th Annual Conference on Distance Teaching, and Learning,* edited by William Winfield, pp. 113–18. Madison, Wis.: University of Wisconsin—Madison, 2002.

Flavell, John H. *Cognitive Development.* Engelwood Cliffs, N.J.: Prentice Hall, 1985.

Gardner, Howard. *The Disciplined Mind.* New York: Penguin Books, 1999.

Gopen, George D., and D. A. Smith. "What's an Assignment like You Doing in a Course like This? Writing to Learn Mathematics." In *Writing to Learn Mathematics and Science,* edited by Paul Connolly and Teresa Vilardi, pp. 209–28. New York: Teachers College Press, 1989.

Grouws, Douglas A., ed. *Handbook of Research on Mathematics Teaching and Learning.* New York: Macmillan Publishing Co., 1992.

Harvey, Stephanie, and Anne Goudvis. *Strategies That Work: Teaching Comprehension to Enhance Understanding.* York, Maine: Stenhouse Publishers, 2000.

Hiebert, James, and Thomas P. Carpenter. "Learning and Teaching with Understanding." In *Handbook of Research on Mathematics Teaching and Learning,* edited by Douglas A. Grouws, pp. 65–97. New York: Macmillan Publishing Co., 1992.

Hiebert, James, Thomas P. Carpenter, Elizabeth Fennema, Karen C. Fuson, Diana Wearne, Hanlie Murray, Alwyn Olivier, and Piet Human. *Making Sense: Teaching and Learning Mathematics with Understanding.* Portsmouth, N.H.: Heinemann, 1997.

Marzano, Robert J. *Designing a New Taxonomy of Educational Objectives.* Thousand Oaks, Calif.: Corwin Press, 2001.

McNiff, Jean, Pamela Lomax, and Jack Whitehead. *You and Your Action Research Project.* London: Routledge, 1996.

Moyer, John C. "Using Research to Guide Mathematics Program Development." In *Planning Curriculum in Mathematics,* edited by Wisconsin Department of Public Instruction (WDPI), pp. 275–84. Madison, Wis.: WDPI, 2001.

National Board for Professional Teaching Standards (NBPTS). *Standards for Adolescent and Young Adulthood Mathematics.* Washington, D.C.: NBPTS, 1999.

National Council of Teachers of Mathematics (NCTM). *Principles and Standards for School Mathematics.* Reston, Va.: NCTM, 2000.

Perkins, David. "What Is Understanding?" In *Teaching for Understanding: Linking Research with Practice,* edited by Martha Stone Wiske, pp. 39–58. San Francisco: Jossey-Bass, 1998.

Reason, Melanie. "Relational, Instrumental and Creative Understanding." *Mathematics Teaching* 184 (September 2003): 5–7.

Schwalbach, Eileen M. *Value and Validity in Action Research: A Guidebook for Practitioners.* Lanham, Md.: Scarecrow Press, 2003.

Skemp, Richard R. *The Psychology of Learning Mathematics.* Harmondsworth, England: Pelican Books, 1971.

———. *The Psychology of Learning Mathematics.* Expanded American ed. Hillsdale, N.J.: Lawrence Erlbaum Associates, 1987.

Stringer, Ernest T. *Action Research in Education.* Upper Saddle River, N.J.: Pearson Education, 2004.

Swan, Karen. "Building Learning Communities in Online Courses: The Importance of Interaction." *Education, Communication and Information* 2 (May 2002): 23–49. (Retrieved on September 7, 2004, from Academic Search Elite database)

Tovani, Cris. *I Read It, but I Don't Get It: Comprehension Strategies for Adolescent Readers.* Portland, Maine: Stenhouse Publishers, 2000.

Weiss, Iris R., Joan D. Pasley, P. Sean Smith, Eric R. Banilower, and Daniel J. Heck. *Looking inside the Classroom: A Study of K–12 Mathematics and Science Education in the United States.* Chapel Hill, N.C.: Horizon Research, 2003.

Wiggins, Grant, and Jay McTighe. *Understanding by Design.* Alexandria, Va.: Association for Supervision and Curriculum Development, 1998.

Wilhelm, Jeffrey D. *Action Strategies for Deepening Comprehension.* New York: Scholastic, 2002.

Zinsser, William. *Writing to Learn.* New York: Harper & Row, 1988.

11

Amplifying Student Learning in Mathematics Using Curriculum-Embedded, Java-Based Software

Eric W. Hart
Christian R. Hirsch
Sabrina A. Keller

THE power and potential of computer technologies for enhancing student learning of mathematics have long been recognized (cf., Conference Board of the Mathematical Sciences 1983; Fey et al. 1984). However, cost and access issues have remained barriers to widespread use of these technologies. More recently, the Internet and World Wide Web are being tapped for their potential to improve the teaching and learning of mathematics. At the interface of the universal World Wide Web and the individual personal computer is the technology of machine-independent software. Java is one example of a machine-independent programming language. Typically, Java software resides on the Web and can be operated through any Web browser, such as Internet Explorer, Firefox, or Safari. Such software applications are called Java applets, or simply applets. Java can also be used to produce applications that can be downloaded and run directly on personal computers outside the environment of a Web browser and without the need for constant Web access. Well-designed, Java-based software offers a way of providing virtually all students access to powerful tools that can help them learn mathematics more effectively.

In this article, we discuss how curriculum-embedded, Java-based software can be used to amplify student learning in high school mathematics. Examples will be drawn from the areas of geometry, algebra, and statistics.

This paper is based on work supported in part by the National Science Foundation under grant no. ESI 0137718. Any opinions, findings, and conclusions or recommendations expressed in this material are those of the authors and do not necessarily reflect the views of the National Science Foundation.

We are grateful to our colleagues James Fey, Harold Schoen, and Ann Watkins for their collaboration in developing the curriculum and shaping the curriculum-embedded software described in this article. We also thank Anna Kruizenga, James Laser, and Beth Ritsema for their helpful comments on improvement of the software.

Amplifying Student Learning

The idea of *amplifying student learning* that we consider stems from Jerome Bruner, who referred to cognitive tools such as written language as *cultural amplifiers* of the intellect, in that they amplify cognitive processes (Bruner 1966). We focus on Java-based technology tools as cognitive tools that amplify students' cognitive processes by allowing them to investigate and solve problems that may be beyond their reach without such tools. The development and use of these tools are embedded in the development and use of a high school mathematics curriculum, as described later in this article. The software tools empower students to learn mathematics and attack problems from several and perhaps new directions (for example, from high-powered numerical, graphical, or symbolic directions) using a variety of possibly new tools (such as simulation or numerical solving routines) that require potentially different capabilities.

Some describe such enhanced cognitive activity as even more than just amplified. For example, Goos, Galbraith, and Renshaw (2004, p. 93) claim, "Cultural tools do not simply amplify cognitive processes—they fundamentally change the nature of the task and the requirements to complete the task. The rapid development of computer technology and its application to classrooms provide numerous examples...." Pea (1985, p. 168) claims that these "cognitive technologies" help "transcend the limitations of the mind ... in activities of thinking, learning, and problem-solving." Similarly, Brown (1984) described technological tools as *idea amplifiers.* For example, early software developed by Brown, AlgebraLand, "freed [students] from hand calculations associated with executing different algebraic operations and allowed [them] to focus on high level problem-solving strategies" (Pea 1987, p. 92).

However, the cognitive power of such technology tools is not automatically realized. Substantially amplifying student learning requires well-designed tools used in pedagogically appropriate ways in a rich curriculum. This is perhaps stronger than, but similar to, what Shaughnessy, Garfield, and Greer (1996, p. 217) note in a review of international research on educational uses of statistics software: "Dorfler (1993) argued that interacting with a computer within a system that includes an individual, some representational and computational tools, and other people within a social context, does not merely amplify cognitive capacities. It in fact brings about a reorganization of cognitive capacity." More generally, in an analysis of student-centered learning environments, Land and Hannafin (2000, p. 15) claim, "Constructivist environments scaffold thinking and actions in order to deepen understanding. They provide opportunities for learners to amplify and extend cognitive capabilities, as well as to reorganize thinking processes."

In this article, we will consider how curriculum-embedded, Java-based software applications can be used as technological tools in a constructivist learning environment to amplify student learning in mathematics. By *curricu-*

lum-embedded software we mean software that is developed from the ground up with specific curriculum applications in mind and is then embedded, in a ready-to-use manner, into the curriculum. A curriculum-embedded software approach using Java applets has been briefly discussed with respect to the middle school Connected Mathematics curriculum in Hart et al. (2005). Here, we will focus on a suite of emerging Java-based software tools called CPMP-Tools (Keller 2006) as incorporated in the high school curriculum, Core-Plus Mathematics.

Developing CPMP-Tools: A Suite of Customizable, Curriculum-Embedded Software Tools

In spite of the considerable promise that computer technology provides for the improvement of school mathematics and student learning, the fulfillment of the promise has been stymied by issues of finance, access, and equity, among others (Heid 1997). We consider the following question:

> What would it take for cognitive technologies like spreadsheets, computer algebra systems (CAS), and interactive geometry, discrete mathematics, and statistics tools to become a more integral part of student learning and work?

Among the answers to this question as reported in Heid (2005), and which we address below, are universal availability, potential to be tailored for particular purposes, and curricula that incorporate these tools as an integral part of the development of mathematics.

Regarding universal availability, recent research on Internet use indicates that 73 percent of American adults use the Internet (Pew Internet and American Life Project 2006), and "most kids, almost all kids, have a place in which they can go online and have gone online" (Vicki Rideout, vice-president of the Henry J. Kaiser Family Foundation, which has studied Internet use by race, ethnicity, and age, as quoted in Marriott 2006, p. A1). As such, increasing availability of Internet access may provide the "tipping point" for software tools to become widely available to support student learning of mathematics.

Since 1992, the Core-Plus Mathematics Project has been involved in the research and development of curriculum materials that interpret and implement the curriculum, teaching, and assessment recommendations in the NCTM *Standards* documents (cf., Schoen and Hirsch 2003). Because of concerns for access and equity, the first-edition materials were based on a modest technology assumption—students would have access to graphing calculators in class and outside school. As work began on the second edition, the contextual and mathematical problems that the curriculum was being organized around, and the learning expectations for students, were such that it was desirable to augment graphing-calculator technology with computer tools. To meet this challenge and maintain our commitment to access and equity, the project systematically explored the development of Java-based software that eventually evolved into CPMP-Tools (see fig. 11.1)—a suite of both general-purpose and custom tools whose development continues to be guided by, and integrated with, the development of the curriculum materials.

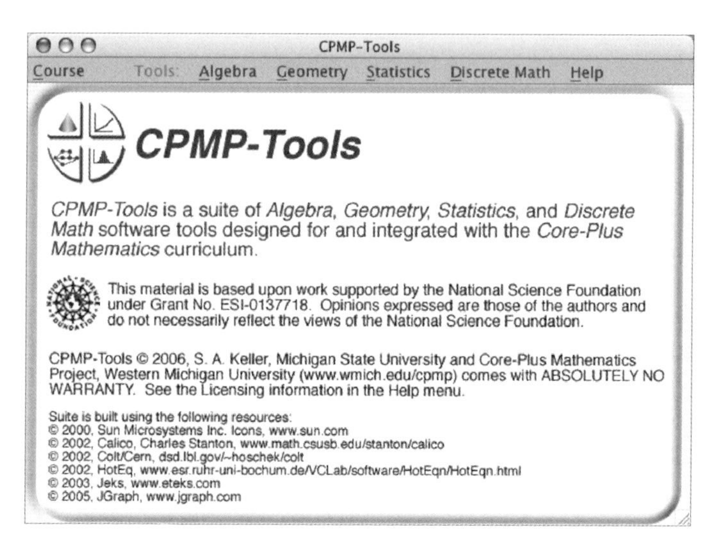

Fig. 11.1. CPMP-Tools opening screen

The CPMP-Tools opening screen suggests several design decisions:

- Tools were developed for each strand of the curriculum—algebra, geometry, statistics, and discrete mathematics. Developing students' disposition and ability to make decisions about what technology tool to use and when was an important consideration. The design of CPMP-Tools keeps tools for all strands up front and readily available.
- Tools and their functionality were organized by course to focus on the intended mathematics and to reduce the steepness of the learning curve. This has allowed the software capabilities to evolve with the mathematics and students' understanding.
- Tools share similar menu screens and interface, promoting learning transfer from one tool to another.
- Tools are built using Java WebStart, which permits safe, easy, and reliable distribution of software and software updates across different types of computers.
- As GNU public license software, CPMP-Tools is available for use by anyone, in particular by teachers and students who are using curricula

other than Core-Plus Mathematics. Equally important, other curriculum developers are welcome to modify and enhance the software and build customized capabilities (much like the Custom Tools in CPMP-Tools) tailored to their curricular goals, subject to the general guidelines of GNU public license software.

In the collection of tools supporting student learning in each strand, the user will have access to a general-purpose tool and custom tools that are focused on a particular mathematical or statistical concept or problem. For example, in the Geometry tools there is a general-purpose, learner-oriented, interactive geometry tool, as well as several custom tools that support student investigation of specialized problems in a particular domain of geometry.

The design of the interactive geometry software illustrates how the functionality of CPMP-Tools is characterized by its intended use in the curriculum. For example, to promote deeper student understanding of the connections between coordinate and matrix representations of geometric ideas, and to support student learning of applications of the mathematics to animation, the interactive geometry software was designed to perform matrix calculations (both automatically and by command) and to offer a simple, easily understood programming language (see fig.11.2). The capability for the learner (and software) to move flexibly between coordinate and matrix representations of figures

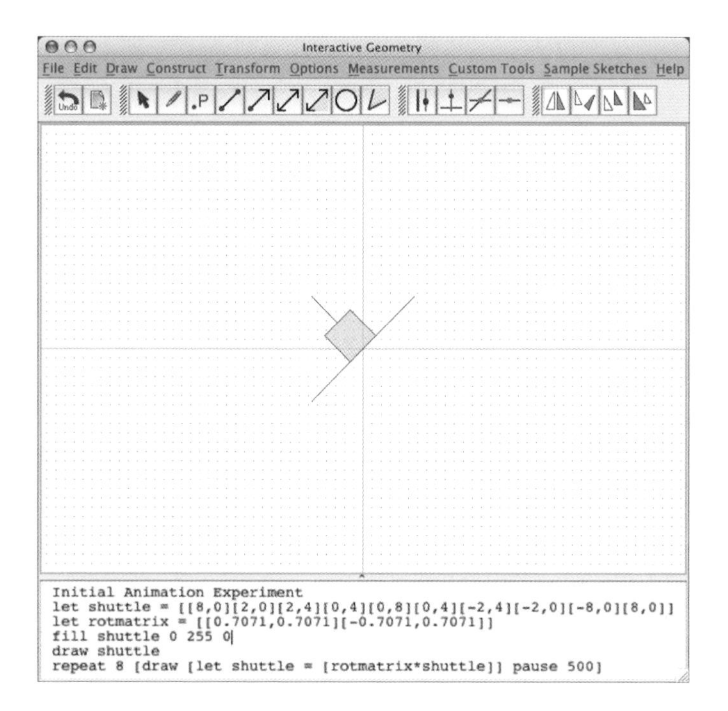

Fig. 11.2. Coordinates, matrices, and animation

and transformations is an important goal of the curriculum. The programming capability further supports the curriculum goals of focusing on contemporary applications of mathematics and developing deeper student understanding of logic, variables, functions, recursion, iteration, and algorithmic problem solving.

For each strand of the curriculum, CPMP-Tools includes custom tools that enable students to investigate particular problems and specific mathematical questions, often with the potential of guiding students to conjecture and mathematical justification. For example, in the Patterns in Shape unit in Course 1, emphasis is on understanding properties of shapes that make them useful in building and design. Using linkage strips and fasteners, students easily discover that, unlike triangles, quadrilaterals and other polygons are not rigid figures. The fact that quadrilaterals are not rigid makes them particularly useful in the design of mechanical linkages. Quadrilateral linkages (see fig. 11.3) have different characteristics depending on the length of the sides and which side is used as a crank.

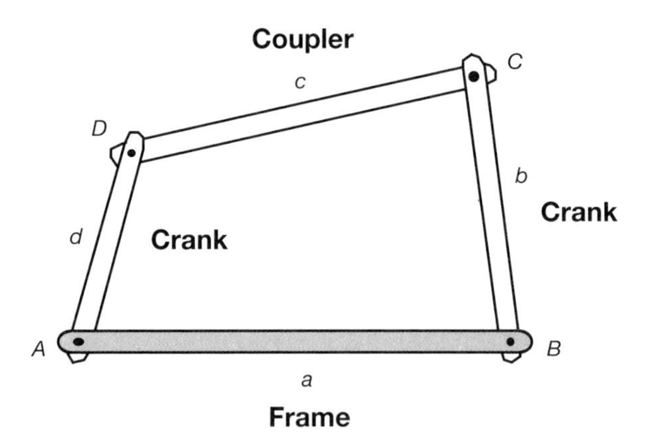

Fig. 11.3. Quadrilateral (4-bar) linkage

Among the questions students consider are the following:

- Working with a partner, make several different quadrilateral linkages so that strip $\overline{AB}$ is the longest side and fixed; strip $\overline{AD}$ is the shortest side and acts as one of the cranks. Investigate how lengths a and d are related to lengths b and c when $\overline{AD}$ can rotate completely. In this instance, how does the follower crank move? The coupler? Write a summary and explanation of your findings.
- How would you design a quadrilateral linkage that has two cranks that rotate completely? How must the side lengths be related?

(Adapted from Hirsch et al. forthcoming [a], Course 1, pp. 367–68)

Investigating properties of quadrilateral linkages with physical linkage strips and fasteners is helpful, but there are drawbacks. These include correctly forming the quadrilaterals, the time-consuming nature of forming many different quadrilaterals, and the visual memory needed to retain movements of linkages. The custom tool "Design a Linkage" (see fig. 11.4) allows students to adjust lengths of already formed quadrilateral linkages to determine when and how properties change. It enables students to experiment quickly with many different configurations of side lengths, trace the motion made by the strips, and print various examples as records to aid in analysis and generalization. For example, in figure 11.4a, point *P*, which is on segment *DC*, has been moved to coincide with point *D*, and the paths traced by both points *P* and *C* when the driver crank (*AD*) is rotated about point *A* are shown. In figure 11.4b, the driver crank has been moved by dragging point *D*, and only the path traced by point *C* is shown.

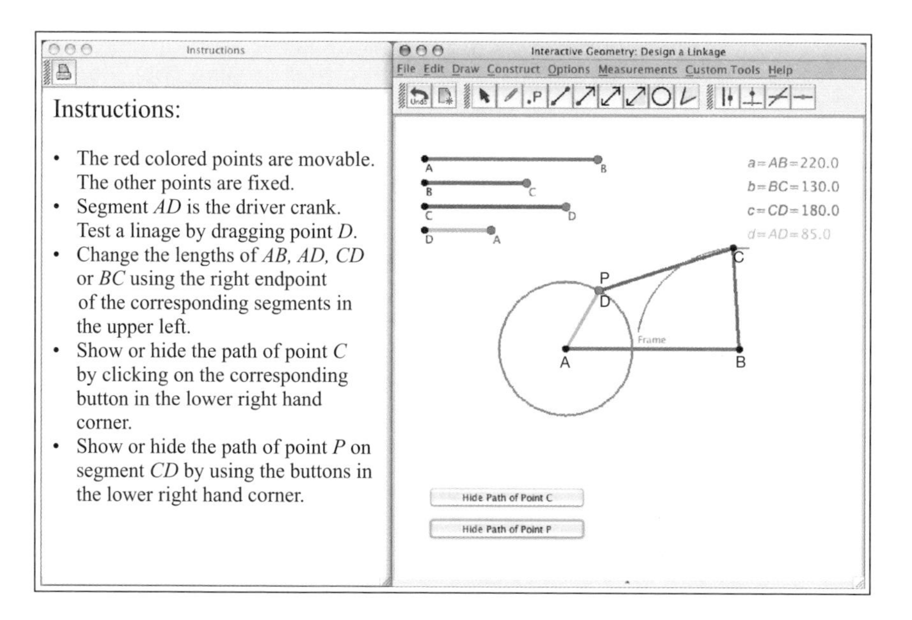

Fig. 11.4a. Design and trace the motion of a quadrilateral linkage.

CPMP-Tools is designed so that it can be used to amplify students' learning in a classroom with one or two computers or a set of laptops, in a computer lab, in a library, or at home. Guidance on how to use these various configurations is provided to teachers in teacher resource materials.

In the next two sections, we illustrate in more detail how CPMP-Tools is being used to amplify student learning in the algebra and statistics strands.

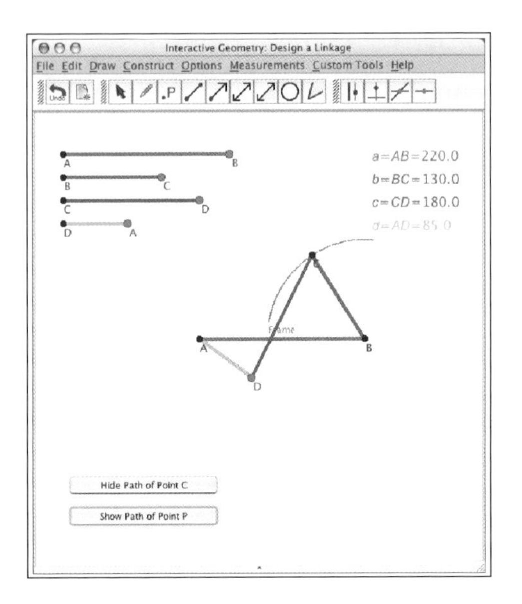

Fig. 11.4b. Investigate 4-bar linkages.

Using Algebra Tools to Amplify Students' Learning of Algebra and Functions

Algebra provides a context in which we can consider what it means to assert, as above, that technology tools can not only amplify cognitive capacities, but they can also fundamentally change the nature of the task and the requirements to complete the task (cf., Goos, Galbraith, and Renshaw 2004; Shaughnessy, Garfield, and Greer 1996). We will consider two curriculum-embedded, Java-based software examples from CPMP-Tools—a spreadsheet and grapher, and a computer algebra system (CAS).

Spreadsheet and Grapher

Spreadsheets and graphing tools are powerful tools to help students learn more mathematics more deeply. Consider the following problem:

> Wildlife management has become an increasingly important issue as modern civilization puts greater demands on wildlife habitat. As an example, consider a fishing pond that is stocked with trout from a Department of National Resources hatchery. Suppose you are in charge of managing the trout population in the pond.... Three factors that affect the changing trout population that you may have listed in your earlier discussion are initial trout population in the pond,

annual growth rate of the population, and annual restocking amount—that is, the number of trout added to the pond each year. For the rest of this investigation, use just the following assumptions:

- There are about 3,000 trout currently in the pond.
- Regardless of restocking, the population decreases by about 20 percent each year because of the combined effect of all causes, including natural deaths and trout being caught.
- 1,000 trout are added at the end of each year.

Using these assumptions, build a mathematical model to analyze the population growth in the pond. Describe the long-term trend in population.
(Adapted from Hirsch et al. 2006, Course 3)

If $P(n)$ is the population at the end of year n, then a closed-form function model for this situation is $P(n) = -2{,}000(0.8)^n + 5{,}000$. Using this formula to find the long-term population requires students to consider, at least informally, $\lim_{n \to \infty}(-2{,}000(0.8)^n + 5{,}000)$. So far this puts a heavy cognitive load on students, especially on their ability to construct and manipulate symbolic representations, so much so that this problem will not be accessible to many students unless they take a precalculus course. However, this is in fact important and useful mathematics, relating to models of population growth and exponential functions, which should be understood and applied by more students.

Appropriate use of technology can amplify students' cognitive capacity to tackle this problem, since it transforms the nature of the problem and its solution. Consider the use of a spreadsheet applet that allows a recursive, numerical, and graphical approach to the problem (see fig. 11.5).

Instead of trying to go straight to the closed-form function model, students can model this situation using a recursive representation, which is quite accessible, sensible, and commonly used in conjunction with computer tools. The population next year will be 80 percent of the population now (because of the 20 percent decrease rate), with an additional 1,000 fish added at the end of each year. Thus, $NEXT = 0.8*NOW + 1{,}000$, with an initial population of 3,000. Students can easily translate from the intuitive *NEXT* and *NOW* language to a more formal recursive formula, $P(n + 1) = 0.8P(n) + 1000$. This formulation also translates directly to spreadsheet language, as illustrated in figure 11.5, by entering 3000 into cell $A1$, then $A2 = 0.8A1{+}1000$, and then fill down to see how the population changes and the long-term trend. Choosing the "Time series" option from the "Graph" menu yields the graph shown in figure 11.6.

Students can now analyze the long-term trend numerically (values in the table are getting close to 5000) and graphically (there appears to be a horizontal asymptote at 5000). Armed with this new knowledge and power, students can explore what-if scenarios. What if the initial population changes? (No change to the long-term population of 5000, which is perhaps quite surprising and

Spreadsheet

File Data Edit Graph Help

Data Sheet 1

	A	B	C	D	E	F	G
1	3000						
2	=.8A1+1000						
3	3720.0						
4	3976.0						
5	4180.8						
6	4344.64						
7	4475.712						
8	4580.569600...						
9	4664.455680...						
10	4731.564544...						
11	4785.2516352						
12	4828.201308...						
13	4862.561046...						
14	4890.048837...						
15	4912.039069...						
16	4929.631255...						
17	4943.705004...						
18	4954.964003...						

Fig. 11.5. Spreadsheet from Algebra Tools

intriguing to students.) What if the restocking rate changes? (Long-term population changes proportionally.) What if the decrease rate changes? (Long-term population changes inverse proportionally.)

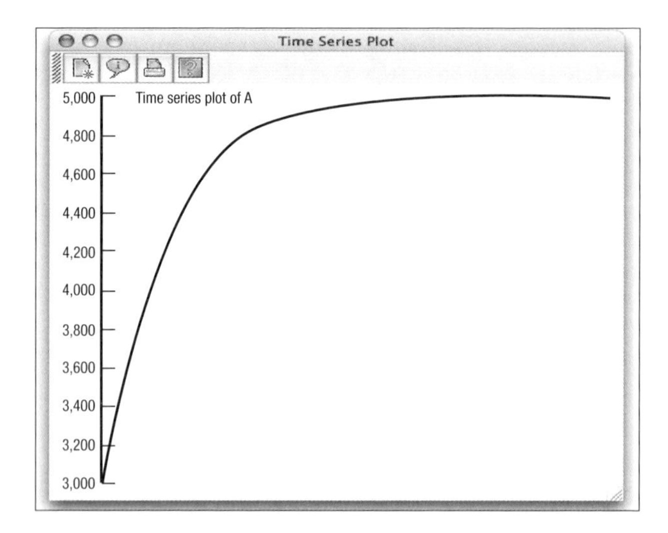

Fig. 11.6. Spreadsheet grapher from Algebra Tools

Using such spreadsheet and graphing tools, especially when easily available as curriculum-embedded, Java-based software, all students now have access to the important ideas in this problem (e.g., shifted exponential functions, recursion, and population growth). On the basis of this initial analysis and understanding, further and more technical analysis can now be pursued.

Many students can and should be encouraged to investigate closed-form algebraic representations. Some students will have already moved in this direction, but now, rather than having been left behind because of an artificial cognitive barrier, many more students may potentially gain an understanding of the problem and mathematical model, and they thus may be more inclined and better able to proceed to other representations. For example, on the basis of the numerical and graphical analysis, students now suspect that the long-term population is 5,000. But can they use symbolic reasoning to support their conjecture? Having gained some understanding and experience of this situation numerically, recursively, and graphically may help students reason that the long-term leveling off must be when the amount lost due to the decrease rate is the same as the amount gained due to the restocking. So, if L is the long-term population, D is the decrease rate as a decimal, and S is the restock amount, $DL = S$. In this problem, $0.2L = 1000$, so L must be 5000.

Getting to the closed-form formula for this situation, $P(n) = -2{,}000(0.8)^n + 5{,}000$, is more challenging, but being able to switch between representations in Algebra Tools can help. For example, they might reason that the graph looks exponential, which makes sense, since the recursive formula clearly shows that we multiply by 0.8 at each step. Thus, starting with $P(n) = 0.8^n$ and using knowledge about families of functions and graph transformations, they might reason about graph transformations and the initial and long-term values to arrive at $P(n) = -2{,}000(0.8)^n + 5{,}000$. Some students might continue the analysis by using the recursive formula $P(n) = 0.8P(n - 1) + 1{,}000$ to iterate expressions for $P(n)$ for several values of n, then do some algebraic recombining, sum a geometric series, and thus arrive at the closed-form formula.

This last approach might not be an appropriate expectation for all students, but that shouldn't preclude all students from learning about the important ideas contained in this problem. The technology-supported approach outlined above transforms the problem and amplifies students' cognitive capacities so that virtually all students can productively tackle this problem and learn much about the underlying mathematics.

Computer Algebra System

A computer algebra system (CAS) is computer software that can do many of the algebraic manipulations of high school mathematics. As such, a CAS presents many promises, challenges, and issues for mathematics education, all of which are compounded by the fact that CASs are now easily accessible to students. For example, the CPMP-Tools software suite has a customized CAS built in as part of the Algebra Tools. The educational value of a CAS can be significant, as noted in an article in a recent international volume on the subject (Kendal, Stacey, and Pierce 2005, p. 105):

> With CAS, students have the opportunity to fulfill their mathematical potential with less computational burden. Using suitable teaching materials, competent teachers can focus student attention on the meaning of the mathematics under consideration. CAS can assist teachers to enhance students' opportunity to acquire insightful problem solving skills, develop deep understanding, develop higher levels of thinking, and gain an understanding of how to validate and interpret solutions. CAS technology can prove to be a powerful mathematical partner.

But there are hard educational questions that must be answered before this great potential is achieved: When and how should a CAS be used? What algebraic manipulations should be done by hand? When can and should algebraic manipulation skill be handed over to a CAS so that students can focus on big ideas and high-level problem-solving strategies (see Brown 1984 and Pea 1987)? One operational answer to these questions is given through the systematic use of a CAS applet in Core-Plus Mathematics.

The use of a CAS is gradually woven into the four courses of Core-Plus Mathematics. In Course 1, a CAS is used as an optional tool mainly to illustrate the existence of such tools and as an occasional check for by-hand manipulation. In Course 2, students are expected to use a CAS to factor some expressions and solve some equations, but then they check the results using by-hand manipulation. Conversely, after solving by hand students will sometimes check using a CAS. In addition, a CAS is used at times to motivate and lead into the development of concepts and skills. Problems that require exclusive use of a CAS are found only in optional homework problems. Thus, in Course 2 students become familiar with using a CAS to solve, factor, and expand, but they also are expected to solve the problems by hand. They do not yet use a CAS exclusively or as a replacement for by-hand manipulation. By the end of Course 2, students have included a CAS as another useful tool in their algebra toolbox. Combined with graphing and spreadsheet tools, appropriate use of a CAS helps students use connected, multiple representations to solve problems as they generate tables of values, draw and trace graphs, manipulate symbols, and use algebraic reasoning to factor, expand, and find approximate and exact solutions.

Below are some examples of the uses of a CAS described above, adapted from Core-Plus Mathematics, Course 2 (Hirsch et al. forthcoming [b]):

- *Example: Use a CAS, then check by hand and by reasoning—Factoring quadratic expressions*

  [This problem follows some previous work on quadratic functions and expressions, including the meaning of factoring, reasoning about factoring, and factoring by hand.]

When the factoring task involves an expression like $-10x^2 + 240x - 950$, things get much more challenging. Fortunately, what is known about factoring quadratic expressions has been converted into routines for calculator and computer algebra systems. Thus, some simple commands will produce the desired factored forms and the insight that comes with them.

For example, the screen display in figure 11.7 below shows how a CAS would produce a factored form of $-10x^2 + 240x - 950$.

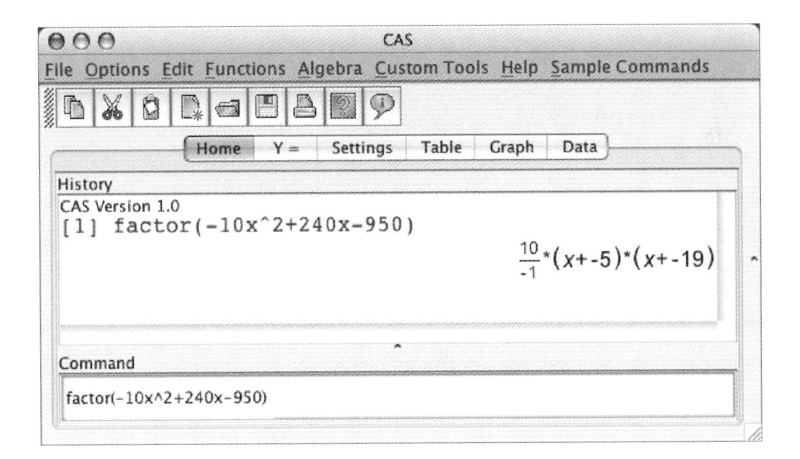

Fig. 11.7

Use a computer algebra system to write each of the following expressions in equivalent form as products of linear factors. Then use your own reasoning to check the reasonableness of the CAS results.

a. $x^2 + 2x - 24$ ... **f.** $3x^2 - 7x - 6$ ...

- *Example: Use a CAS to motivate and lead into the development of concepts and skills—The quadratic formula*

[This problem follows some previous work on quadratic functions, expressions, and equations, including some initial work with the quadratic formula.]

In addition to expanding and factoring expressions, you can also use a computer algebra system to solve quadratic equations. The screen display in figure 11.8 below shows how a CAS would solve the equation $2x^2 - 9x - 5 = 0$.

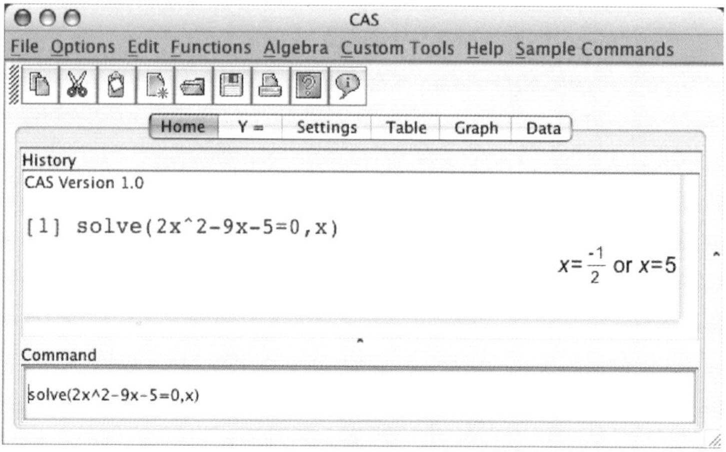

CAS
File Options Edit Functions Algebra Custom Tools Help Sample Commands
Home Y = Settings Table Graph Data
History
CAS Version 1.0
[1] solve(2x^2-9x-5=0,x)
$x=\frac{-1}{2}$ or $x=5$
Command
solve(2x^2-9x-5=0,x)

Fig. 11.8

Use a computer algebra system to solve the following equations.

a. $x^2 + 5x + 3 = 0$ **b.** $2x^2 - 5x - 12 = 0$

c. $-10x^2 + 240x - 950 = 0$ **d.** $x^2 - 6x + 10 = 0$

In solving the equations above, you discovered that some equations that look fairly complex can actually produce simple whole number solutions, while equations that look quite simple end up with irrational number solutions or even no real-number solutions at all. When quadratic expressions involve fractions or decimals, mental factoring methods are not often easy to use. In those cases and in situations where you might not have access to a calculator or computer algebra system program, there is a quadratic formula for finding solutions.... [continues with investigation and application of the quadratic formula]

- *Example: Use multiple connected representations and multiple tools, including a CAS, to solve problems—Solving equations of the form* $mx + d = ax^2 + bx + c$

[This summary section follows the development of concepts, methods, and applications for solving equations that relate linear and quadratic functions.]

Summarize the Mathematics

In this investigation, you developed strategies for solving problems that involved combinations of linear functions and quadratic functions.

a. What strategy would you use to solve a system of equations of the form $y = mx + d$ and $y = ax^2 + bx + c$?

b. What are the possible numbers of solutions for equations in the form $mx + d = ax^2 + bx + c$?

c. How can you estimate the solutions for equations like those in part b by inspecting tables and graphs of functions? What will graphs look like for the various solution possibilities?

d. How can exact values of the solutions be found by reasoning with the symbolic expressions involved? By using a computer algebra system *solve* command?

Be prepared to explain your ideas to the class.

In Course 3, a CAS becomes a more powerful cognitive amplifier. That is, students now use a CAS to find solutions to complex problems that are not easily approachable without technology tools. Students use a CAS to help solve two general classes of problems for which technology is helpful—problems involving complex calculations, in this example, complex symbolic calculations; and what-if analyses in which many instances are examined, in this example what-if situations involving patterns and results of symbol manipulation. Students also consider the limitations of a CAS and carefully interpret the output produced.

This increased use of CAS—along with the previous multiple-tool, multiple-representation use—not only helps students develop deeper conceptual understanding and more powerful problem-solving ability but also helps them develop "deep procedural knowledge," as discussed, for example, in Star (2005). Procedural knowledge needs to be deep knowledge, not just superficial knowledge of "the syntax, steps, conventions, and rules for manipulating symbols" (p. 407). Deep procedural knowledge is "knowledge of procedures that is associated with comprehension, flexibility, and critical judgment" (p. 408). The use of a CAS can be instrumental in helping students develop deep procedural knowledge. This potential benefit of using technology tools is also noted by Kieran (2004, p. 31): "As has been emphasized in some recent research, a deeper conceptualization of the manipulative processes of algebra can be obtained from stressing the mathematically relevant aspects of techniques of manipulation within technological environments."

Some students in Course 3—those planning to go on to major in mathematics and science in college—are expected to solve all problems by hand as well as with a CAS. Other students, who would otherwise be left behind, use a CAS exclusively for some problems. This allows all students to learn important mathematical ideas in advanced algebra, rather than denying access to this powerful mathematics because of the complexity of symbol manipulation. In this way, borrowing a slogan from the college calculus reform efforts, using a CAS helps algebra become *a pump, not a filter,* for high school mathematics education.

Thomas, Monaghan, and Pierce (2004, p. 159) comment on these potential benefits of a CAS in a recent book on the future of algebra:

> CAS can provide a stimulus for, and access to, more sophisticated algebra.... In current curricula, students often become fluent at performing mathematical processes that are, in fact, meaningless to them. The opportunities that CAS can provide to choose, and work with, one or more representations may help students to build a conceptual understanding of mathematically complex systems. CAS affords the opportunity to tackle harder and more realistic problems, to extend topics, and to connect topics.

Below are some examples of the uses of a CAS described above, adapted from *Core-Plus Mathematics*, Course 3 (Hirsch et al., 2006)

- *Example: Limitations and interpretations of a CAS—Quadratic formula*

  After suitable development, the quadratic formula is eventually derived in the form

$$x = \frac{-b}{2a} + \frac{\sqrt{b^2 - 4ac}}{2a} \text{ or } x = \frac{-b}{2a} - \frac{\sqrt{b^2 - 4ac}}{2a}.$$

  Shortly following the derivation, students are asked to use a CAS to solve $ax^2 + bx + c = 0$ for x, compare the CAS result to the quadratic formula derived above, and then resolve and explain any apparent differences. For example, the result using the CAS in CPMP-Tools is shown in figure 11.9. To complete this task, students need to use symbol sense and symbolic reasoning to interpret the CAS output and reconcile the different forms of the quadratic formula given by the CAS and by their by-hand symbolic derivation.

- *Example: Limitations and interpretations of a CAS—Simplify rational expressions*

  Students are asked to do the following task:

$$\text{Simplify } \frac{x-2}{x^2-25} \bullet \frac{2x^2+11x+5}{6x-12}$$

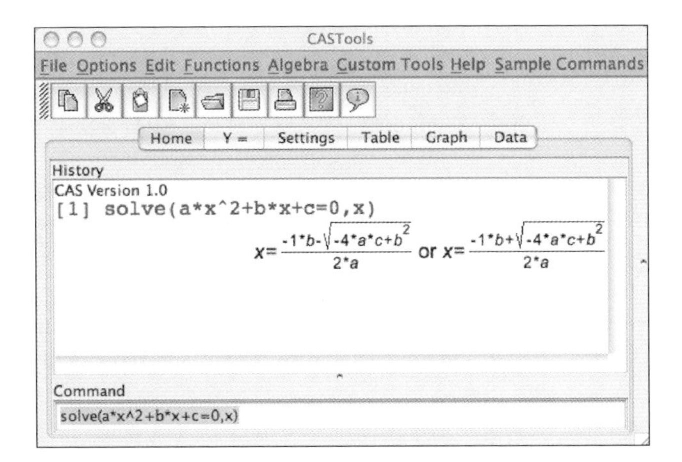

Fig. 11.9. Quadratic formula on a CAS

> by factoring and canceling factors before expanding the products in numerator and denominator. Check your answer with a CAS, but be careful to explain the values of x for which your answer is invalid.

To complete the task, students must first determine how to use the particular features of their CAS. For example, there may be a "Simplify" command, but perhaps only "Factor" and "Expand" commands. Then they must interpret the output. For example, the output may be

$$\frac{2x+1}{6(x-5)},$$

in which case they must be able to state and explain the restrictions that $x \neq 5, -5$, or 2.

- *Example: Use CAS for what-if analyses—Properties of rational expressions and functions*

  Why is it, students often ask or wonder, that it is okay (usually) to multiply the top and bottom of a rational expression by the same quantity, but it is not okay to add the same quantity top and bottom? And of course this question has implications for incorrect "canceling." Use of a CAS can help students explore this situation and understand the relevant properties of rational expressions. For example, students are given the following task, in which they are encouraged to use a CAS:

  > Consider next the rational function $f(x) = \dfrac{x+3}{x-5}$. You might find it helpful to use a CAS as you work through the following problems.

a. Multiply both the numerator and the denominator by $(x + 4)$ to yield the function $g(x)$. Expand the expressions in the numerator and denominator to eliminate all parentheses.

b. Graph both $f(x)$ and $g(x)$ and explain what the result suggests about the relationship of the two functions.

c. Compare tables of values for $f(x)$ and $g(x)$ for x in the interval $[-10, 10]$. Find any values of x for which $f(x) \neq g(x)$ and explain why those inputs produce different outputs for the two functions.

d. Now see what happens if you *add* $(x + 4)$ to both the numerator and the denominator of $f(x)$ and simplify to yield the function $h(x)$.

i. Compare graphs and tables of values for $f(x)$ and $h(x)$ for x in the interval $[-10, 10]$. Find any values of x for which $f(x) \neq h(x)$ and explain why those inputs produce different outputs for the two functions.

ii. Then explain why adding $(x + 4)$ to both numerator and denominator of a rational expression has a different effect than multiplying by that common factor.

A CAS like the one in CPMP-Tools is particularly useful here, since when you define the functions in the Home window, they are automatically entered into the "Y=" window, and thus the functions are immediately available to view as represented by tables (in the Table window) or graphs (in the Graph window), simply by clicking on the appropriate button. Thus, the symbolic, numeric, and graphical representations can all be brought to bear in a connected and coordinated manner to help students make sense of the sometimes mysterious ways of manipulating rational expressions.

In this section we have explored how a curriculum-embedded, Java-based CAS can be used to amplify students' learning of algebra. The particular importance of the notion of "curriculum embedded" is highlighted by Heid et al. (2002, p. 590) in their closing thoughts in a review of research on the use of CASs:

> As we begin to use CASs in our classrooms, we need to be aware that the combination of the capabilities and underlying mathematics of the CAS, our students' understandings of the CAS, our use of CASs in the classroom, and the curriculum that we use affects students' understandings.

The examples and discussion above offer one view of how this combination of factors is addressed by embedding a customizable, Java-based CAS into a specific curriculum.

It is now time to consider how curriculum-embedded, Java-based tools can be used to enhance the learning and teaching of statistics and probability.

Using Statistics Tools to Amplify Student Learning of Statistics and Probability

Following the publication of *Curriculum and Evaluation Standards for School Mathematics* (NCTM 1989), data analysis and descriptive statistics have become a more prominent component of school mathematics. However, Jere Confrey, as reported in Heid (2005, p. 357), points to the need for developing a kind of reasoning that often is not yet an integral part of many school mathematics courses:

> I think the idea of probabilistic reasoning and reasoning that is much more contingent is going to be of major proportions in the future and people are going to have to begin to grapple with complexity in the sense of not just what is, in a descriptive sense. I think many more people are going to have to begin to understand inferential reasoning and what you can say and not say relative to these complex systems.

The representational and computational capabilities of cognitive technologies like CPMP-Tools statistics software enable students to deepen and extend their understanding of making inferences from data. Consider the following problem that launches the Patterns in Data unit in Course 1 of Core-Plus Mathematics and is revisited in the Reasoning and Proof unit in Course 3.

> A teacher in Traverse City, Michigan, was interested in whether eye-hand coordination is better when students use their dominant hand (the hand they write with) than when they use their nondominant hand. She asked students in her first two classes to perform a penny-stacking experiment. In each of two classes, students were told, "You can touch pennies only with the one hand you are using; you have to place each penny on the stack without touching others; and once you let go of a penny, it cannot be moved. Your score is the number of pennies you had stacked before a penny falls."
>
> Each class was randomly divided into two groups of about equal size. In one group, each student stacked pennies using only their dominant hand. The students in the other group used their nondominant hand. Each student's score was the number of pennies stacked before a penny falls.
>
> (Adapted from Hirsch et al. 2006)

Following is a description of a teaching episode related to this problem in field-test classes using interactive statistics software. Once the data for the two classes are collected and pooled, the question becomes how to analyze them. In keeping with what students learned in previous Core-Plus Mathematics courses, they produce and interpret graphical displays and summary statistics (see figs. 11.10 and 11.11). The displays show considerable overlap in these two distributions. The mean number of pennies stacked by those using their dominant hand is 32.89, whereas the mean for those using their nondominant hand is 27.33. The difference of the dominant hand mean minus the nondominant hand mean is 5.56. Can we conclude from this difference in means that more

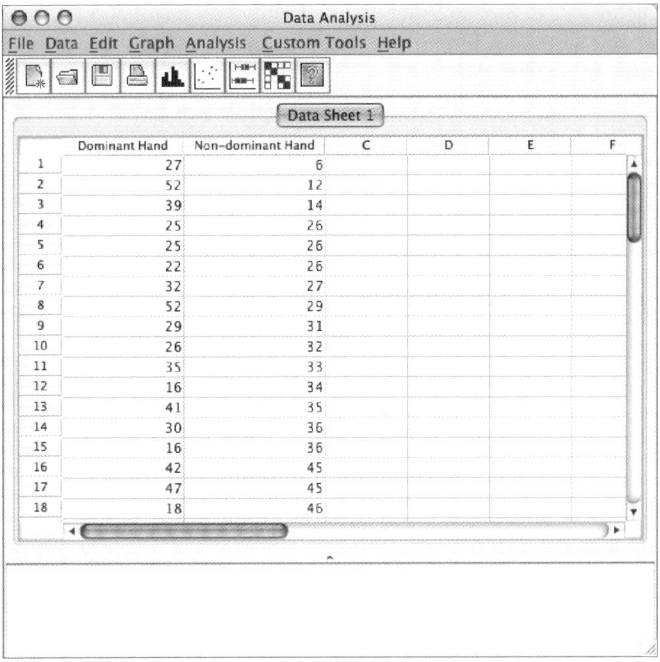

Data Analysis

File Data Edit Graph Analysis Custom Tools Help

Data Sheet 1

	Dominant Hand	Non-dominant Hand	C	D	E	F
1	27	6				
2	52	12				
3	39	14				
4	25	26				
5	25	26				
6	22	26				
7	32	27				
8	52	29				
9	29	31				
10	26	32				
11	35	33				
12	16	34				
13	41	35				
14	30	36				
15	16	36				
16	42	45				
17	47	45				
18	18	46				

Fig. 11.10. Penny-stacking data

pennies can be stacked with the dominant hand?

Through a series of scaffolded tasks, that is, where the difficulty level is gradually increased, students reason to the following preliminary conclusion. If the hand that people use makes absolutely no difference in how many pennies they can stack, there would still almost always be a nonzero difference in every class that does this experiment. This is because of two sources of vari-

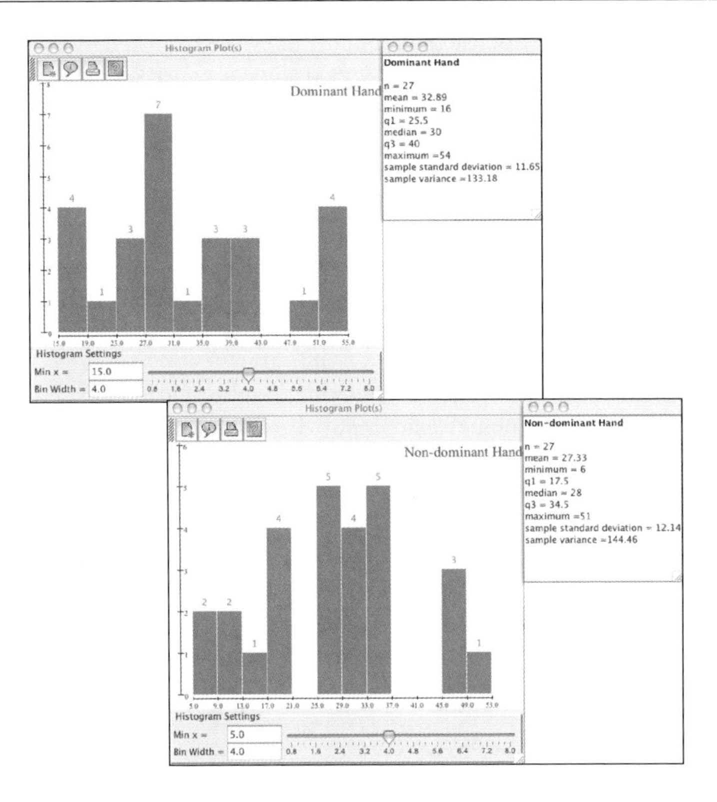

Fig. 11.11. Number of pennies stacked by dominant and nondominant hand

ability—variability in the actual difference due to effect of hand dominance and variability due to chance. The difference between the means of the two groups in this instance is 5.56. It is possible that a difference this large or larger could occur simply by chance even if no actual difference exists between using the dominant hand or the nondominant hand.

So the questions for the students are as follows:

> What is the probability of obtaining a difference of at least 5.56 just by chance, assuming that no actual difference exists in what hand people use to stack pennies?
>
> On the basis of this probability, do you think that people can stack more pennies with their dominant hand than with their nondominant hand? Explain.

Most statistics books address questions like this using a t test for the difference between two means. However, there is a more accessible approach that is easily understood by high school students. Students in field-test classrooms proceeded as follows. Suppose the numbers of pennies stacked by the students in the two classes were written on identical slips of paper, mixed up well, and then half of them are drawn out to represent the students who used their dominant hand. Compute the mean number of pennies stacked for the dominant hand on the basis of these slips of paper. Then compute the mean number of pennies stacked for the nondominant hand, using the remaining slips of paper. Subtract the *nondominant hand mean* from the *dominant hand mean.* If this process is repeated, say, 100 times, you can examine how many times a difference as large as, or larger than, 5.56 is obtained. The result of that simulation can be used to estimate the answer to the first question above. To get a better estimate, you could increase the number of trials.

Because this method (called a *randomization test*) is tedious, it is seldom taught. Computing technology can amplify this simple and powerful idea and make it an integral part of how students think about statistical inference and statistical significance. Figures 11.12–11.15 show results of using the Randomization Distribution custom tool within CPMP-Tools for 250, 500, 750, 1000 runs. The learner orientation of the software enables students to control the number of runs, observe the samples as they are generated, and see "in the mind's eye" how, as the number of runs increases, the distribution begins to stabilize.

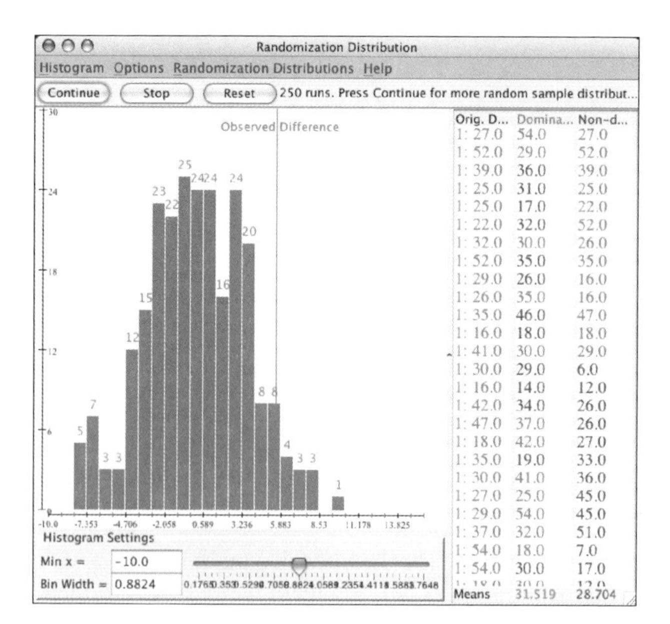

Fig. 11.12. Randomization distribution based on 250 runs

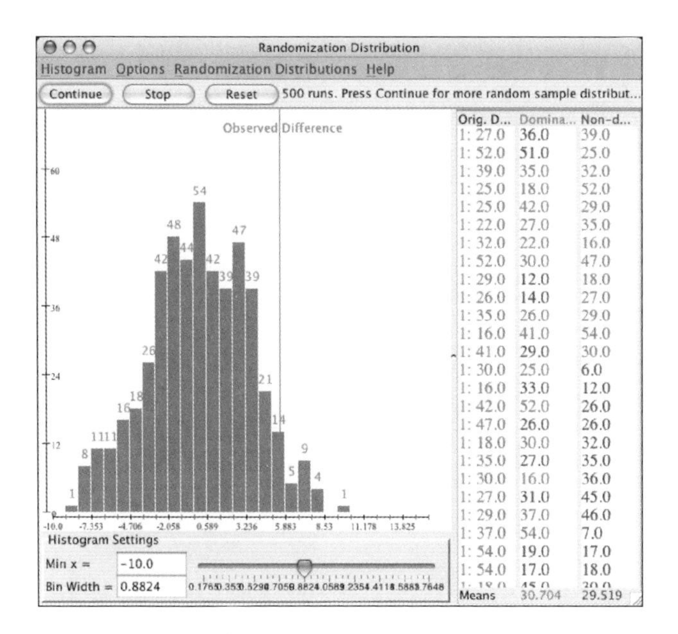

Fig. 11.13. Randomization distribution based on 500 runs

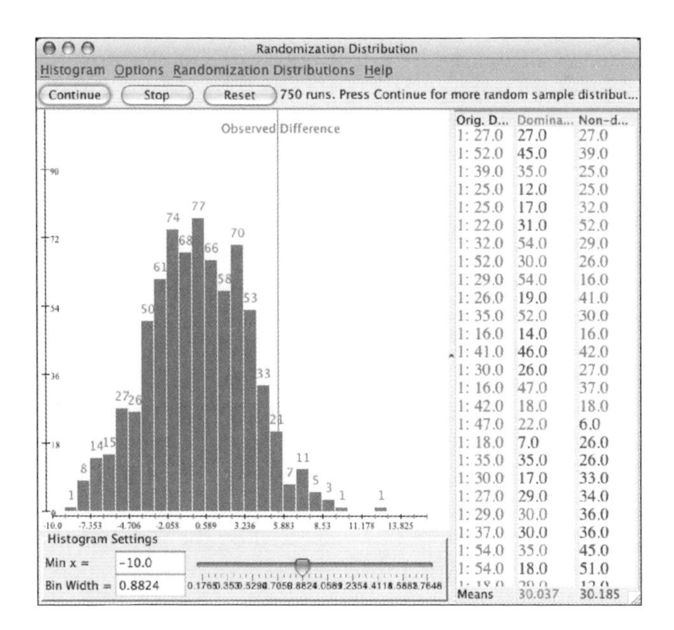

Fig. 11.14. Randomization distribution based on 750 runs

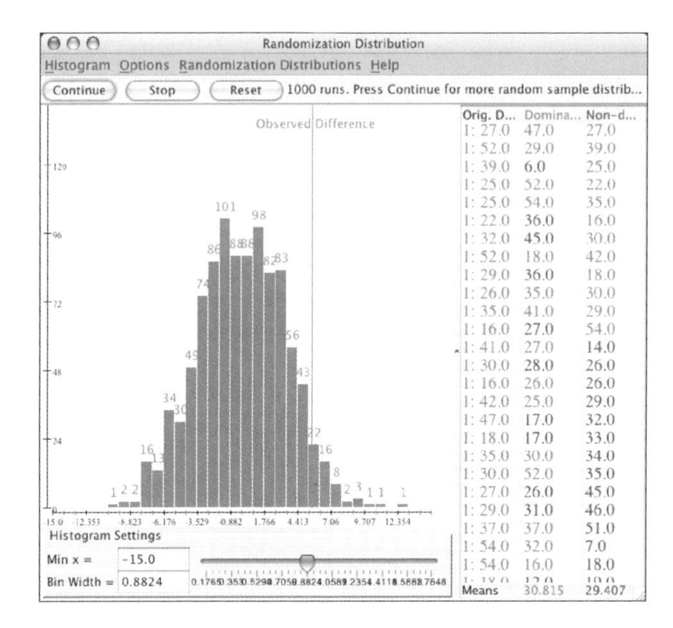

Fig. 11.15. Randomization distribution based on 1000 runs

In the instance of the random distribution of 1,000 runs, the probability of getting a difference as large or larger than 5.56 just by random chance is 0.04. Since this is a small probability, we conclude that the difference of 5.56 is unusually large; it's a "rare event." (More formal analysis occurs later in the curriculum.) Thus, we conclude that a person can probably stack more pennies with his or her dominant hand.

The randomization test, developed and first described by R. A. Fisher in his 1935 book *The Design of Experiments,* makes no assumption of normality or minimum sample size—only that treatments are randomly assigned to subjects (Fisher 1960). This test, enabled by technology, amplifies student learning so that many more students can gain access to, carry out, and understand hypothesis testing. Moreover, student learning is amplified beyond just testing the difference between the means of two groups. Students are also able to use this method with other statistics—median, interquartile range, and standard deviation—for comparing two groups (cf. Barbella, Denby, and Landwehr 1990).

With the cognitive amplification provided by the Randomization Distribution feature built into CPMP-Tools, high school students in heterogeneous classrooms are now able to address questions involving inferences from data that were previously reserved for those who had studied standard hypothesis testing based on a t or normal distribution. The conceptual understanding of statistical inference developed by students through the use of such technology is invaluable in a world filled with data and claims about data. And it will furnish a solid foundation for those students who go on to further study of statistical tests.

Closing Comments

In this article, we have described features of emerging curriculum-embedded, Java-based software and how that software has the potential to amplify student learning of important mathematics. Java-based software is becoming increasingly available. Examples of such software available in spring 2006 are given in the Resources section below, including interactive geometry tools, interactive data-analysis tools, and computer algebra systems.

CPMP-Tools is learner-oriented software in that its design and capabilities are shaped by goals of the Core-Plus Mathematics curriculum for student learning and application of that learning. Embedding this software in the curriculum and making it available over the Web and free to users of the curriculum (and of other curricula) should enable these cognitive tools to become a more integral part of student learning and work.

Some Java-Based Software Resources

Interactive Geometry Software

- Cabri® Geometre (can create Java Applets), www-cabri.imag.fr
- Cinderella (can create Java Applets), cinderella.de/tiki-index.php
- CPMP-Tools, www.wmich.edu/cpmp
- GeoGebra, www.geogebra.at
- Geometer's Sketchpad® (can export to Java Sketchpad), www.keypress.com/sketchpad

Computer Algebra Systems

- CPMP-Tools, www.wmich.edu/cpmp
- Hartmath, sourceforge.net/projects/hartmath
- Maple (includes Java-Server Application), www.maplesoft.com
- Mathematica (can create Web Demonstrations), www.wolfram.com
- MuPad (Java version in development), www.mupad.com
- XCAS, sourceforge.net/projects/xcas

Interactive Data Analysis Software

- Calico, www.math.csusb.edu/faculty/stanton/calico/index.html
- CPMP-Tools, www.wmich.edu/cpmp
- StatCrunch, www.statcrunch.com
- StatGraphics, www.statgraphics.com
- Statistics101, www.statistics101.net

Discrete Mathematics Software

- CPMP-Tools, www.wmich.edu/cpmp
- JGraph (Java-based library), www.jgraph.com
- JGraphEd, www.jharris.ca/JGraphEd
- JGraphPad, www.jgraph.com/jgraphpad.html
- JGraphT, jgrapht.sourceforge.net/
- JUNG, jung.sourceforge.net

REFERENCES

Barbella, Peter, Lorraine Denby, and James M. Landwehr. "Beyond Exploratory Data Analysis: The Randomization Test." *Mathematics Teacher* (February 1990): 144–49.

Brown, John Seely. *Idea Amplifiers: New Kinds of Electronic Learning Environments.* Palo Alto, Calif.: Xerox Palo Alto Research Center, Intelligent Systems Laboratory, 1984.

Bruner, Jerome S. "On Cognitive Growth." In *Studies in Cognitive Growth*, edited by Jerome S. Bruner, Rose R. Oliver, and Patricia M. Greenfield, pp. 30–67. New York: John Wiley & Sons, 1966.

Conference Board of the Mathematical Sciences (CBMS). *The Mathematical Sciences Curriculum K–12: What Is Still Fundamental and What Is Not.* Report to the National Science Board Commission on Precollege Education in Mathematics, Science, and Technology. Washington, D.C.: CBMS, 1983.

Fey, James T., William F. Atchison, Richard A. Good, M. Kathleen Heid, Jerry Johnson, Mary G. Kantowski, and Linda P. Rosen. *Computing and Mathematics: The Impact on Secondary School Curricula.* College Park, Md.: National Council of Teachers of Mathematics and the University of Maryland, 1984.

Fisher, Ronald A. *The Design of Experiments.* London: Oliver & Boyd, 1960.

Goos, Marilyn, Peter Galbraith, and Peter Renshaw. "Establishing a Community of Practice in a Secondary Mathematics Classroom." In *Mathematics Education: Exploring the Culture of Learning*, edited by Barbara Allen and Sue Johnston-Wilder, pp. 36–61. New York: Routledge Falmer, 2004.

Hart, Eric W., Sabrina Keller, W. Gary Martin, Carol Midgett, and S. Thomas Gorski. "Using the Internet to Illuminate NCTM's *Principles and Standards for School Mathematics*." In *Technology-Supported Mathematics Learning Environments*, Sixty-seventh Yearbook of the National Council of Teachers of Mathematics (NCTM), edited by William J. Masalski, pp. 221–40. Reston, Va.: NCTM, 2005.

Heid, M. Kathleen. "The Technological Revolution and the Reform of School Mathematics." *American Journal of Education* 106 (November 1997): 5–61.

———. "Technology in Mathematics Education: Tapping into Visions of the Future." In *Technology-Supported Mathematics Learning Environments,* Sixty-seventh Yearbook of the National Council of Teachers of Mathematics (NCTM), edited by William J. Masalski, pp. 345–66. Reston, Va.: NCTM, 2005.

Heid, M. Kathleen, Glendon W. Blume, Karen Hollebrands, and Cynthia Piez. "Computer Algebra Systems in Mathematics Instruction: Implications from Research." *Mathematics Teacher* 95 (November 2002): 586–91.

Hirsch, Christian R., James T. Fey, Eric W. Hart, Harold L. Schoen, and Ann E. Watkins, with Beth Ritsema, Rebecca Walker, Sabrina Keller, Robin Marcus, Arthur F. Coxford, and Gail Burrill. *Core-Plus Mathematics,* Course 1. Columbus, Ohio: Glencoe/McGraw-Hill, forthcoming (a).

———. *Core-Plus Mathematics,* Course 2. Columbus, Ohio: Glencoe/McGraw-Hill, forthcoming (b).

———. *Core-Plus Mathematics,* Course 3. Field-test version. Kalamazoo, Mich.: Western Michigan University, 2006.

Keller, Sabrina. CPMP-Tools. Version 1.1, trademark pending. Computer software. East Lansing, Mich.: Michigan State University and Core-Plus Mathematics Project, 2006.

Kendal, Margaret, Kaye Stacey, and Robyn Pierce. "The Influence of a Computer Algebra Environment on Teachers' Practice." In *The Didactical Challenge of Symbolic Calculators: Turning a Computational Device into a Mathematical Instrument,* edited by Dominique Guin, Kenneth Ruthven, and Luc Trouche, pp. 83–112. New York: Springer, 2005.

Kieran, Carolyn. "The Core of Algebra: Reflections on Its Main Activities." In *The Future of the Teaching and Learning of Algebra: The 12th ICMI Study,* edited by Kaye Stacey, Helen Chick, and Margaret Kendal, pp. 21–34. Norwell, Mass.: Kluwer Academic Publishers, 2004.

Land, Susan M., and Michael J. Hannafin. "Student-Centered Learning Environments." In *Theoretical Foundations of Learning Environments,* edited by David H. Jonassen and Susan M. Land, pp. 1–24. Mahwah, N.J.: Lawrence Erlbaum Associates, 2000.

Marriott, Michel. "Blacks Turn to Internet Highway, and Digital Divide Starts to Close," *New York Times,* March 31, 2006.

National Council of Teachers of Mathematics (NCTM). *Curriculum and Evaluation Standards for School Mathematics.* Reston, Va.: NCTM, 1989.

Pea, Roy D. "Beyond Amplification: Using the Computer to Reorganize Mental Functioning." *Educational Psychologist* 20 (1985): 167–82.

———. "Cognitive Technologies for Mathematics Education." In *Cognitive Science and Mathematics Education*, edited by Alan H. Schoenfeld, pp. 89–122. Hillsdale, N.J.: Lawrence Erlbaum Associates, 1987.

Pew Internet and American Life Project. "73% of Americans Go Online." Press release. March 31, 2006.

Schoen, Harold L., and Christian R. Hirsch. "The *Core-Plus Mathematics Project:* Perspectives and Student Achievement." In *Standards-Based School Mathematics Curricula: What Are They? What Do Students Learn?*, edited by Sharon L. Senk and Denisse R. Thompson, pp. 311–43. Mahwah, N.J.: Lawrence Erlbaum Associates, 2003.

Shaughnessy, J. Michael, Joan Garfield, and Brian Greer. "Data Handling." In *International Handbook of Mathematics Education*, Part 1, edited by Alan J. Bishop, M. A. (Ken) Clements, Christine Keitel, Jeremy Kilpatrick, and Colette Laborde, pp. 205–38. Dordrecht, Netherlands: Kluwer Academic Publishers, 1996.

Star, Jon R. "Reconceptualizing Procedural Knowledge." *Journal for Research in Mathematics Education* (November 2005): 404–11.

Thomas, Michael O. J., John Monaghan, and Robyn Pierce. "Computer Algebra Systems and Algebra: Curriculum, Assessment, Teaching, and Learning." In *The Future of the Teaching and Learning of Algebra: The 12th ICMI Study*, edited by Kaye Stacey, Helen Chick, and Margaret Kendal, pp. 155–186. Norwell, Mass.: Kluwer Academic Publishers, 2004.

ADDITIONAL READING

Fey, James T., Al Cuoco, Carolyn Kieran, Lin McMullin, and Rose Mary Zbiek, eds. *Computer Algebra Systems in Secondary School Mathematics Education*. Reston, Va.: National Council of Teachers of Mathematics, 2003.

PART THREE

Measuring and Interpreting Students' Learning

Marilyn E. Strutchens

There are a variety of methods for gathering information about what students have learned. It is important to recognize that the choice of measurement tools determines what is known about what students have learned; moreover, the inferences drawn from assessments are incomplete and inexact. In this era of high-stakes accountability, the perspective of "what is measured is what is learned" has implications for both administrative and instructional decision making. Moreover, according to the Assessment Principle from NCTM's *Principles and Standards for School Mathematics* (NCTM 2000, p. 22):

> [T]he tasks used in an assessment can convey a message to students about what kinds of mathematical knowledge and performance are valued. That message can in turn influence the decisions students make—for example, whether or where to apply effort in studying. Thus, it is important that assessment tasks be worthy of students' time and attention. Activities that are consistent with (and sometimes the same as) the activities used in instruction should be included.

Also, a constant theme of the *Assessment Standards* is that decisions regarding students' achievement should be made on the basis of a convergence of information from a variety of balanced and equitable sources (NCTM 1995). In this section authors use a variety of assessment items and instructional tasks to examine students' understanding of fractions and geometric shapes.

In their article, Cramer and Wyberg illustrate the importance of bringing out the different strategies that children use as they solve problems and how these strategies can be highlighted to help children make connections among the topics they learn. Cramer and Wyberg show that although there are many reasons for assessing students in a mathematics classroom, assessing to understand their thinking is one of the most important reasons. They illuminate the fact that "it is not enough to know if students' answers are right or wrong; you also have to investigate the strategies that led to their answers." They present an example of what happened when they questioned students orally about the students' written responses. By examining thinking among three students, the authors gained insight into the reasons that, although the three were equally successful on fraction-order tasks, only one student was successful on a related, but more complex, fraction task—fraction estimation.

Saxe, Shaughnessy, Shannon, Langer-Osuna, Chinn, and Gearhart exploit the potential that number lines have for supporting students' understanding of fractions in the upper elementary and middle school grades. Their research interweaves developmental studies of children's understandings of fractions with classroom studies on students' learning and the pedagogical strategies that support their learning.

Wilson describes the findings of research conducted in developing a learning trajectory for children's shape-composition abilities. He discusses what can be learned about students' understanding of geometric shapes and their properties from observing children while they solve puzzles of one form or another, match letters or shapes to a cutout form, assemble multiple pieces to make a picture, and play with wooden blocks or pattern blocks. The developmental progression and complete learning trajectory were created in collaboration with teachers; and the description, assessment, and labeling of the levels was designed to maintain its comprehensibility.

REFERENCES

National Council of Teachers of Mathematics (NCTM). *Assessment Standards for School Mathematics.* Reston, Va.: NCTM, 1995.

———. *Principles and Standards for School Mathematics.* Reston, Va.: NCTM, 2000.

12

When Getting the Right Answer Is Not Always Enough: Connecting How Students Order Fractions and Estimate Sums and Differences

Kathleen Cramer
Terry Wyberg

We begin this article by asking you to solve two problems involving fractions. The first problem is as follows: Determine which fraction is larger, 2/3 or 1/4. Remember the strategy you used to solve this problem. The right answer is 2/3. The second problem is as follows: Estimate the sum, 2/3 + 1/4. Is this sum larger or smaller than 1/2? Is the sum smaller or larger than 1? Remember how you thought about this problem. The right answer to the second problem is a little less than 1. Think about the relationships between the strategies you used to solve both problems. We asked a large group of fifth graders to solve similar problems involving fractions. We found that most of the children were able to get the right answer when ordering fractions similar to the first problem but were not very successful when estimating sums and differences similar to the second problem.

A common answer children find when calculating the sum of 2/3 + 1/4 is 3/7, and you may see this answer when you ask your students to add fractions. One strategy that may help children see that this answer is not reasonable is to ask them to estimate the sum. They should realize that because 2/3 is larger than 1/2 and more is being added, the result has to be more than 1/2; 3/7 is clearly less than 1/2. For children to be successful in estimating sums of fractions, they need to be able to judge the relative size of the fractions being added. Why is it that the children we assessed were able to judge the size of the fractions easily when ordering but not when they are adding? There may be many possible answers for this question. One possibility may lie with the different strategies that children use when ordering fractions.

We first look at the different strategies that you could use to find the right answer when asked to determine the larger fraction between 2/3 and 1/4. One possible strategy is to convert 2/3 to 8/12 and 1/4 to 3/12. Once there is a common denominator, it is easier to see that 8/12 is larger than 3/12. Another strategy is to convert both fractions to percents, 2/3 is about 66 percent and 1/4 is

25 percent. Another possible strategy is to realize that 2/3 is larger than 1/2 and 1/4 is smaller than 1/2. Some of you may have cross-multiplied to determine the larger fraction. Although all the four strategies mentioned above (common denominator, percent, compare to one-half, cross-multiplication) would help you get the right answer, which of the strategies would help you the most when you estimate the sum 2/3 + 1/4?

The purpose of this article is to illustrate the importance of bringing out the different strategies that children use as they solve problems and how these strategies can be highlighted to help children make connections among the topics they learn.

Although there are many reasons for assessing students in a mathematics classroom, assessing to understand their thinking is one of the most important reasons. It is not enough to know if students' answers are right or wrong; you also have to investigate the strategies that led to their answers. This may be difficult to do, but in this article we present an example of what happened when we did this using data from written tests where students described in writing their strategies, and from interviews of students. By examining students' thinking among three students, we gained insight into why, when the three were equally successful on fraction-order tasks, only one student was successful on a related, but more complex, fraction task—fraction estimation.

The three fifth graders presented in this article were among the fifth graders in a district using one of the NSF-funded, *Standards*-based curricula. This district was interested in investigating what impact the new curriculum was having on students' fraction learning. Nearly all fifth graders in the district took a written assessment to measure fraction learning; 26 fifth graders were randomly selected for one-hour interviews. On the written test, students were asked to show their work, and during the interviews, students were asked to explain their thinking. We examined in detail the fraction-order and estimation tasks to gain insights into students' number sense for fractions. We wanted to know (1) whether students could order fractions, (2) how students solved these order tasks, (3) whether they used conceptual or procedural strategies, (4) whether students could estimate fraction addition and subtraction tasks, (5) if there was a connection between how students ordered fractions and their ability to estimate, and (6) on the basis of this information on students' learning, what curriculum adjustments might be needed.

Indicators of Fraction Number Sense

The NCTM Number and Operations Standards for grades 3–5 recommend that students in this age span be able to use models, benchmarks, and equivalent forms to judge the size of fractions and to develop and use strategies to estimate computation involving fractions (NCTM 2000). Understanding the relative size of fractions is an essential prerequisite to estimating sums and differences involving fractions, and both skills underlie what it means to have number sense for fractions. For example, the Rational Number Project (RNP) group reported the following student's reasoning as an example of a student with number sense (Cramer, Post, and delMas 2002). In response to the question "Estimate where the answer to 11/12 – 4/6 would be on a number line from 0 to 2 with 0, ½, 1, 1½, and 2 noted," a student responded by putting an X between 0 and 1/2. Her explanation was as follows: "11/12 is more than 4/6. I know that 11/12 is nearly one whole thing and 4/6 is way over 1/2. So when I take 4/6 away from 11/12 I will get way less than 1/2" (p. 137). Notice the order ideas embedded in this student's thinking and how she used her understanding of the relative size of the fractions to estimate a reasonable range for the answer.

Although it is reasonable to assume that to estimate sums and differences with fractions students need to understand the relative size and magnitude of the fractions involved, how students have constructed that understanding also is important. Consider the following fraction pairs: 3/4 and 2/3; 1/2 and 5/8; 3/12 and 7/12; 4/9 and 4/11; 6/14 and 5/9. A student could order each pair in a variety of ways: she or he could find common denominators, calculate cross products, or find percents. In addition, it is possible to order these fraction pairs correctly but use incorrect whole-number thinking. For example, a student could conclude 3/4 is greater than 2/3 (or 5/8 >1/2 or 7/12 >3/12) because 3/4 includes larger numbers; often when students order fractions by focusing on the size of individual parts they will conclude incorrectly that 4/11 is larger than 4/9 or 6/14 is larger than 5/9. Another common, incorrect strategy that leads to a correct answer at times is to look at the difference between the numerator and denominator—for example, "4/9 is greater than 4/11 because 4 is closer to 9 than 11."

Another strategy students could use to order the fraction pairs is to construct mental images for the fractions that are related to concrete representations they may have used while learning about fractions. For example when ordering 3/4 and 2/3, a student might note that 3/4 is only 1/4 away from the whole, whereas 2/3 is 1/3 away. Since 1/3 is greater than 1/4, 2/3 must be less than 3/4. Using 1/2 as a benchmark, a student could reason that 5/8 is larger than 1/2 because 4/8 equals 1/2. This benchmark approach works for the last pair, too: 6/14 is less than 1/2, whereas 5/9 is greater than 1/2. Seven-twelfths is greater than 3/12 because seven of the same-sized parts is more than three of the same-sized parts. The type of reasoning described above has been noted by the RNP group and represents conceptually based reasoning as opposed to procedural or whole-number thinking (Bezuk and Cramer 1989; Cramer, Post, and delMas 2002).

As we see, there are different strategies for judging the size of fractions. How a student thinks about fraction size will play a role in his or her understanding of other fraction ideas, in particular, operating on fractions in a meaningful way. In the following section we present examples showing how students' understanding of one mathematical idea affects their learning of a connected idea.

How Kevin, Natalie, and Ben Ordered Fractions and Estimated on a Written Test

On a fraction test, Kevin, Natalie, and Ben all correctly ordered at least four of the five fraction-ordering tasks shown in table 12.1. Because students were asked to show how they ordered the fraction pairs, that information is available and also is shown in table 12.1. Notice the differences in how each student approached the problem. Kevin was adept at finding common denominators and used that strategy to find the larger fraction in each pair. He had difficulty finding the common denominator for the last fraction pair; he selected the correct answer, but since his work is incomplete, we are not sure how he determined the final answer. Natalie's thinking seemed to be based on pictures of fractions and the insight that with fractions, the larger the denominator, the smaller each fraction part. There are some limitations to relying on the size of the denominator only, but in the case of 6/14 and 5/9, the strategy works; but it would not have worked if it was 13/14 instead of 6/14. Ben used a percent strategy; given that calculators were available, we can assume he adeptly used the calculator to find the percent for each fraction pair. (Note that he made an incorrect conclusion for problem 3 even though he did find the correct percents for 4/9 and 4/11.)

These students were less successful on the test's three estimation items shown in table 12.2. The teacher presented each item at the overhead separately for thirty seconds and asked students to record the whole number the answer was closest to. Ben and Kevin answered one of three correctly, whereas Natalie answered two of three correctly.

Students were not asked to explain their reasoning, so we do not have insights into their thinking based on these written test items. But we can uncover their thinking strategies by looking at students' interviews on similar items to see if how they ordered fractions pairs might explain differences in estimation ability.

How Kevin, Natalie, and Ben Ordered Fractions and Estimated in Interviews

The eight order tasks students were asked to solve on the interview are shown in table 12.3, along with the students' explanations. Consider Kevin's responses. He answered all eight correctly. With the exception of ordering 1/5, 1/4, and 1/3 where he used a percent strategy, he solved the others correctly using a common-denominator strategy. He also used the common-denominator strategy on the written test. He had to ponder how to order 1/17 and 1/20 at first because the common denominator is not easily calculated, but he did conclude that 1/20 was less than 1/17 because "20 goes into 200 a less amount of time than 17 goes into 200." What Kevin did not do was to construct a men-

Table 12.1
Written Test: Order Items

Order Items	Kevin	Natalie	Ben
$\frac{3}{4}$ $\frac{2}{3}$	$\frac{3}{4}=\frac{9}{12}$ $\frac{2}{3}=\frac{8}{12}$	Drew pictures of circles	$\frac{3}{4}=75\%$ $\frac{2}{3}=66\%$
$\frac{1}{2}$ $\frac{5}{8}$	$\frac{5}{8}=\frac{10}{16}$ $\frac{1}{2}=\frac{8}{16}$	Drew pictures of circles	$\frac{1}{2}=50\%$ $\frac{5}{8}=62\%$
$\frac{4}{9}$ $\frac{4}{11}$	$\frac{4}{9}=\frac{44}{99}$ $\frac{4}{11}=\frac{36}{99}$	Nine is smaller numbers so it has bigger pieces because it has less to divide	$\frac{4}{9}=46\%$ $\frac{4}{11}=36\%$ (In this instance only, he concluded that 4/11 > 4/9.)
$\frac{3}{12}$ $\frac{7}{12}$	$\frac{7}{12}=\frac{7}{12}$ $\frac{3}{12}=\frac{3}{12}$	Both have the same denominator and 7 is bigger than 3	$\frac{3}{12}=25\%$ $\frac{7}{12}=58\%$
$\frac{6}{14}$ $\frac{5}{9}$	$\frac{6}{14}=\frac{}{108}$ 14 × 12 = 18, 140, 158; 14 × 6 = ; 14)108 = 00	Nine has bigger pieces so therefore brings it bigger than 14	$\frac{6}{14}=42\%$ $\frac{5}{9}=55\%$

tal representation for both fractions and use the "size of piece" strategy many children use to order fractions with the same numerator. Since twentieths are smaller than seventeenths, then 1/20 would be less than 1/17. He learned to order fractions using a procedural strategy and was quite adept at it.

Table 12.2
Written Test: Responses for Estimation Items

Estimation Item / (Correct Answer)	Kevin	Natalie	Ben
$\frac{7}{8} + \frac{12}{13}$ (2)	1	2	2
$\frac{3}{8} + \frac{5}{12}$ (1)	1	0	8
$\frac{8}{9} - \frac{7}{8}$ (0)	1	0	2

Natalie took a different approach to the order tasks. She talked of "pieces."The fraction 1/20 is less than 1/17 because "denominator is bigger so each piece is smaller." She used 1/2 as a benchmark when she ordered 3/9 and 3/4. She reflected on the amount left over to order 11/12 and 4/5: "a lot bigger piece left from 5 pieces than 12." She also demonstrated procedural skill in simplifying 6/8 to 3/4 to compare to 3/5. Her thinking was conceptually based; she judged the size of the fractions by relying on mental representations connected to a concrete model that involved fraction pieces. Her strategies on the interview are similar to the explanations she gave on the written test.

Ben's strategies for ordering fractions on the students' interview depended on whole-number thinking. This contrasts with his percent strategy that he used on the written test. During the interview students did not have access to a calculator, so we could determine if they could approach the ordering tasks in a conceptual way.

For 11/12 and 4/5, Ben noted that 11/12 had bigger numbers so it was the bigger fraction—a correct answer but one based on incorrect reasoning. This same strategy led him to conclude incorrectly that 4/15 was greater than 4/10. When ordering 3/4 and 5/12, he tried to use a percent strategy but abandoned that strategy to look at the difference between the numerator and denominator. Three-fourths is larger because "there is one away from that (pointed to the 3 and 4) and there is a lot away from 12." Once he started to think about the fractions from this difference perspective, he continued to do so for the last five questions. This strategy does give a correct answer for many fraction pairs; but when Ben compared 6/4 and 6/5, this difference strategy led to an incorrect answer, as it did when he compared 3/5 and 6/8. Ben resorted to whole-number thinking as a way to make sense of fractions when he was unable to rely on mental constructis for the fractions or when a calculator was not available. Relying on whole-number thinking is a common error among students lacking mental representations for fractions (Cramer, Post, and delMas 2002).

Table 12.3
Comparing Ordering Strategies

Ordering Task	Kevin	Natalie	Ben
Which fraction is larger? 4/5 11/12	11/12, because 12 and 5 go into the same number, that's 60, because 12 times 5 is 60. So what I did is 11 times 5 is 55… then I did 12 times 4 and that is 48. That's my whole new fraction, 48/60 and 55/60.	11/12. 12 pieces are smaller than 5 pieces. A lot bigger piece left from 5 pieces than 12.	This one (pointing to 11/12) … is larger. "Why is that?" Because 4/5 is a smaller number and this is a higher number (pointing to the 11/12). 'Cause, see, they are both one number apart. "What do you mean that they are both one apart?" See, 11 is one less than 12, and this (pointing to the 4) is one less than 5. "Even though both are one apart the 11/12ths is still bigger?" But this is still a bigger number (pointing to the 11/12) than that. So this is still bigger.
Which fraction is larger? 4/15 4/10	15 and 10 go into 30. (After correctly converting both fractions to thirtieths, he states that 4/10 is larger.)	4/10 pieces bigger. Splitting between 10 people or 15. Four tells the number.	Umm. This one (pointing to 4/10). "Why is that?" Because these both numbers are the same (pointing to the 4's). But I do not know. 'Cause this (10) is less than that (15). Actually, this (4/15) is bigger. "Why is that?" Cause these are both the same (fours) but this (15) is bigger than that (10). The 15 is bigger than the 10.

Continued on next page

Table 12.3—*Continued*

Ordering Task	**Kevin**	**Natalie**	**Ben**
5/12 3/4 Are these fractions equal or is one less?	Fifths is less, because if you change the 3/4 into the same denominator, twelfths, 9/12 is bigger than 5/12.	I think 5/12 is smaller. There are 3/12 in 1/4 … 3 + 3 is 2/4 and that's 1/2. Would need to be 9/12 to equal 3/4.	This (pointing to the 3/4) is 75. "The 3/4 is 75?" Yeah, and this is … I think this one is bigger (pointing to the three fourths). "The three fourths?" Yeah. "OK, why is that?" 'Cause this is 1 away from that (the 3 and 4) and this is a lot away from 12.
Order from smallest to largest: 1/5 1/3 1/4	1/5 is 20 percent; 1/4 is 25 percent, and 1/3 is 33 1/3 percent.	1/5, 1/4, 1/3. Bigger the denominator the smaller the pieces because that's what you have to divide them up into.	This (1/5) is the smallest, this (1/4) is the second one, and this (1/3) is the third. "Why is 1/5 the smallest?" Because 1 is closest to 3 and for the medium one because 1 is closer to 4 than 5.
Which fraction is smaller? 1/20 1/17	(Long pause). 1/17 is smaller … no, 1/20 is because 20 goes into 200 a less amount of times than 17 goes into 200.	1/20. Denominator is bigger, each piece is smaller.	Smaller? This one (pointing to 1/20). "Why is that?" Because 17 is closer to 1 than 20. Yeah, this one (1/17) is bigger. "The 1/17 is bigger?" Yeah, and the 1/20 is smaller. "Because the 17 is closer to 1?" Yeah.

Continued on next page

Table 12.3—*Continued*

Ordering Task	Kevin	Natalie	Ben
Which fraction is smaller? 3/4 3/9	(Long pause.) This one, 3/9, because 4 goes into 36 nine times and then you do 9 times 3 and that is 27 so that is 27/36. 9 goes into 36 four times so you do 4 times 3 and that is 12. 27/36 is greater than 12/36.	3/9 smaller. 3/4 is greater than 1/2. 3/9 is a third and that's less than 1/2.	3/9 is smaller. "Why is that?" Because 9 is farther away from 3 than 4 is to 3.
Which fraction is smaller? 6/4 6/5	6/5 because 4 and 5 go into 20. 4 goes into 20 five times so 5 times 6 is 30 so 30/20ths. In other words, that is one and one-half. It's 1 and 10/20, which is also one and one-half. And then um 6, um 5 goes into 20 four times. Then you do 6 times 4 and that is 24 so that would be one and one-**sixth** yeah one and one-**sixth** and that would be one. So one-half equals the bigger one.	6/4. Bigger since bigger pieces. 6/5 has one extra piece and 6/4 has 2 extra pieces, bigger pieces.	6/4. Because 4 takes longer to get to 6 than 5 does.
Which fraction is smaller or are they equal? 3/5 6/8	3/5 is smaller. Because 5 and 8 both go into 40 and then 5 goes into 40 eight times. So 8 times 3 is 24. And then 8 goes into 40 five times. And 5 times 6 is 30 so 30/40ths is bigger than 24/40ths.	6/8 bigger. Not in lowest terms. Divide by 1/2 and get 3/4 and 3/4 is greater than 3/5.	They are equal. "Why is that?" Because this (pointing to the 3) is 3 away from this (pointing to the 6) and this (pointing to the 5) is 3 away from this (pointing to the 8). So they are equal.

Kevin and Natalie correctly answered all the ordering tasks on a written test, whereas Ben correctly answered four of the five questions. The test and the interviews show that these students had different ways of judging the relative magnitude of fractions. Natalie's thinking was conceptual, whereas Kevin's approach was procedurally based. Ben relied on incorrect whole-number strategies when a calculator was not available; the incorrect whole-number strategies did lead in many instances to correct answers. We should not be surprised that, given their different ways of determining fraction size, the three students approached estimation items on the interview in different ways and with different levels of success.

Simple fraction estimation requires an understanding of the relative size of the numbers involved and insight into what happens to these numbers when added or subtracted. We now look at how the three students estimated fraction addition and subtraction tasks in an interview situation to see if how they ordered fractions influenced their ability to estimate. Table 12.4 shows the problems and students' responses. Kevin, who ordered fractions using a common-denominator procedure, approached the first problem in the same manner. He found the exact answer using 24 as a common denominator. But in the second two problems, where the interviewer pushed him more for an estimate, we can uncover the limits of his procedural understanding for judging the relative size of two fractions. He reverted to whole-number strategy, adding numerators and denominators to find an estimate. He was unable to label the number line with his incorrect estimate as well. (See fig. 12.1.) Notice how he misjudged the unit on the number line.

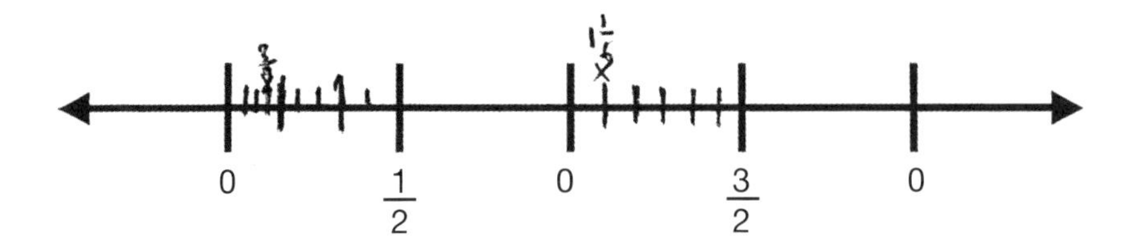

Fig. 12.1. Kevin's estimate for 2/3 + 1/6

Natalie, who demonstrated conceptual strategies for ordering fractions, put her understanding of the relative magnitude for fractions to work on the estimation tasks. In the first problem, she knew that the sum would be greater than 3/4 but less than 1 because 1/6 < 1/4 and 1/4 is needed to construct one whole cup. She could solve a problem procedurally as demonstrated in task 2 when she quickly calculated the exact answer, but when asked, she could estimate in a more conceptual manner by judging that 2/3 > 1/2 and 1/6 is 1/2 of 1/3.

Her last response was her most complete one. She used a common-de-

Table 12.4
Students' Responses to Addition Estimation Interview Items

Interview Item	Kevin	Natalie	Ben
"Marty was making two types of cookies. He used 1/6 cup of flour for one recipe and 3/4 cup of flour for the other. How much flour did he use altogether? Without working out the exact answer, give an estimate that is reasonable. Ask: "Is it greater than 1/2? Less than 1/2? Greater than 1? Less than 1?"	Used algorithm to get correct answer. Did not estimate. Wrote 18/24 + 4/24 = 22/24. When asked to explain: six and four go into 24. Six goes into 24 four times and 4 times 1 is 4. Four goes into 24 six times and 6 times 3 is 18. 18 + 4 = 22.	A little more than 3/4. One-sixth isn't very big, just estimating. If you add it to 3/4 only end up being maybe another 1/2 of a 1/4 or less. When prompted as to why she was sure it was less than one: 1/6 isn't equivalent to a 1/4 and 1/4 is needed to make whole.	I know that this (pointing to 3/4) is equal to 75. I think this (the 1/6) might be like 15. So this would be 90. This does not have to be exact right? "Right. So this would be about 90 cups of flour?" (Hesitation) Yeah. So does this have to be in cups? (I read the problem again) He hesitates then says yes cups. Yeah well. Didn't this ask for how many cups? "Yes, so about 90 cups?" Yeah.

Continued on next page

Table 12.4—*Continued*

Interview Item	Kevin	Natalie	Ben
"Tell me about where $\frac{2}{3} + \frac{1}{6}$ would be on this number line". 	Divides the number line into thirds (between 0 and 1/2), then (incorrectly) puts three dashes between the 0 and the first dash. Puts three ninths on the third dash. When asked, "How did you get that?" Because 3 plus 6 is 9 and 2 plus 1 is 3 so that is three ninths. You know that is the same as 1/3.	(Calculated 5/6 mentally.) OK. Right here between 1/2 and 1. You have to make the denominators the same. (When asked not to use common denominators, she said the answer was between 1/2 and 1. She pointed that 2/3 was between 1/2 and 1 and stated that 1/6 was one-half of 1/3.)	(Writes 2/3 + 1/6 = 3/9.) 3/9. "Where would 3/9 go on the number line?" (Puts x between 1 and 3/2.)
Jon calculated $\frac{2}{3} + \frac{1}{4} = \frac{3}{7}$. Do you agree?	Yes. Because he just added the 3 plus 4 and that is 7 and 2 plus 1 is 3 so this is the same thing as 3/7.	NO, just adding those. Adding numbers not fractions. Mathematically find same denominators and then times tops. Multiply 4 × 3 = 12. 3/12 and 8/12 so 11/12. (When asked to estimate and not get the exact answer, she said that 2/3 + 1/4 is greater than 1/2 because 2/3 is greater than 1/2. 2/3 + 1/4 is less than 1 because 1/4 is less than 1/3. When asked about 3/7, she said, 3/7 $<$ 1/2 because 1/2 of 7 is 3 1/2.)	Yes because 2/3rds plus 1/4th is 3/7ths.

nominator strategy but then went into detail on why the answer was greater than 1/2 but less than 1. Notice how her conceptually based order strategies played a role in her thinking. She used 1/2 as a benchmark, and she ordered two fractions with same numerators.

On the written test, Ben answered all but one ordering question correctly by changing each fraction to a percent. Insights into how he understood fraction ordering and how these misunderstandings affected his ability to estimate were revealed in the interviews. Ben attempted a percent strategy on the story problem but did not have an understanding of what a percent is and used 75 percent as the whole number 75. For the other two interview problems, he reverted to whole-number strategies, adding numerators and denominators, to estimate a sum of two fractions.

When looking at the written test, each student was successful on the ordering tasks. Ben and Kevin answered one of the three estimation tasks correctly on the written test, whereas Natalie answered two out of three correctly. The work they showed on the written test hinted at different ways of ordering fractions. It was the interviews, though, that uncovered how deep the differences were and the impact these different ways of thinking had on their ability to estimate. Conceptual understanding for judging relative size of fractions as evidenced by Natalie's way of knowing enabled her to estimate a reasonable sum of two fractions. Even though Kevin had a correct procedure for ordering fractions, his way of knowing did not provide him with the type of understanding needed for more complex number-sense tasks. Ben's incorrect whole-number thinking about fraction order may lead to correct answers sometimes but, of course, will provide him no support for more difficult fraction tasks. For him, the percent strategy that he used on the written test also proved ineffective when asked to estimate the sum to two fractions.

Reflecting on Students' Learning

As mentioned in the beginning of this article, these students were among all the fifth graders in a district who took the fraction written test and among the randomly selected group of 26 students who participated in interviews. But their thinking was representative of the type of students' thinking we found overall. Some had good procedural skill, whereas others demonstrated more conceptual thinking. And, among a good number of students, whole-number thinking was still prevalent.

A goal for this article was to describe how we measured students' understanding of ordering and estimation items using a written test and an interview. If we were interested in learning how successful Ben, Natalie, and Kevin were on the ordering items only, we might be satisfied with our findings from the written test. All three children answered at least four of the five order tasks correctly and were able to describe correct procedures. It could be argued that Ben and Kevin used more advanced strategies than Natalie on the written test.

But one of the reasons we want children to be able to order fractions is so they can develop mental pictures of the size of the numbers and use these mental pictures when they operate on these numbers. The formal strategies that Ben and Kevin used to order fractions did not help them when they were asked to estimate fraction sums like 2/3 + 1/4, whereas Natalie was able to use her mental images to provide a reasonable estimate and a correct answer.

Lesson Learned

This article shows how important it is for teachers to recognize students' thinking behind the answers they give. Kevin's thinking documents how procedural skill detached from conceptual understanding is a limited form of knowledge. Ben's thinking shows that a student can obtain a correct answer based on incorrect thinking. In both instances, how they ordered fractions affected their ability to estimate, a skill that requires conceptual understanding of the numbers being operated on.

Because getting the right answer is not always enough, teachers should implement classroom strategies to gather information on how students obtained the right answer (or wrong answer). One strategy is to use problems that effectively uncover students' thinking. The fraction tasks presented in this article have been used widely in RNP research studies. They have consistently shown if children's thinking is conceptually or procedurally based and have revealed children's misunderstandings as well.

The problems used here can be presented to students on written assessments where students explain their thinking in writing. They also can be used to initiate classroom discussions on how to order fractions and estimate sums and differences. Students can share their thinking strategies for ordering and estimating; the teacher can lead a discussion on differences and similarities among strategies and if incorrect ones arise, help students see why the strategy is ineffective.

This article also has shown how the interviews clarified for us students' understanding or misunderstanding more clearly than the written test; these insights garnered from the interviews helped to explain Ben's and Kevin's lack of success on the fraction estimation tasks. A teacher might consider doing selected interviews with students using similar problems Although it would be difficult to interview all students, a teacher could find time to interview a select number of students across a range of skills to gather insights into how students in her class solve problems like the ones used in this study.

Whatever ways a teacher chooses, we have seen how important it is to understand children's thinking. When a teacher has information on students' thinking, then she can make more effective instructional decisions. With Kevin and Ben, a teacher would likely conclude that they need more-concrete experiences so they can develop more conceptually based understandings of fraction size and then see if that new knowledge supports their estimation skills.

We conclude this article first with a list of questions that teachers in Kevin's, Natalie's, and Ben's—and perhaps your—school might consider as they reflect on students' performance on fraction order and estimation tasks and then with suggestions for possible solutions to the issues raised.

1. Since Natalie's way of knowing was more productive than Ben's and Kevin's strategies for ordering fractions, how can teachers ensure that all students develop conceptual understanding for fraction size as seen in Natalie's thinking?
2. Does the curriculum provide enough concrete models for students to use early in their work with fractions, and throughout the elementary school years, to develop these mental representations? If so, are the teachers using them?
3. How can teachers address incorrect whole-number thinking among students?
4. Since the least common denominator strategy is not introduced in the curriculum but was used by many children, should the curriculum continue to be supplemented with this procedural strategy?
5. Does the curriculum provide enough opportunities for developing and using fraction number sense?

RNP research has shown the impact that extended use of manipulative models has on students' thinking about fractions (Cramer, Post, and delMas 2002; Cramer and Henry 2002). Students who use fraction circles and other models construct for themselves ordering strategies similar to those shown by Natalie. They develop proficiency in estimation by using these strategies. Students who use concrete models extensively come to understand the limitations of whole-number thinking as applied to fractions.

If the district curriculum does not provide adequate experiences with concrete models, then teachers will need to supplement the curriculum. The RNP Web site, education.umn.edu/rationalnumberproject/, includes a set of some twenty-three lessons that use fraction circles, chips, and paper-folding models to develop fraction concepts, order, equivalence ideas, fraction estimation, and addition and subtraction with fractions. This is one possible resource for teachers to address the issues raised in this article.

REFERENCES

Bezuk, Nadine, and Kathleen Cramer. "Teaching about Fractions: What, When, and How?" In *New Directions for Elementary School Mathematics,* 1989 Yearbook of the National Council of Teachers of Mathematics (NCTM), edited by Paul Trafton, pp. 156–67. Reston, Va.: NCTM, 1989.

Cramer, Kathleen, and Apryl Henry. "Using Manipulative Models to Build Number Sense for Addition of Fractions." In *Making Sense of Fractions, Ratios, and Proportions,* 2002 Yearbook of the National Council of Teachers of Mathematics (NCTM), edited by Bonnie Litwiller, pp. 41–48. Reston, Va.: NCTM, 2002.

Cramer, Kathleen, Thomas Post, and Robert delMas. "Initial Fraction Learning by Fourth- and Fifth-Grade Students: A Comparison of the Effects of Using Commercial Curricula with the Effects of Using the Rational Number Project Curriculum." *Journal for Research in Mathematics Education* 33 (March 2002): 111–44.

National Council of Teachers of Mathematics (NCTM). *Principles and Standards for School Mathematics.* Reston, Va.: NCTM, 2000.

13

Learning about Fractions as Points on a Number Line

Geoffrey B. Saxe
Meghan M. Shaughnessy
Ann Shannon
Jennifer M. Langer-Osuna
Ryan Chinn
Maryl Gearhart

IN THIS article we consider the use of number lines for supporting students' understanding of fractions in the upper elementary and middle school grades. Although the number line is introduced to students in elementary school textbooks, its potential for students' learning has not been exploited by educators or researchers. We are engaged in a research program to address the gap, focusing on urban students whose needs are often underserved in mathematics classrooms. Our research interweaves developmental studies of children's understandings of fractions with classroom studies on students' learning and the pedagogical strategies that support their learning.

We argue that number lines can support students' understanding of important properties of fractions. Fifth- and sixth-grade students can use the number line as a vehicle for understanding ideas like numerical unit, relations between whole numbers and fractions, the density of the rational numbers (there are infinitely many rational numbers between any two), and although every number is unique, the number can be named in infinitely many ways (equivalence). The understandings students construct about rational number as well as their familiarity with number line conventions will provide them important resources for algebra in the middle and high school grades. Before describing our studies, we present background on students' learning in the domain of fractions and introduce our research framework.

This paper was made possible by support from the Spencer Foundation (#200100026) and a National Science Foundation consortium grant, Diversity in Mathematics Education (DiME, #406F766). The data presented, the statements made, and the views expressed are solely the responsibility of the authors. Appreciation is extended to the members of the Fractions Research Group at UC Berkeley, including Amanda Arendtsz, Rachel Coben, Charles Hammond, Maxine McKinney, Julie McNamara, and Edd Taylor.

Background and Research Framework

Many students have difficulty developing a conceptual understanding of fractions, and one way that teachers and curriculum developers help students develop this understanding is to engage them with representational models. To date, the most widely used is the area model, a representation that allows students to build on their informal understandings of fair sharing as they partition pizzas, brownies, and other shapes into fractional parts of wholes. But area models have certain limitations, as we have learned in two recent studies (Saxe, Gearhart, and Seltzer 1999; Saxe et al. 2005). For example, when elementary school students were asked to represent the fraction shown in an area model like that shown in figure 13.1, many labeled the fraction as "1/3," using whole-number reasoning to count pieces without considering the relationship between the size of the piece and the size of the whole. Despite repeated experiences with area model tasks, these students were not distinguishing the idea of a fractional part of a set (a discrete quantity) with a fractional part of a continuous quantity (area).

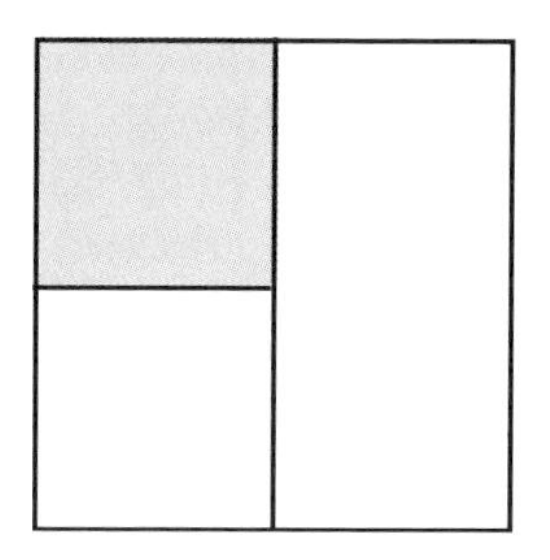

Fig. 13.1. A square partitioned into unequal parts. Many elementary school students call the gray area one-third.

Our goal is to understand how number lines can help students develop understandings of the fraction concepts that are often obscured by area models. Our number line studies are founded on two important postulates prevalent in research on students' mathematical learning. The first is the constructivist view that students are "sense makers" (National Council of Teachers of Mathematics [NCTM] 1989, 2000); in the classroom, students use their prior understandings to make sense of mathematical tasks, mathematical conventions, and mathematical communications. The second derives from Vygotsky's idea of the "zone of proximal development" (Vygotsky 1986), the idea that instruction can "scaffold"—or gradually build the difficulty level of—the development of concepts and problem-solving strategies as students make efforts to understand new material. Building on these two ideas, we conducted three coordinated studies on students' learning (see fig. 13.2) to capture students' sense making and the ways that instruction can support learning. First, we interviewed students of different grade levels to understand how students make sense of whole numbers and fractions represented on the number line. Second, we tutored students and investigated the ways students used our representations and prompts to guide their learning. Third, we designed and implemented whole-class lessons to investigate processes of students' learning in classroom contexts. Each study built on findings from the prior study.

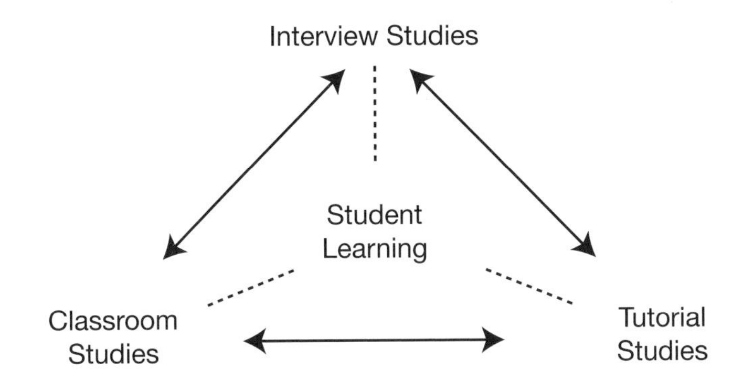

Fig. 13.2. Exploring students' learning—interrelated strands of research

How Do Elementary School Students Make Sense of Points on Number Lines?

Our initial study was designed to reveal the mathematical intuitions and approaches that students bring to instructional contexts. We focused on two grade levels—third grade, when students are first exposed to fractions in formal instruction, and fifth grade, after students have had considerable instruction in fractions. We asked students to identify points on the number line depicting whole numbers and fractions and to give us additional expressions for these points—that is, equivalent fractions—and to explain their reasoning. Figure 13.3 illustrates four number lines used in our assessments.

The number lines in these assessments varied in the use of hash marks and the points that students were required to label. How did students make sense of the points on the lines? Our findings showed that students, regardless of grade level, had little difficulty identifying whole numbers on number lines, but many were challenged when the points required fraction and equivalent fraction names. Difficulties in identifying fraction names were varied. Some students offered nonnumerical answers (e.g., "a tiny line," "a dot"); others were unsure what hash marks or spaces to count for the denominator; and others

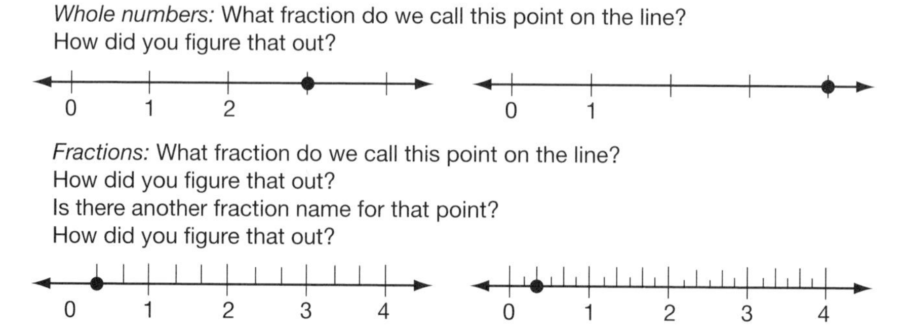

Whole numbers: What fraction do we call this point on the line?
How did you figure that out?
0 1 2
0 1
Fractions: What fraction do we call this point on the line?
How did you figure that out?
Is there another fraction name for that point?
How did you figure that out?
0 1 2 3 4
0 1 2 3 4

Fig. 13.3. Four number lines used in our assessments of students' ability to identify points on number lines

simply did not know what to call the fraction. The few students who correctly identified points typically used a two-count strategy, as depicted in figure 13.4: (1) they determined the denominator of the fraction by counting the number of equally partitioned intervals for the unit distance, and (2) they determined the numerator by counting the number of equally partitioned intervals from the beginning of the unit to the target point. One student offered the following explanation for his answer "two-sixths" for the point depicted in figure 13.4: it is "six spaces and that's two."

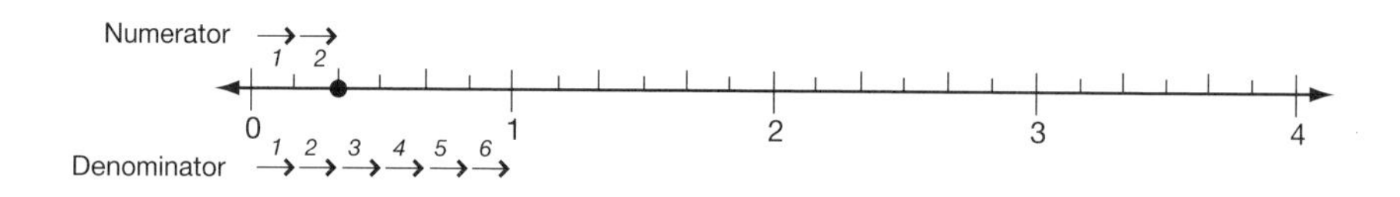

Fig. 13.4. Students' successful two-count strategy

We also queried students about additional names for the same point. Students' difficulties in addressing this question were varied. Some indicated that the point had a single numerical expression (for example, another name for "one-third" is "three pieces and one thing"); others argued that an equivalent fraction name did not exist; others did not know how to answer. In sum, though students were familiar with number lines and could identify points to represent whole numbers, many had difficulty identifying fractional points, and most had difficulty generating equivalent names for the same point.

In the tutorial studies we describe next, we guided students using the strategies that the most successful students employed in our interview study to increase the likelihood that our tutoring was within the students' "zone of proximal development" (fig. 13.4). The tutorial focused on the values of points in between whole numbers and equivalent representations of the same points.

By engaging students with the mathematical thinking of peers who were slightly more advanced in their understandings of fractions on number lines, we were able to investigate how students took advantage of our support.

How Can Tutorials Support Students' Learning of Fractions?

We tutored twenty-one third-grade and twenty-three fifth-grade students individually and evaluated their learning. Our tutorial was designed to help students develop strategies for two aspects of fractions on a number line—(1) conceptualizing and labeling a point as a fraction, and (2) naming equivalent fractions. Since strategies for identifying fractions and equivalent representations of the same point require students to understand number line conventions (the crucial role of 0, and directionality from 0 to whole numbers), the tutorial began with an orientation to these fundamental properties of number lines.

The tutor first administered preassessments used in our prior study, and then, using stick puppets, the tutor told each student a story about a fast rabbit and a slow turtle racing each other on number lines like those depicted in table 13.1. We modeled a two-count strategy to name fraction points on a number line—the strategy used by the most successful students in our interview study. To encourage students to make sense of the tutorial ideas, we asked them to explain their reasoning during each task; whenever a student offered an inadequate response, the tutor commented that "another student" had produced a different answer (the correct solution) and explanation and asked the student what he or she thought about this answer. These "hints" offered students opportunities to reconcile their intuitions with other students' conceptions, and the ways that they made use of hints furnished us additional information on students' learning. We concluded the session with postassessments of two types: (*a*) problems similar to those in the tutorial, and (*b*) nonroutine problems designed to probe students' conceptual understandings of fractions.

Students' Learning on Tutorial Problems

When we coded students' preassessment and postassessment responses, we credited students for the following: *Identification* (giving an appropriate fraction name for a point), *Strategy* (making a separate determination of the numerator and the denominator as modeled in the tutorial), and *Equivalence*[1] (ensuring that the numerator and denominator were coordinated multiples of their original numerators and denominators).

1. In order to ensure that all students were engaged in an identical equivalence task, students needed to produce an adequate identification for the point in order to receive credit for equivalence.

Table 13.1
Tutorial on Number Lines

Tutorial Target	Number Line Representation	Tutorial
Naming whole numbers	0 1 2 3	Let's start with the rabbit. We start at 0 because the rabbit hasn't moved yet. Now look at the rabbit hopping from 0 to 1. If the rabbit hops to here [interviewer stops rabbit on 4], what number is the rabbit on now? Good. We call this point "four." How did you figure that out?
Naming fractions	0 1 2 3 4	Now it's the turtle's turn. We also start on 0 because the turtle hasn't moved yet. Now look at the turtle move to here [interviewer moves the turtle to 2/3]. Since the turtle is so slow, the turtle didn't even make it to 1, so we divided the distance between 0 and 1 to see how far the turtle went. As you can see, we divided the distance into three equal parts, called thirds. Here is zero-thirds because the turtle has not moved yet. Here the turtle moved to one-third. Okay, now the turtle moved again. What point is the turtle on now?
Equivalent expressions	0 1 2 3 4	On another day, the turtle moved here [interviewer moves turtle to the one-third mark]. What fraction would you call the point where the turtle stopped? Is there anything else you can call it?

As shown in figure 13.5, the tutorial resulted in improvement in students' learning for students in both the third and the fifth grades, but their patterns of learning differed by grade level. Some third graders appropriated the tutor's strategy for Identification of points, but did not use that Strategy effectively because of counting errors. The third graders' errors were often conceptual—for example, when students counted only internal hash marks to determine the denominator. Third graders showed no improvement at all with Equivalent fractions. In contrast, fifth graders brought more knowledge to the tutorial, as shown by greater scores on all measures at the pretest. Though fifth graders showed improvement in fraction Identification, Strategies for naming points, and Equivalence, their progress was limited, and many students still had much to learn. A promising finding was that when fifth graders appropriated the strategy modeled in our tutorial, this gain was reflected in their accuracy for identifying points.

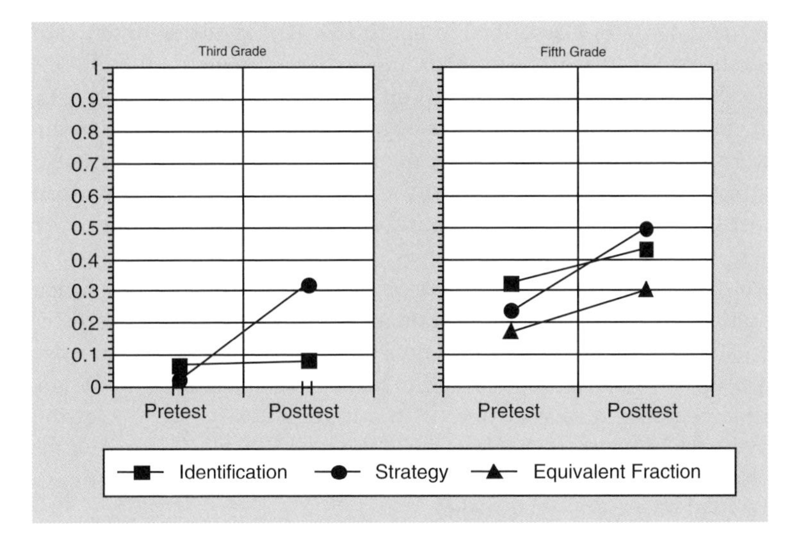

Fig. 13.5. Students' improvement in identifying fractions on the number line, using appropriate fraction-identification strategies, and naming equivalent fractions

Students' Understanding of Nonroutine Number Lines: Probing Conceptual Understanding of Fractions

Our postassessment included nonroutine problems like the one shown in figure 13.6. In this task, students identify a point on a number line divided into intervals of unequal lengths. Nonroutine tasks like these help reveal students' thinking, and the students' responses were fascinating glimpses into their developing understandings of the necessity of equal units in fractions involving a continuous quantity, like length. We found these results particularly useful when designing our classroom study.

As we summarize in table 13.2, some students viewed the figure as an in-

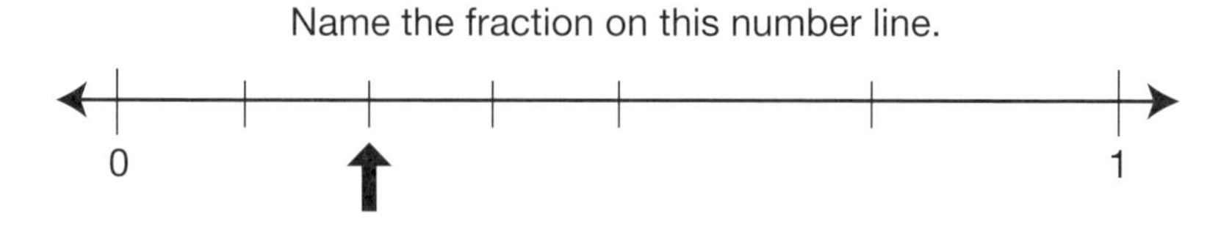
Name the fraction on this number line.
0
1

Fig. 13.6. Nonroutine number line problem with unequal intervals

completely partitioned number line, and they added or removed hash marks to create intervals of equal length, providing correct fraction names such as "2/8" or "1/4" (the shaded rows in table 13.2). These students typically made use of the two-count strategy described in figure 13.4 (and modeled in our tutorial). But others made sense of the number line with fraction names like "2/6," "2/7," "2," and "2/4," responses suggesting whole-number reasoning (counting marks) and unconventional understandings of number line conventions (confusions over the role of the zero, hash marks, or arrows). These patterns persisted after the tutorial, so clearly our very short tutorial was not a sufficient opportunity for most students to develop conceptual understanding of fractions on a number line.

We decided that a classroom lesson had greater potential to engage students with the big ideas of fractions on a number line if we could create ways to help students make sense of the nonroutine problem through whole-class discussions and small-group work. The lesson was challenging to design, and we implemented it a second time with a different class after reviewing and modifying the initial plan. We describe the second lesson below, but we include some contrasts with the first to share what we learned about classroom practices that support students' learning.

How Can Teachers Help Students Learn the Big Ideas of Fractions on Number Lines?

We undertook this third phase of our research in collaboration with Ryan Chinn, one of our team members and a sixth-grade teacher at a local urban middle school. Sixth grade is a crucial transition point from elementary to middle school mathematics, yet many students still struggle with elementary school mathematics. Mr. Chinn knew that many of his students would respond just like the fifth graders did in our tutorial study and that even students who could give correct solutions would benefit from explaining their understandings of fractions—understandings that were based on the conventions of the number line. Mr. Chinn was eager to collaborate in designing number line lessons if we could create lessons that would give all his students opportunities to learn.

We designed this study guided by the principles we used in the design of

Table 13.2
Common Answers for the Nonroutine Number Line Problem (Adequate Answers Appear in the Shaded Rows)

Fraction Names	Typical Explanations
2/6	*a*) There are 6 spaces (intervals) and the arrow is pointed to the end of the second space. *b*) There are 6 hash marks (disregarding the hash mark at 0 while counting) and the arrow is pointed at the second hash mark.
2/7	*a*) Add 2 hash marks to create equal-sized sections (intervals). Then, count the internal hash marks (7). The point is the second hash mark. *b*) There are 7 hash marks (disregarding the need for equal-sized intervals), and the arrow is pointed to the second hash mark.
2	The arrow is pointed to the second hash mark.
2/4	Move the "1" marker to "1/2." The point is 2/4 of the distance between 0 and 1.
2/8	Add 2 hash marks to divide the number line into eighths. The arrow is pointed to the second hash mark.
1/4	Simplify 2/8; cross out hash marks so that the number line is divided into fourths. The point is at the end of the first section.

our tutorial but adapted for the classroom context. We focused on techniques that would encourage students to exchange and reconcile mathematical ideas about number lines—teacher-orchestrated whole-class discussion and small-group work. For Mr. Chinn, the challenge was to encourage students to express their mathematical ideas about number lines and to devise ways to honor students' ideas while honoring the mathematics that students ultimately need to understand (Ball 1993). Our plan incorporated several pedagogical techniques:

- The first was our choice of the number line shown in figure 13.6 as the task. A nonroutine fraction problem could create pedagogically useful difficulties for students, since they are unable to use counting as a strategy for naming fraction points. A nonroutine representation of a number line could engage students with the big ideas about unit length and equal partitions that are at the crux of rational numbers.

- A second technique was to create mathematical dilemmas and the need for resolution. We decided that students should first grapple with the problem independently and then sort out their disagreements in whole-class and peer discussion. For students who produce inadequate solutions that they believe are sound, disagreements should help them discover problems with their reasoning. For other students who are confused, reflections on other students' explanations should open up new solution paths to explore. Students whose solutions are correct would have to defend and clarify their reasoning.
- Our third technique was the teacher's role in orchestrating discussion and introducing important ideas about number lines at particular teaching moments. If students are actively grappling with a mathematical dilemma, a teacher's explanation within their "zone of proximal development" can provide a powerful moment of insight.

The Japanese Itakura method of instruction offered us a model of these techniques in action (Itakura 1967, cited in Inagaki, Hatano, and Morita 1998). The crux of an Itakura lesson is the resolution of mathematical disagreements through student discussion. Students are presented with a problem and several possible answers, each representing ways that students are likely to reason according to their understandings of the underlying concepts. We modified the traditional Itakura lesson structure to allow Mr. Chinn to play a primary role in orchestrating the discussion and establishing mathematical conventions. Mr. Chinn posed the nonroutine unequal-interval problem, and the answer choices we provided were some of the most common responses of the younger students in our tutorial study. Figure 13.7 contains a schematic of our six-phase lesson structure.

In the first phase, students work independently to choose one of five solutions and justify their choice (Phase 1). Their responses are collected and used by the teacher as an initial assessment of students' thinking. In the second phase, on the basis of his quick review and observations of the students' work, the teacher asks several students to present solutions for their options, making sure that justifications for all options are presented to the class; the teacher guides students to agree that some options can be eliminated. In Phase 3, the teacher sends students to their small work groups to present their thinking to one another and try to reach consensus on a single solution and justification. In Phase 4, students are re-presented with the problem of the day but now with the opportunity to make use of the whole-class and small-group discussions as they reconceptualize the problem. In Phase 5, the class reassembles, and the teacher orchestrates the discussion toward the correct solution with an extended explanation. In the final phase, the class concludes as it started, with independent work on two extension problems that the teacher uses as evidence to evaluate the lesson's effectiveness. Appendix A contains a more detailed lesson plan with a focus on the function of each phase for students' learning and the roles of both the teacher and the students.

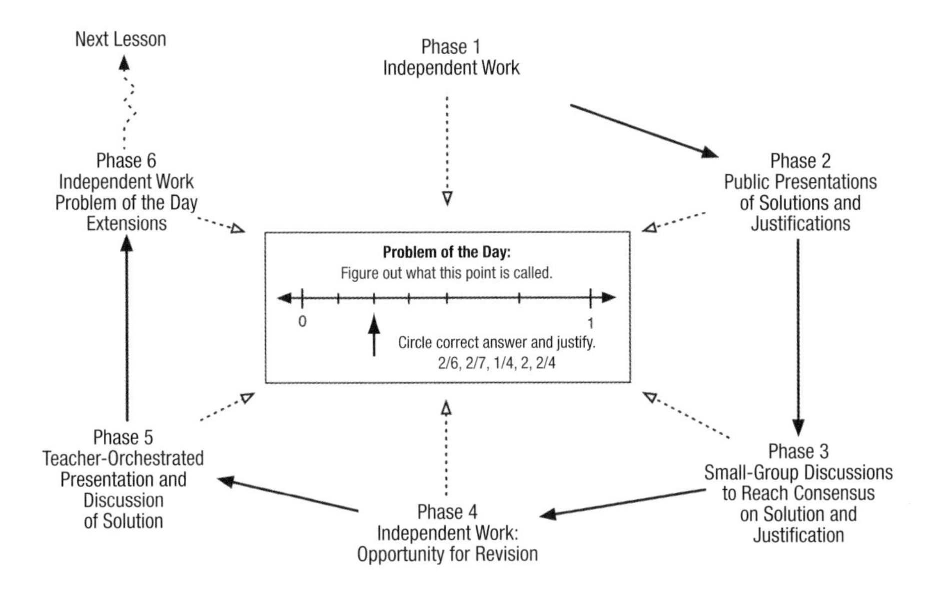

Fig. 13.7. The six-phase lesson structure used in the classroom study

Mr. Chinn implemented the lesson twice, once with each of his two math classes. Between lessons, we reviewed the students' learning and modified the design for the second cycle. For each implementation, we analyzed students' work from three assessments during the course of the lesson: Phase 1, students' independent work; Phase 4, students' independent work following small-group discussions; and Phase 6, students' independent work on the extension problems.

The Lesson: Students' Learning

Though the lessons were virtually identical in overall structure, in the second cycle Mr. Chinn took on a more central role in orchestrating student discussion. Although we focus primarily on findings from the second lesson, we will conclude with comparative findings from the first lesson to draw attention to the teacher's role in guiding the students' learning.

Mr. Chinn implemented the second lesson essentially as it had been planned. He presented his sixth graders with the Problem of the Day (see fig. 13.8) as a worksheet, and after the students had completed the task, he polled them to determine how many had selected each value, challenging them to make sense of their disagreements. To focus students' attention on several important issues, Mr. Chinn guided the class to eliminate two of the five answer choices. He had two goals for this phase. On the one hand, he wanted to help students consolidate their understanding of number line conventions: the answer "2" was eliminated because the point is between 0 and 1 and therefore cannot be 2—a discussion that drew students' attention to the 0 and the meaning of the hash marks to the right of 0. On the other hand, he wanted to end up with three choices—"2/6," "2/7," or "1/4"—that would focus students' attention on the more common solutions, solutions that were most likely to spark interesting discussion and reflection. In Phase 3, Mr. Chinn asked small groups to reach consensus on which of these three responses was the correct solution.

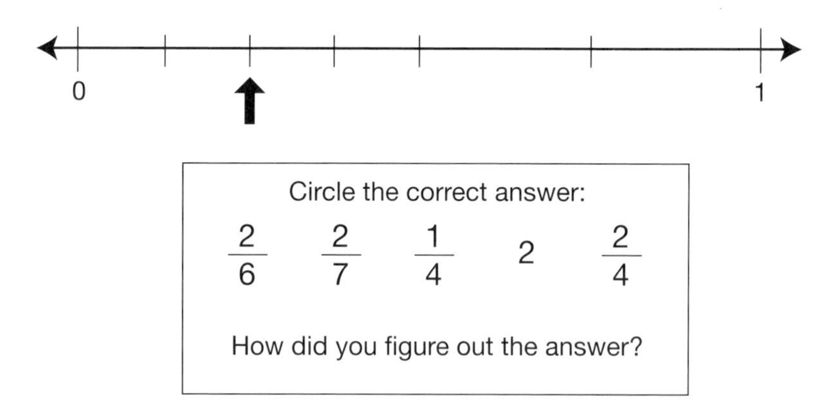

Fig. 13.8. Problem of the Day

Consider the opportunities for learning that occurred during Phase 3 as one lively group of four engaged in a debate about (1) strategies for naming a point on a number line and (2) number line conventions. Jorge incorrectly asserted that the answer was 2/6, arguing that it was wrong to add hash marks to divide the interval into equal distances (a strategy that other students were advocating); he reasoned that the test developers had purposely made the problem in this way, and therefore the correct answer must be "2/6" because there are 6 intervals and the arrow is pointed at the end of the second one. This is a form of reasoning that treats the points or intervals on the line as a discrete quantity. Kelsey countered that Jorge's explanation did not make mathematical sense, because fractions need equal-sized sections (intervals). Jorge looked back at the whiteboard where Mr. Chinn had written something a student had said—"pieces have to be equal size"—and tells the group, "Oh, that's what he said, huh—they're supposed to be equal." Jorge was convinced, and he soon joined Ethan and Kathy in attempting to convince Kelsey that the answer was "1/4." Jorge's performance on the second worksheet suggests that he gained new insights about fractions on number lines in this interaction; on a repeat of the same nonroutine task, Jorge circled "1/4" as his answer and wrote, "Well, first of all I think the number line should have the same length [use intervals of the same lengths]. So that means that if you add the two marks [in the middle of the longer intervals], then you would have eight segments. So you simplify 2/8, which equals 1/4. So it would be the correct answer." This episode illustrates the way a student shifted his approach to conceptualizing a problem, since he coordinated his thinking about point identification with his peers as well as with sources of authority (the perceived motives of the test developers, the principle written on the board). The coordination opened up new problems and new insights for Jorge.

Another example of students' learning comes from Phase 5 of the lesson, when Mr. Chinn's task was to guide students to an adequate solution and justification. During this phase, Mr. Chinn introduced a new version of the mathematical dilemma to focus on strategies for naming points on a number line. David, known to the class as a high-achieving student, had told Mr. Chinn privately that the answer could be either 1/4 *or* 2/7, and Mr. Chinn now asked David to explain his reasoning to the whole class. In his explanation, David argued that if hash marks were removed to create fourths, then the point was 1/4, but if hash marks were added to create equal intervals, then the point was called 2/7 (he counted only the internal hash marks).

As students considered David's argument, they referred to prior whole-class and small-group discussions. Jason argued that David's solution was problematic on the basis of the premise that the unit length between 0 and 1 must be fixed. Jason showed on the overhead (fig. 13.8) that the answer could not be 2/7, because the unit distance would need to be altered. To show this, he counted intervals by sevenths, beginning with the leftmost interval and ending with the equivalent of 1, or 7/7; if the point were 2/7, the unit distance would need to be moved from its position on the number line, something that the class agreed to as fixed. To support his students' learning, Mr. Chinn used the 1/4 vs. 2/7 discussion as a context for a teachable moment. He crossed out hash marks on figure 13.9 so that the number line was divided into fourths; he also added hash marks to divide the number line into eighths, named the point "2/8," which simplifies to 1/4, and stated the principle of equal intervals.

We have shared some examples of students' learning, but to what extent did all students learn in the class? To answer this question, we contrasted stu-

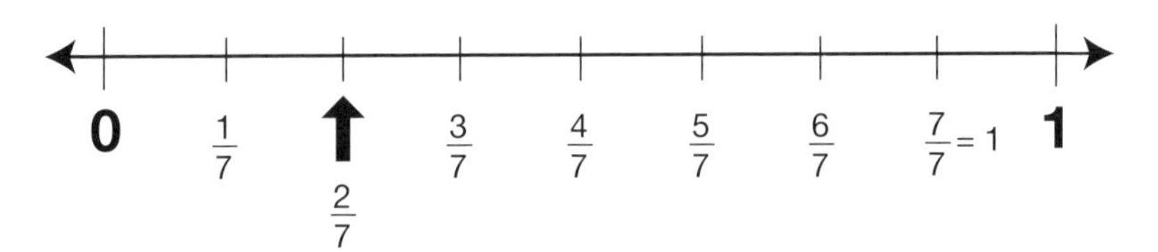

Fig. 13.9. Jason's identification of a contradiction for the "two-sevenths" solution

dents' learning in the first and second cycles at different points in the lessons. As indicated in figure 13.10, only students in Lesson Cycle 2 showed gains in their performance from Worksheet #1 to #2 as a result of student whole-class presentations and small-group discussions. Furthermore, students in Lesson Cycle 2 sustained their high level of performance on the first Extension Problem A (the same fractions family) and could extend their learning to Extension Problem B (a new fractions family). In contrast, although Lesson Cycle 1 students showed gains on Extension Problem A after they were presented with the correct answer to the Problem of the Day, their gains dropped precipitously with an extension problem that differed in fraction family (Extension B) from the Problem of the Day.

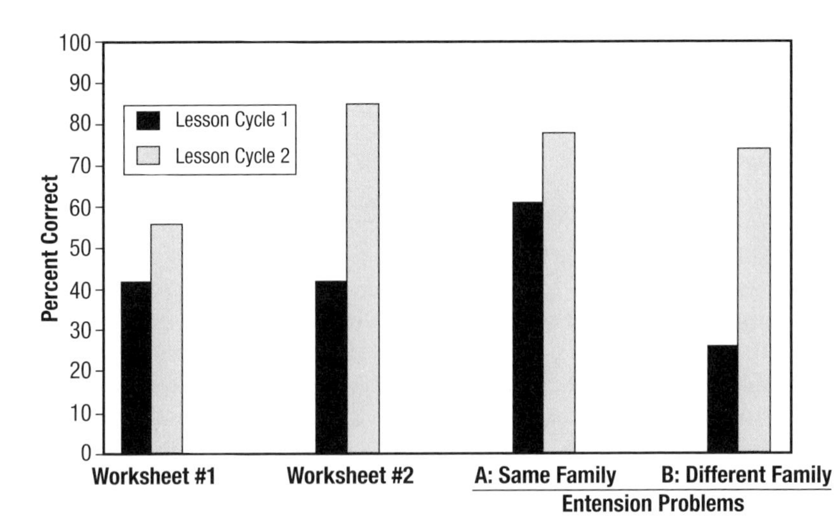

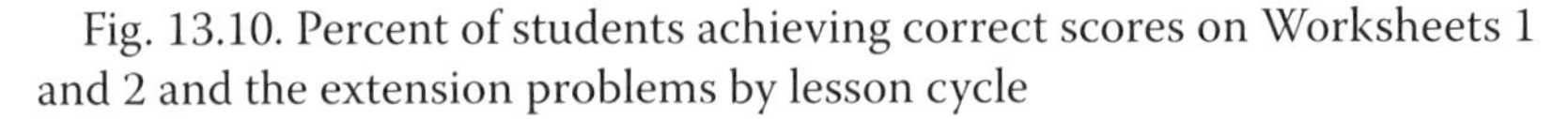

Fig. 13.10. Percent of students achieving correct scores on Worksheets 1 and 2 and the extension problems by lesson cycle

What accounted for the differences in students' learning between the two lesson cycles? Of course, we cannot isolate a single factor when two classes are compared. Even when the same teacher is engaging a different class with a similar lesson, many elements differ that can plausibly affect learning outcomes. But what we do know is that Mr. Chinn shifted his role from being primarily a facilitator of discussion (Lesson Cycle 1) toward a more directive role in guiding discussion toward the "big mathematical ideas" targeted in the lesson (Lesson Cycle 2). For example, in Phase 2, Mr. Chinn offered a more explicit focus on number line conventions as he and the class eliminated the unpopular choices, and he showed how the number line conventions rendered these alternatives inappropriate. Further, he used the simplified answer set to motivate the students in their small-group work, charging the groups with the task of winnowing down the possibilities even further. Finally, Mr. Chinn presented a more detailed explanation of the correct solution, drawing on students' prior arguments.

Conclusions and Next Steps

Our findings demonstrate the value of coordinating developmental and classroom-based research, particularly when investigating learning processes in hard-to-teach and hard-to-learn domains of mathematics. We are currently using the approach to investigate other important arenas of students' learning involving the number line, including properties of number related to the density of points (where to locate the point 1 49/200, and how many points are between any two points on a number line?) and operations on fractions and integers using number lines. Our goal is to design studies that reveal important processes of students' learning that the number line can support and ways of efficiently supporting these processes in classroom contexts. We believe our efforts will better prepare upper elementary and early middle school students for the mathematics of middle school and beyond.

REFERENCES

Ball, Deborah Loewenberg. "With an Eye on the Mathematical Horizon: Dilemmas of Teaching Elementary School Mathematics." *Elementary School Journal* 93 (March 1993): 373–97.

Inagaki, Kayoko, Giyoo Hatano, and Eiji Morita. "Construction of Mathematical Knowledge through Whole Class Discussions." *Learning and Instruction* 8 (December 1998): 503–26.

Itakura, K. "Instruction and Learning of Concept 'Force' in Static Based on Kasetsu–Jikken–Jigyo (Hypothesis–Experiment–Instruction): A New Method of Science Teaching (in Japanese)." *Bulletin of National Institute for Educational Research* 52 (1967): 1–121.

National Council of Teachers of Mathematics (NCTM). *Curriculum and Evaluation Standards for School Mathematics.* Reston, Va.: NCTM, 1989.

———. *Principles and Standards for School Mathematics.* Reston, Va.: NCTM, 2000.

Saxe, Geoffrey B., Maryl Gearhart, and Michael Seltzer. "Relations between Classroom Practices and Student Learning in the Domain of Fractions." *Cognition and Instruction* 17 (1999): 1–24.

Saxe, Geoffrey B., Edd V. Taylor, Clifton McIntosh, and Maryl Gearhart. "Representing Fractions with Standard Notation: A Developmental Analysis." *Journal for Research in Mathematics Education* 36 (March 2005): 137–57.

Vygotsky, Lev. *Thought and Language.* Cambridge, Mass.: MIT Press, 1986.

Appendix A: Whole-Class Lesson Plan

Phase of Lesson	Function for Students' Learning	Students' Role	Teacher's Role
Phase 1: Independent work: Problem of the Day worksheet	Make sense of a point on a number line.	Listen carefully to the Problem of the Day, and work on the problem independently.	Present the problem and monitor students' solutions by walking around to students' desks.
Phase 2: Students' public presentations of solutions and justifications	Consider alternative solutions and rationales.	Present a solution or listen carefully to presenters' solutions, asking questions and thinking about why an answer does or does not make good mathematical sense.	Ask student presenters to explain their reasoning; establish norms for whole-class discussion, steering the lesson toward big mathematical ideas.
Phase 3: Small-group discussions	Debate solutions in active discussion.	Explain which answer makes good mathematical sense and why. Listen to group members present their answers. With your group, decide which answer makes the most mathematical sense.	Establish norms for the small-group discussions.
Phase 4: Independent work: Problem of the Day worksheet	Revise the interpretation of a point on a number line.	Independent work with the option to take a different approach to the Problem of the Day.	Observe students' solutions as they work.
Phase 5: "The Correct Solution"	Make sense of the teacher's answer and rationale, integrating it with prior thinking.	Listen carefully and think about why the answer makes the most mathematical sense.	Present the correct answer, building on students' thinking.
Phase 6: Independent work: extension problems	Extend thinking to a new set of nonroutine number line tasks.	Independently figure out what the numbers are called and explain why.	Observe students' solutions as they work.

14

Beyond Puzzles: Young Children's Shape-Composition Abilities

David C. Wilson

We have all watched children solve puzzles of one form or another, from matching letters or shapes to a cutout form, to assembling multiple pieces to make a picture, to engaging in more freestyle activities, such as play with wooden blocks or pattern blocks. What can we learn as we observe children engaged in puzzle-solving activities? The mental processes underlying such activities are complex of course, but recent research has revealed some interesting developmental patterns that can help teachers recognize those abilities present in their students and offer specific instructional tasks to help develop and extend those abilities.

As children manipulate shapes during play and activities, they are developing their concept of shape (including length and angle) through their physical interactions with the objects. As the tasks and their interactions increase in sophistication, children further develop fundamental geometric and spatial concepts, including orientation, transformations (rotations and reflections), parallelism, congruence, and part-whole relationships; and more significantly, they begin to develop mental representations that allow them to engage in increasingly complex tasks. Thus, the abilities to define, use, and visualize the effects of composing (putting together) and decomposing (taking apart) geometric shapes constitute a major conceptual field, as well as a set of competences, within the domain of geometry. Furthermore, the acts of creating composed units in the contexts of patterns, measurement, and calculations are foundational building blocks of mathematical understanding (Clements et al. 1997; Reynolds and Wheatley 1996; Steffe and Cobb 1988). Clements and his colleagues (1996) suggest that children's abilities to compose and decompose shapes correspond with and support their abilities to compose and decompose numbers.

The following pages describe the findings of research conducted in developing a learning trajectory for children's shape-composition abilities.[1] In general, a learning trajectory includes "conjectures about both a possible

1. For a complete description of the development and research supporting the hypothesized learning trajectory, see Clements, Wilson, and Sarama (2004) and Wilson (2002).

learning route that aims at significant mathematical ideas and a specific means that might be used to support and organize learning along this route" (Clements 2002, p. 605). Thus, a hypothetical developmental progression, or learning route, was first created and then refined through the cooperative efforts of teachers and researchers. Once the developmental progression was assessed and found to be valid, it then became the basis for activities and computer software that support children's developing abilities in composing shapes. The verbal descriptions of children's actions stem from the research on the developmental progression.

The Developmental Progression

The hypothesized developmental progression and learning trajectory for children's shape-composition abilities were developed as a part of a larger project aimed at developing software and print materials for prekindergarten to second-grade mathematics.[2] The developmental progression was developed as a basis for the creation of the print activities and software, which in turn assist in moving children along the progression. Thus, the working model for the project involved constructing a developmental progression, conducting research to assess the validity of the progression, and finally developing both on- and off-computer activities that move a child along the developmental progression. Each phase of the project involved classroom teachers and students as well as the researchers.

The theoretical assumption underlying the developmental progression of shape composition is that to solve composition tasks effectively and efficiently, children must build a mental image of a shape and then match that image to the goal shape by superposition, performing mental rotations as necessary to match those images (Clements, Sarama, and Wilson 2001). This development is related to their evolving knowledge of shapes, which Pierre and Dina van Hiele have theorized develops from a generic visual impression to property recognition and hierarchical classifications (van Hiele 1986). The learning trajectory goes beyond existing van Hielian thought in adding the composition process as an essential element of geometric knowledge. That is, in addition to children's increasing understanding of shapes through their properties (and eventual recognition of properties that exist between and within shapes), the competences within composition and decomposition are an additional developmental progression that may parallel and support development within the

2. This article is supported in part by National Science Foundation Grants No. ESI-9730804, "Building Blocks—Foundations for Mathematical Thinking, Pre-Kindergarten to Grade 2: Research-Based Materials Development" (Douglas H. Clements and Julie Sarama, Co-PIs) and REC-9903409, "Technology-Enhanced Learning of Geometry in Elementary Schools" (Daniel Watt, Douglas H. Clements, and Richard Lehrer, Co-PIs). Any opinions, findings, and conclusions or recommendations expressed in this material are those of the author and do not necessarily reflect the views of the National Science Foundation.

van Hiele levels. Similar to the van Hiele levels of development, children's initial abilities to combine shapes are based on trial-and-error combinations of whole shapes as they form pictures and attempt puzzle tasks. Later, through an increasing ability to combine shapes on the basis of their attributes (e.g., side lengths, angle size), they solve increasingly complex puzzles and begin to foster mental images of shapes and their respective attributes. The following paragraphs elaborate on these levels and detail the developmental progression and significant aspects of children's developing abilities at each level.

The developmental progression and complete learning trajectory were created in collaboration with teachers, and the description, assessment, and labeling of the levels were designed for comprehensibility. The following levels constitute the initial hypothesized developmental progression for the composition of shapes.[3]

1. *Precomposer.* Children manipulate shapes individually but are unable to combine them to compose a larger shape. For example, in free-form "make a picture" tasks, children might use a single shape for a sun, a separate shape for a tree, and another separate shape for a person.
2. *Piece assembler.* Children at this level are similar to precomposers, but they can concatenate shapes (i.e., place shapes adjacent to one another with shared sides) to form pictures. In free-form "make a picture" tasks, each shape used represents a unique role or function in the picture (e.g., one shape for one leg). In puzzle tasks, children can fill simple frames using trial and error (Mansfield and Scott 1990; Sales 1994) but have limited ability to use turns or flips to do so; they cannot use geometric motions to see shapes from different perspectives (Sarama et al. 1996). Thus, children at the first two levels view shapes only as wholes and see few geometric relationships between shapes or between parts of shapes (i.e., a property of the shape).
3. *Picture maker.* Children can concatenate shapes to form pictures in which several shapes play a single role (e.g., a leg might be created from three contiguous squares), but they use trial and error and do not anticipate creation of new geometric shapes. Children at this level choose shapes using gestalt configuration or one component, such as side length (Sarama et al. 1996). If several sides of the existing arrangement form a partial boundary of a shape (instantiating a schema for it), the child can find and place that shape. If such cues are not present, the child matches by a side length. The child may attempt to match corners but does not possess angle as a quantitative entity, so he or she tries to match shapes into corners of existing arrangements in which their angles do not fit

3. The hypothesized developmental progression has two additional levels that relate to older children and are more fully described in the previously referenced works.

(i.e., uses a "picking and discarding" strategy). The child uses rotating and flipping, usually by trial and error, to try different arrangements.

4. *Shape composer.* Children combine shapes to make new shapes or fill puzzles with growing intentionality and anticipation ("I know what will fit"). They choose shapes using angles as well as side lengths. Eventually, children consider several alternative shapes with angles equal to the existing arrangement. They use rotation and flipping intentionally and mentally (i.e., with anticipation) to select and place shapes (Sarama et al. 1996). They can fill complex frames (frames whose filling requires multiple shapes) (Sales 1994) or cover regions (Mansfield and Scott 1990). Their imagery and systematicity grow within this and the following levels. In summary, children exhibit intentionality and anticipation based on the shapes' attributes, and thus they exhibit imagery of the component shapes, although imagery of the composite shape develops within this level.
5. *Substitution composer.* Children form composite units of shapes (Clements et al. 1997) and recognize and deliberately use substitution relationships among those shapes (e.g., two trapezoid pattern blocks can make a hexagon).

Children's Actions and the Developmental Progression

Two Puzzle Tasks

The levels composing the developmental progression are best understood through examining children's responses to specific tasks. The research conducted to assess the validity of the developmental progression involved administering a set of seventeen tasks to approximately twenty children at each of four grade levels—prekindergarten, kindergarten, first grade, and second grade. The tasks were completed on paper and frequently employed shape manipulatives—primarily pattern blocks. Many of the tasks later became part of the set of print activities and computer software that were the outcome of the project.

One of the tasks, the "dog" puzzle, shown in figure 14.1, was successfully completed by 70 percent of the children; however, unlike for some tasks, all children attempted to complete the puzzle and thus gave the researchers an opportunity to observe different strategies at many levels. The dog puzzle elicited different responses, in part because it is composed of simple frames and complex frames.[4]

4. The term *simple* is used to refer to a frame that defines a single pattern block piece, such as the "legs" in figure 14.1, whereas *complex* refers to frames that require the concatenation of several pieces to fill the frame, such as the "head" and "body."

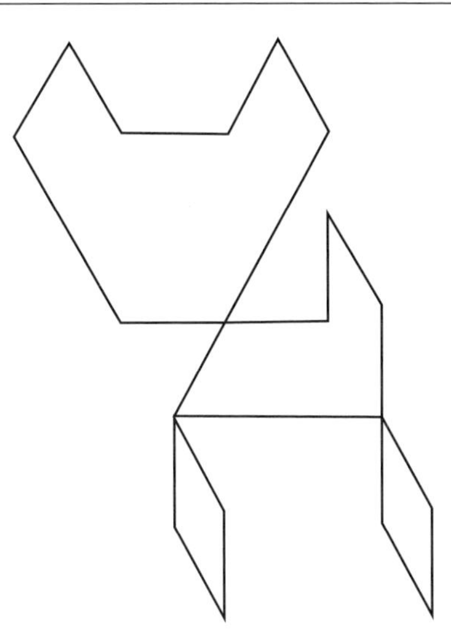

Fig. 14.1. The "dog" puzzle

A second task that illustrates the developmental progression levels is a complex frame that was designed to assess some of the actions of children at the higher levels. The puzzle shown in figure 14.2, the "dog bone," prompted a variety of solution paths; it was accompanied by limited sets of pattern blocks and challenged children to imagine the completed figure through mental constructions involving the pattern blocks.

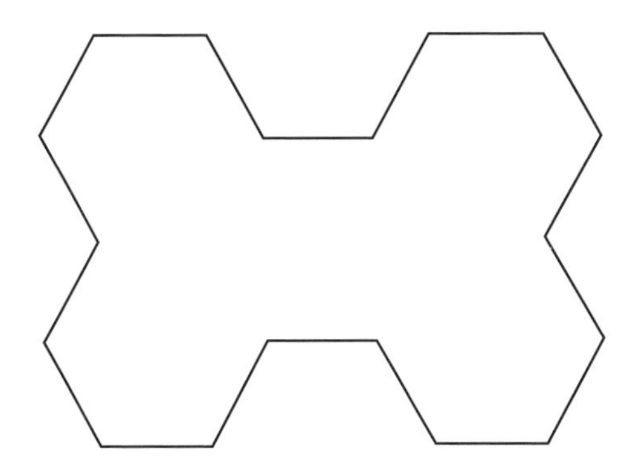

Fig. 14.2. The "dog bone" task

The dog-bone task was composed of three parts. The children were initially presented with eight hexagons and asked whether the puzzle could be filled using only hexagons. Once the children had ascertained the solution with hexagons, the hexagons were removed and the children were presented with four trapezoids and asked whether the puzzle could be filled using only trapezoids if they had as many as they wanted. That is, the children were asked to imagine that they had as many trapezoids as they would like and to determine if the trapezoids would then fill the puzzle. Similarly, they were then presented with four triangles and asked whether the puzzle could be filled using only triangles, again, if they had as many as they wanted. Thus, children were provided with an abundance of hexagons followed by too few trapezoids and too few triangles to fill the puzzle frame. In order to assess the children's thinking, following each child's response to whether only trapezoids could fill the puzzle, they were asked to state how many trapezoids would be required. The same question followed their response regarding whether triangles could fill the puzzle. Thus, the task offered an opportunity for children to recognize and discuss the relationship between the pattern blocks (e.g., two trapezoids can form a hexagon) as well as for teachers to assess their ability to conceptualize the concatenation of shapes in the absence of the actual pieces.

The dog puzzle and the dog-bone tasks form the basis for the following discussion of several children's behaviors as they attempted to complete the tasks in ways that illustrated their level of ability within the developmental progression.

Developmental Level Responses

1. A precomposer

Anna,[5] a three-year, eleven-month-old, was the only child whose behaviors exclusively exemplified the precomposer level. Figure 14.3 displays Anna's solution to the dog puzzle. Anna began by placing a square on each "ear" and then moved to the "legs" and placed a square on each leg. Those actions suggest that Anna was able to see some symmetrical aspects of the puzzle and attend to them in the selection and placement of pattern blocks but was not able or motivated to attend to the frame and shape simultaneously. However, as she attempted to fill the remaining complex frames that required her to concatenate the pattern blocks, Anna exemplified precomposer-level behaviors.

Anna did not attempt to concatenate pieces to fill, or even partially fill, the complex frames. The placement of the triangles indicated she was able to see that the section of the puzzle required multiple pieces to fill, but she visualized the pieces as separate entities, not as parts of a larger shape or figure. Anna was consistent in that behavior throughout the entire set of tasks. In every opportunity presented to her, she allowed a maximum of vertex-to-side contact, never side-to-side concatenation except when incidental contact occurred through placement of additional pattern blocks.

5. All proper names are pseudonyms.

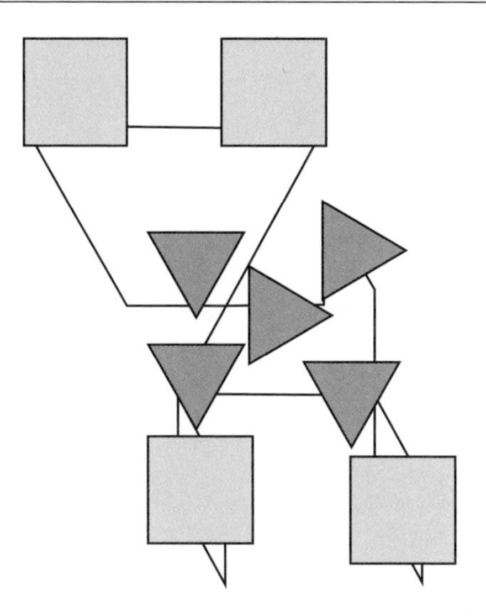

Fig. 14.3. A precomposer's completed "dog" puzzle

2. A piece assembler

Christopher, a four-year, ten-month-old kindergartener, was able to concatenate shapes to attempt to fill frames. Christopher's response to the puzzle dog is shown in figure 14.4. Christopher correctly filled the simple frame legs with the rhombus but had difficulty filling the complex frame of the head. Although he began by correctly placing a trapezoid at the top of the head, he selected squares to fill the remaining space despite the mismatch with the frame. Typical of piece assemblers, he chose squares to fill the frame perhaps because they were the easiest shape to concatenate with the trapezoid. That is, placing two squares adjacent to one another results in the same length as the long base of the trapezoid.

Once the complex frame had been somewhat filled, Christopher believed that the task was completed. His visual image of the frame had been replaced by the visual presence of the concatenated pieces, creating a unique solution that satisfied the gestalt image of the puzzle; that is, to him his solution represented a "dog."

A second item that illustrates Christopher's piece-assembler level of thinking was his response to the dog-bone task. Christopher was able to correctly complete the puzzle using five hexagons to begin the task. However, when he was given the four trapezoids, he placed them in separate parts of the puzzle, as shown in figure 14.5.

When asked whether he would be able to complete the puzzle if he had as many trapezoids as he wanted, he said no. The question was repeated with the emphasis on "as many as you need," but he still shook his head no. Chris-

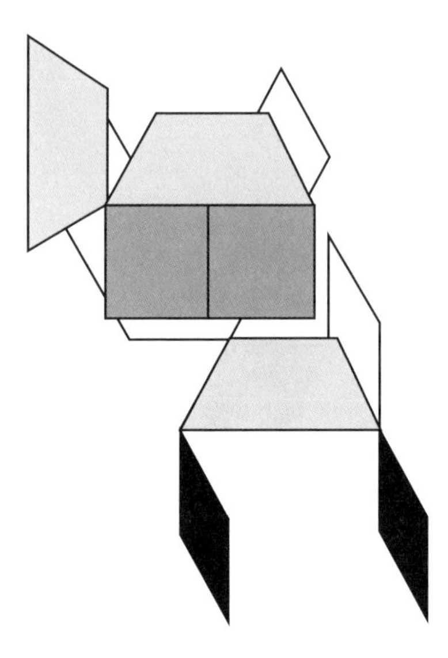

Fig. 14.4. A piece assembler's completed "dog" puzzle

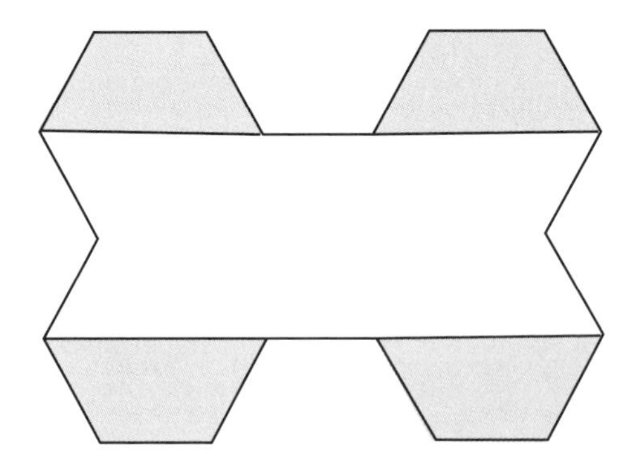

Fig. 14.5. A piece assembler's response to the "dog bone" task

topher's negative response reflects his focus on the shape as an independent entity and the lack of ability to see interrelationships between the shapes and the frame (and other shapes). That is, his difficulty was not that he was unable to see the interrelationship between the trapezoid and the hexagon, which does not occur until much later in the trajectory, but rather that he was unable to see that the remaining spaces could be filled with trapezoids even though the portions of the frame remaining after the pieces were in place were highly indicative of the trapezoid. Additionally, he did not concatenate the trapezoids to fill regions of the puzzle, nor move them about the puzzle, thus reinforcing his view of them as independent pieces, without relationship to one another or to the frame, which is typical of piece-assembler thinking.

3. A picture maker

Kevin, a six-year-old first grader, exhibited behaviors reflective of the picture-maker level. The two puzzles shown in figure 14.6 depict his progress as he attempted to complete the puzzle dog. Kevin used a "picking and discarding" strategy repeatedly as he tried to fill the puzzle's complex frames. He selected and "tried out" shapes for a fit through placement and manipulation of the shapes directly on the puzzle—in apparent absence of anticipation and visualization. However, unlike a piece assembler, Kevin maintained the image of the frame and rejected those pieces that did not match it.

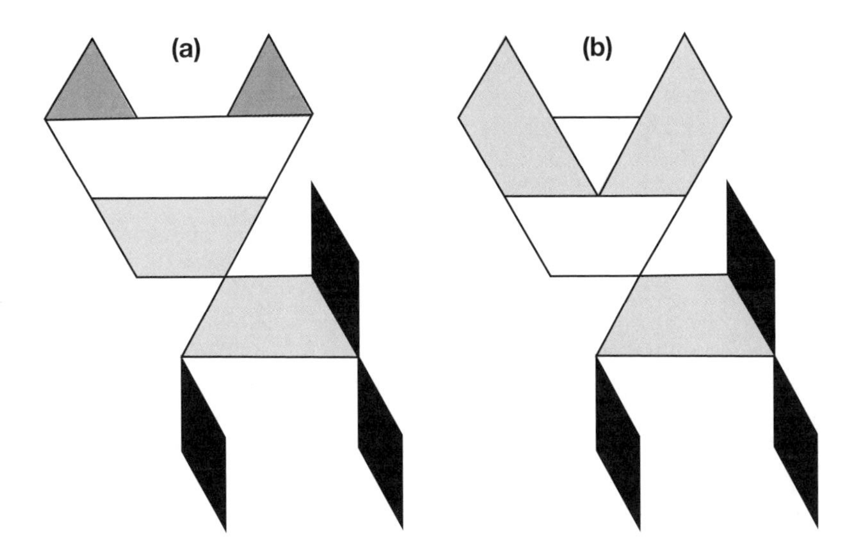

Fig. 14.6. A picture maker's initial (a) and later attempt (b) to complete the "dog" puzzle

Figure 14.6a displays the puzzle after Kevin initially placed the triangles in the ears. The complex frame that remained once he had placed the trapezoid at the base was difficult for him to complete. Kevin attempted to fit a rhombus into the open space, then a square, and eventually, being unable to fill the complex frame, he rejected the arrangement and cleared away the shapes in the head. Figure 14.6b depicts his new attempt. His placement of a trapezoid along the left side of the head and then similarly on the right created two remaining simple frames that allowed him to complete the puzzle. Thus, typical of a picture maker, Kevin was able to maintain an image of the individual shapes and recognize those shapes when simple frames were created, although he was unable to see relationships between shapes or to mentally compose shapes to fill complex frames.

Kevin's work on the dog-bone puzzle further distinguishes his thinking as typical of the picture-maker level. He answered correctly that five hexagons would be needed to fill the puzzle once he placed them on the puzzle, but, unlike Christopher, a piece assembler, Kevin concatenated the trapezoids to cover a region in his attempt to answer part (b), as shown in figure 14.7.

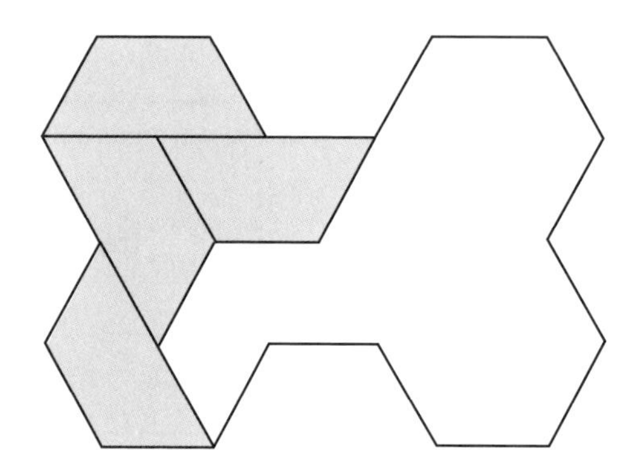

Fig. 14.7. A picture maker's response to the "dog bone" task

Kevin responded correctly that the frame could be filled if he had more trapezoids. However, he was less certain when responding to the number of trapezoids that would be required. His uncertainty was evident in his incorrect response that nine trapezoids would be needed to fill the dog bone and in his admission that "I don't know if it's really going to work" as he wrote his response on the paper.

Kevin's responses reflect the developmental nature of mental images within the picture-maker level. When manipulatives were provided, he was able to complete frames through trial and error, often using side length as a cue to find shapes. However, the less definition supplied by the frame (i.e., fewer sides defining the individual shapes), the less ability he demonstrated in using his developing mental image. In addition, unlike the following levels, he did not recognize the relationship between the trapezoid and the hexagon.

4. A shape composer

Marissa, a seven-year, seven-month-old second grader, was one of two children who were categorized as shape composers. This level is characterized by the emergence of well-developed mental images and the ability to operate on them. Marissa completed the dog puzzle quickly and efficiently. She was able to look at the puzzle frame, look at the shapes available, and select and correctly place each shape on the frame. Trial and error was completely absent in her work. She made deliberate selections with anticipation of the precise placement of the shape, and she used a systematic method to complete the task.

Marissa completed the dog-bone task with similar efficiency and anticipation. Unsurprisingly, she was able to state the correct number of hexagons required to fill the puzzle after placing two of them on the frame. To determine whether the trapezoids could fill the frame, she placed two of the four trapezoids together to form a hexagon and did the same with the other two. The following dialogue occurred as Marissa attempted to answer whether trapezoids could fill the puzzle.

M:	Of course I could if I had as many as I wanted [as she was about to place the fourth trapezoid in place].
DW:	How do you know for sure?
M:	Because if I could have as many as I wanted, I could have as many as there are in the puzzle.
DW:	But how do you know for sure that they would cover it evenly—like they wouldn't stick out, or you'd be missing a piece and maybe you'd need a triangle. How do you know they'd fill it exactly?

Marissa then made a hexagon with two trapezoids in the lower right and again in the upper right, and then slid the pair from the lower right to the middle and nodded her head yes.

DW:	Okay, and how did you decide?
M:	Because I took these four and made a hexagon and another hexagon [the upper right], and then I took it like that [as she slid it to the middle position], and then I could move it here [as she showed how the hexagon would fill both hexagonal spaces on the left].

Marissa had to think about why she knew that the trapezoids would be able to fill the frame. The fact that she needed to slide the concatenated trapezoids on the puzzle to determine whether they unquestionably could fill the frame is significant. A shape composer has a strong mental image of shapes, and thus, Marissa was easily able to make a hexagon and then another, because that shape is the logical and most efficient way to fill the space. However, as her response indicated, she had not yet begun to think of united shapes as a whole, nor reason that since five hexagons would cover the puzzle, and that two trapezoids can make a hexagon, then the trapezoids could cover the puzzle.

Marissa completed the last part of the dog-bone task by correctly responding that triangles could indeed fill the puzzle and that thirty would be needed to do so. She was able to visualize the two triangles needed to complete a hexagon in addition to the four given triangles. She then used methods similar to her analysis of the trapezoids as she moved the group of four triangles to the middle hexagonal space to be certain they could fit there.

Worth noting is the fact that she answered the question of filling the frame separately from the question of how many of a shape would be needed to do so. That is, she first did the work required to verify that the triangles would fill the frame, wrote that answer in the space provided, and then said, "Okay, now, how many" and replaced the triangles in the lower right corner to begin counting. Her summary of her arithmetic methods yielding a total of thirty triangles reflected her strong mental imagery at this point. She explained as follows:

> Because, if I had 6 over here and 6 over here [pointing to the two hexagonal spaces on the right], that would be 12 ..., and then 12 plus 12 [as she points to the left side] is, um, 24 ..., and then 24 plus 6 [in the middle] is 30.

Although Marissa may clearly appear to be uniting the triangles into a hexagon and using that composite shape to calculate the total number required, an important aspect to recognize is that she had to place the triangles on the puzzle and count the unfilled spaces with her finger to know that six triangles made up the hexagon; that knowledge was *not* part of her mental image. Her strong mental image of the individual triangle, not the image of the composite shape made up of six triangles, was what allowed her to know that two triangles were missing.

5. A substitution composer

A substitution composer extends the strong mental image of a shape composer to include recognizing and using substitution relationships between shapes. Evidence of this level was seldom encountered, in part due to the ages of the children targeted in the research as well as the limited number of tasks employed that were designed to elicit responses reflective of this level.

Evan, a seven-year, ten-month-old second grader, revealed thinking indicative of his developing understanding of substitution relationships between pattern blocks in his response to the dog-bone task. In the first part, he was easily able to identify five as the number of hexagons required to fill the frame. He then answered the trapezoid part of the question without using the four trapezoid pieces he was given. He wrote "yes" in response to the question of whether the trapezoid shapes could fill the frame while simply looking at the trapezoids. He followed that response by writing "10" for the number required to fill the frame, again without touching the shapes. When asked how he had arrived at his answers, he said, "I pointed my pencil to half of the shape at each part, and then I knew it was going to be ten." His response indicated he was using his knowledge that a trapezoid was half of a hexagon to arrive at his conclusions. When responding to the third part, Evan did move the triangles around on the frame, but his explanation as to how he knew that the puzzle could be completed with only triangles was again based on the substitution relationship between triangles and the hexagon. He explained, "I knew that each hex—each of this shape would be six, so then I counted six [*sic*] groups of six to equal thirty."

Evan was recognizing and using the fact that six triangles make one hexagon to respond to the task. However, his reliance on moving the group of triangles on the puzzle, and the absence of the realization that three triangles also make a trapezoid, reflect his developing abilities within this level. That is, as with many of the children involved in the assessment of the developmental progression, he was in transition between levels.

Children in Transition

A particularly large number of children were in transition from picture maker to shape composer. This preponderance may not be surprising, since the previously discussed items clearly show that the ability to maintain a mental image and perform operations on that image is a significant advancement from the picture maker level. This development occurs over an extended time period and depends on numerous varied experiences and tasks involving shapes.

The figures and descriptions above, detailing the response of children engaged in the two tasks, allow for further understanding of the levels and help visualize the children's actions-on-objects that characterize each level. They may also help teachers assess their own students' composition abilities and make instructional decisions, as well as devise appropriate activities aimed at developing their students' composition abilities. The outcome of the project described at the beginning of this section and some suggestions for classroom implementation are given in the following paragraphs.

Classroom Activities and Practices

The developmental progression laid the foundation for both the development of activities within the learning trajectory and, following pilot tests and field tests of the activities, the print materials and software that were outcomes of the project (Clements and Sarama 2003; DLM Math Software 2003). In computer activities, such as "shape puzzles," shown in figure 14.8a, children complete puzzle frames by performing actions on geometric shapes reflective of some common manipulative sets. The computer presents different shape sets (e.g., pattern blocks and tangrams) depending on the frame presented, which in turn depends on the developmental level at which the individual child is working. The actions available include duplication (to get more of the shape), sliding, rotation, and reflection. The software assesses children's performance in completing puzzles (similar to those in the figures discussed above) and rewards the successful completion of a puzzle by the appearance of a creative background, as shown in figure 14.8b.

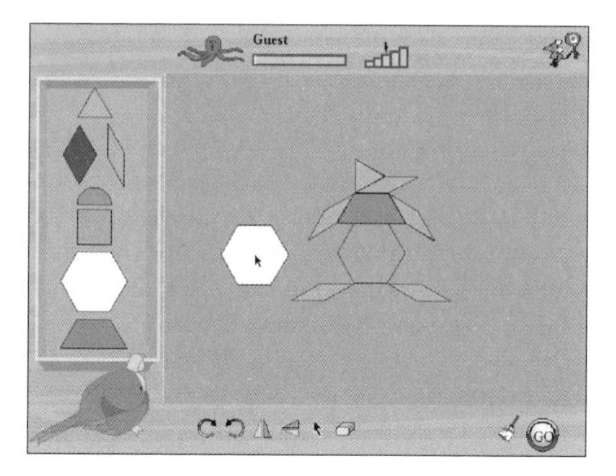

(a) A puzzle in progress.

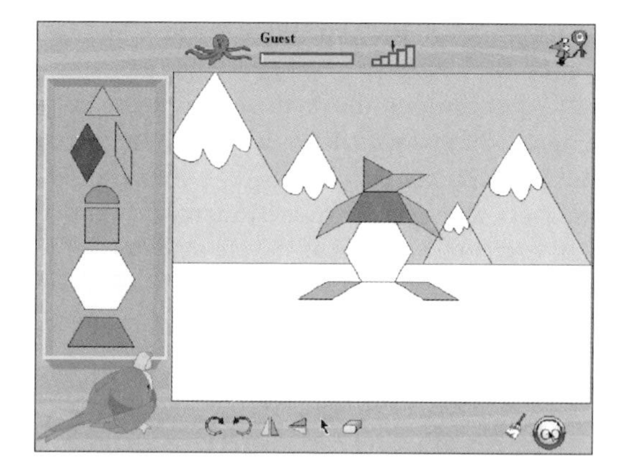

(b) Feedback after a successful solution.

Fig. 14.8. The DLM Early Childhood Math Software (Clements and Sarama 2003), which automatically presents puzzles at the children's current level of thinking.

The software moves children up (or down) through the trajectory by presenting puzzles appropriate to each level of the progression. For example, the precomposing child is presented puzzles in which a frame suggests the placement of each shape; further, each shape plays a separate role in the picture. For the child who is at the piece-assembler level, the software encourages the development of thinking at the picture-maker level by presenting puzzles that suggest the placement of each shape but in which several shapes together may play a single semantic role in the picture. To encourage shape-composer thinking, the software presents puzzles in which the child must completely fill a region that consists of multiple corners, requiring selecting and placing shapes to match angles. In addition, the software includes tangram puzzles that, unlike pattern block puzzles, can require reflections of figures to complete frames.

Although the software has clear advantages, puzzles are an integral part of most children's learning experiences, and many classrooms have puzzles involving tangrams or pattern blocks already available. The children's abilities, or lack thereof, in finding and placing shapes on such puzzles can be supported through appropriate selection of puzzles or modification of existing puzzles. Puzzles that require few shapes and that have lines to define the placement of each shape within the larger frame are ideal for children in transition from the piece-assembler level. The level of support is high, but children are attending to the individual shapes as independent entities. The additional frame structure helps develop their mental image of the individual shapes and of the geometric motions required to match a shape to a frame. With appropriate teacher interaction and support, it also helps them attend to the larger composite shape they have constructed.

The children working at, and in transition from, the picture-maker level often attend to side length as the primary characteristic. Thus, puzzles should offer hints about what shape to select through well-defined frames. The "dog" puzzle provides a fair mix of challenge and support for a child at this level. The number of pieces is limited, and the frame defines all but one side of most shapes that could be placed. Less-defined puzzles can be modified by drawing additional lines and creating additional simple frames, which then support the developing abilities rather than cause frustration. The child can also be challenged within this level and above to try to make the puzzle in a different way following initial success. This modification will facilitate their image of the large shape as being a whole composed of parts.

Children in transition from the picture-maker level and at the shape-composer level should be given increasingly challenging puzzles. The number of pieces can increase substantially at these levels, and the frames can gradually become less defined as children's mental images become increasingly flexible. Picture makers conduct geometric motions by hand (i.e., not mentally), and thus, puzzles that require geometric motions help develop the cognitive structures that will eventually allow children to operate mentally on the shapes. Challenging children at this level to make puzzles in different ways, but perhaps with such additional goals as all one color, will help develop awareness of the interrelationships of the shapes as well.

In light of the preceding paragraphs, puzzles may appear to be foundational to developing shape-composition abilities in children. However, free exploration of shapes should not be inferred to be unimportant. At all levels of the developmental progression, the creativity and geometric thinking that are enhanced through such free-explore activities are essential. Puzzles can be used to challenge children's abilities and support their growth, but they should not replace the creative designs and enjoyment that stem naturally from children's exploration of shapes in many environments, including geoboards, toothpicks, play dough, straws and connectors, drawing, and of course, the shapes and compositions of shapes found everywhere in the world around them. All these types of experiences are necessary for children's well-developed shape-composition abilities and will foster their increased geometric knowledge and abilities.

REFERENCES

Clements, Douglas H. "Linking Research and Curriculum Development." In *Handbook of International Research in Mathematics Education,* edited by Lynn D. English, pp. 599–630. Mahwah, N.J.: Lawrence Erlbaum Associates, 2002.

Clements, Douglas H., Michael T. Battista, Julie Sarama, and Sudha Swaminathan. "Development of Students' Spatial Thinking in a Unit on Geometric Motions and Area." *Elementary School Journal* 98 (2) (1997): 171–86.

Clements, Douglas H., and Julie Sarama. *DLM Early Childhood Express Math Resource Guide.* Columbus, Ohio: SRA/McGraw-Hill, 2003.

Clements, Douglas H., Julie Sarama, Michael T. Battista, and Sudha Swaminathan. "Development of Students' Spatial Thinking in a Curriculum Unit on Geometric Motions and Area." In *Proceedings of the Eighteenth Annual Meeting of the North America Chapter of the International Group for the Psychology of Mathematics Education,* edited by Elizabeth Jakubowski, Dierdre Watkins, and Harry Biske, pp. 217–22. Columbus, Ohio: ERIC Clearinghouse for Science, Mathematics, and Environmental Education, 1996.

Clements, Douglas H., Julie Sarama, and David C. Wilson. "Composition of Geometric Figures." In *Proceedings of the 25th Conference of the International Group for the Psychology of Mathematics Education,* vol. 2, edited by Marja van den Heuvel-Panhuizen, pp. 273–80. Utrecht, Netherlands: Freudenthal Institute, Utrecht University, 2001.

Clements, Douglas H., David C. Wilson, and Julie Sarama. "Young Children's Composition of Geometric Figures: A Learning Trajectory." *Mathematical Thinking and Learning* (6) (2004): 163–84.

DLM Math Software. Columbus, Ohio: SRA/McGraw-Hill, 2003.

Mansfield, Helen M., and Joy Scott. "Young Children Solving Spatial Problems." In *Proceedings of the 14th Annual Conference of the International Group for the Psychology of Mathematics Education (PME).* Oaxlepec, Mexico: PME, 1990.

Reynolds, Anne, and Grayson H. Wheatley. "Elementary Students' Construction and Coordination of Units in an Area Setting." *Journal for Research in Mathematics Education* 27 (November 1996): 564–81.

Sales, Christie. "A Constructivist Instructional Project on Developing Geometric Problem Solving Abilities Using Pattern Blocks and Tangrams with Young Children." Master's thesis, University of Northern Iowa, 1994.

Sarama, Julie, Douglas H. Clements, Julie Jacobs Henry, and Sudha Swaminathan. "Multidisciplinary Research Perspectives on an Implementation of a Computer-Based Mathematics Innovation." In *Proceedings of the Eighteenth Annual Meeting of the North America Chapter of the International Group for the Psychology of Mathematics Education,* edited by Elizabeth Jakubowski, Dierdre Watkins, and Harry Biske, pp. 560–65. Columbus, Ohio: ERIC Clearinghouse for Science, Mathematics, and Environmental Education, 1996.

Steffe, Leslie P., and Paul Cobb. *Construction of Arithmetical Meanings and Strategies.* New York: Springer-Verlag, 1988.

van Hiele, Pierre M. *Structure and Insight: A Theory of Mathematics Education.* Orlando, Fla.: Academic Press, 1986.

Wilson, David C. "Young Children's Composition of Geometric Figures: A Learning Trajectory." Ph.D. diss., State University of New York at Buffalo, 2002.

PART FOUR

Teachers' Learning of Mathematics

Marilyn E. Strutchens

Teachers are also learners of mathematics, both as students preparing for careers in education and subsequently as professionals committed to lifelong learning. *Principles and Standards for School Mathematics* (National Council of Teachers of Mathematics 2000, p. 17) states:

> To be effective, teachers must know and understand deeply the mathematics they are teaching and be able to draw on that knowledge with flexibility in their teaching tasks. They need to understand and be committed to their students as learners of mathematics and as human beings and be skillful in choosing from and using a variety of pedagogical and assessment strategies (National Commission on Teaching and America's Future 1996). In addition, effective teaching requires reflection and continual efforts to seek improvement. Teachers must have frequent and ample opportunities and resources to enhance and refresh their knowledge.

In this section, authors discuss the impact on teachers' practice of how they learn mathematics.

Mewborn and Cross present an argument for how teachers' beliefs about the nature of mathematics can affect (1) how they view their roles as teachers and the role of the students, (2) their choice of classroom activities, and (3) the instructional approaches they use in the classroom. Mewborn and Cross also assert that teachers' beliefs about what counts as mathematical understanding and knowledge, about how students learn mathematics, and about the purposes of schools in general also affect their mathematics teaching practices. They conclude that teachers' beliefs have an intimate link to students' opportunities to learn and therefore to students' beliefs. In the remainder of their article they furnish information to assist teachers and teacher developers in addressing the issue of teachers' beliefs. First, Mewborn and Cross describe the manner in which beliefs are held and changed. Then they supply some examples of tasks they have used with in-service and preservice teachers to raise their awareness of their beliefs about the nature of mathematics and about mathematics learning.

DeBellis and Rosenstein present a program in which teachers are exposed to *new* mathematics topics and are provided with an environment that enables all of them to learn the topics. In this program teachers are amazed with what they are able to learn, and they are constantly encouraged to reflect on how they are learning the new material. It is hoped that through the teachers' own experiences with learning in such a risk-free and empowering environment that the teachers will do the same with their own students.

McAdam offers a glimpse into how prospective teachers in an elementary methods course make meaning of algorithms using base-ten blocks in their procedural and conceptual study of addition and subtraction algorithms. While manipulating these concrete materials, prospective teachers were required to move back and forth between concrete and symbolic representations to explore the meaning of number, place value, and mathematical operations using whole-number algorithms. Constructing algorithms with base-ten blocks provided the context and opportunity for conceptual growth and understanding for the teaching and learning of mathematics for students in kindergarten through grade 8.

Albert and McAdam present a vignette of prospective teachers as learners of mathematics in an elementary mathematics methods course as they learn to apply decimal fraction algorithms using base-ten blocks. A description of a classroom episode is included to give a sense of how the prospective teachers' knowledge of place value regarding whole numbers served as a "scaffold" for learning and thinking about decimals. Albert and McAdam also include a discussion of some of the essential components and underlying principles for teaching and learning decimal fractions, such as the use of precise mathematical language.

REFERENCE

National Council of Teachers of Mathematics (NCTM). *Principles and Standards for School Mathematics.* Reston, Va.: NCTM, 2000.

15

Mathematics Teachers' Beliefs about Mathematics and Links to Students' Learning

Denise S. Mewborn
Dionne I. Cross

THE beliefs that students and teachers hold about mathematics have been well documented in the research literature (e.g., Cooney 1985; Frank 1988, 1990; Garofalo 1989a, 1989b; Schoenfeld 1987; Thompson 1984, 1985). This research has shown that some beliefs are commonly held and quite salient across several populations. They include the following (Frank 1988):

- Mathematics is computation.
- Mathematics problems should be solved in less than five minutes, or else there is something wrong with either the problem or the student.
- The goal of doing a mathematics problem is to obtain *the* correct answer.
- In the teaching-learning process, the student is passive and the teacher is active.

It is generally agreed that these beliefs are not "healthy" in that they are not conducive to the type of mathematics teaching and learning envisioned in *Principles and Standards for School Mathematics* (National Council of Teachers of Mathematics {NCTM] 2000). In contrast, beliefs that align more closely with *Principles and Standards* might look like this:

- Mathematics is problem solving.
- Mathematics problems come in different types. Some can be solved quickly by recall. Other mathematics problems require a significant amount of time to understand the task, experiment with possible solution methods, reach an answer, and check to see that the answer makes sense.
- The goal of doing a mathematics problem is to make sense of the problem, the solution process, and the answer.
- In the teaching-learning process, the student and teacher are both active in making sense of the mathematics and of students' reasoning.

This disparity between the two sets of beliefs has been of concern in the mathematics education community, specifically as it relates to teachers, because of the influence beliefs tend to have on individual thought and action. Before we continue the discussion, however, it is necessary to talk about what we mean by beliefs and how we see them as intimately related to a person's decisions and behavior.

The definition of beliefs has been much debated, with attention being placed on differentiating the concept of beliefs from both attitudes and knowledge, and also on making it distinct from alternative labels, such as "conceptions." (See Pajares 1992 for an extensive discussion of this issue.) For purposes of this article, we are using a broad conception of beliefs—one coined by Tann (1993) as "personal theories." Personal theories encompass "a person's set of beliefs, values, understandings, assumptions—the ways of thinking about the teaching profession" (p. 55).

Beliefs are personal, stable, and often reside at a level beyond the individual's immediate control or knowledge. They are more influential than knowledge in determining how individuals frame problems and structure tasks and are strong predictors of behavior (Nespor 1987; Rimm-Kaufman and Sawyer 2004). Beliefs are stable across time and are unaffected by educational attainment or teaching experience (Torff and Warburton 2005). They are often held implicitly and can be difficult to articulate (Tann 1993). Since these beliefs are often tacit, individuals can hold many beliefs, some of which may be incoherent or contradictory (Tann 1993). Beliefs in general are very powerful; they exert enormous influence over an individual's actions and are highly resistant to change.

Teachers' beliefs about the nature of mathematics influence their beliefs about what it means to learn and do mathematics. These beliefs, in turn, influence their instructional practices, which dictate the opportunities that students have to learn mathematics. Ultimately, students' learning experiences affect their beliefs about the nature of mathematics and how an individual learns and engages in the subject. According to NCTM's *Curriculum and Evaluation Standards for School Mathematics,* "[Students'] beliefs exert a powerful influence on students' evaluation of their own ability, on their willingness to engage in mathematical tasks, and on their ultimate mathematical disposition" (NCTM 1989, p. 233).

Consider, for example, a classroom in which a teacher holds the first set of beliefs outlined above. This teacher's classroom practice is likely to be similar to that described by Romberg and Carpenter (1986), consisting of homework review (in which single-word answers prevail), the introduction of a new skill or procedure, time for students to practice the new material, and then assigned homework over the new material. Dialogue will be minimal and will consist of answers from students and verification of the correctness of the answer by the teacher, followed by another question. Students' learning in this classroom will be characterized by memorizing and mimicking rules and procedures, and students are likely to see mathematics as an odd collection of facts and rules rather than as a way to think critically about problem situations. They are also likely to see mathematics as something that is done by very smart and quick-thinking people, which may lead them to conclude that they are not good at and do not like mathematics.

In contrast, consider a classroom in which a teacher holds the second set of beliefs described above. This teacher's classroom practice might consist of students working together to generate a number of examples and then examining those examples for patterns that might lead to generalizations. There might be a lot of student-to-student dialogue, and the teacher's role would be to ask questions to encourage or clarify students' thinking, to introduce mathematical terminology and symbols as appropriate, and to help students reach closure. (See NCTM 1991 for a description of the teacher's role in discourse in this type of classroom.) Students in this classroom are likely to view mathematics as exploratory and sensible and view themselves as capable of creating new mathematics.

It should be clear from these examples that teachers' beliefs about the nature of mathematics can affect how they view their roles as teachers, the role of the students, their choice of classroom activities, and the instructional approaches they use in the classroom. Teachers' beliefs about what counts as mathematical understanding and knowledge, how students learn mathematics, and the purposes of schools in general also affect their mathematics teaching practices. Thus, teachers' beliefs have an intimate link to students' opportunities to learn and therefore to students' beliefs.

It is important to note, however, that the link between teachers' beliefs and instructional practices is not a direct one, since it is affected by a variety of contextual factors (Skott 2001). Other factors that influence the pedagogical decisions teachers make are the social context in which learning takes place, the constraints and affordances of this environment, the beliefs and expectations of others involved in the educational process (including students, parents, administrators, and policymakers), and the philosophical structure of the educational system (Thompson 1992).

This apparent relationship between teachers' beliefs and students' learning raises the issue of how teachers' beliefs can be shaped to lead to more productive learning experiences for students. Thus, in the remainder of this article we provide information for teachers and teacher developers to assist them in addressing the issue of teachers' beliefs. First, we describe the manner in which beliefs are held and changed. Then we provide some examples of tasks we have used with in-service and preservice teachers to raise their awareness of their beliefs about the nature of mathematics and about mathematics learning. Many of these same activities can be used with students to elicit conversations about mathematics learning and the discipline of mathematics itself.

Raising Awareness of Beliefs

Helping teachers (preservice and in-service) become aware of their beliefs is a significant step toward improving students' opportunities to learn mathematics. If teachers become aware of their beliefs, they can begin to question the evidence for those beliefs, the compatibility of different sets of beliefs, and the relationship between their beliefs and their instructional practices. Below we describe three activities that can help teachers become aware of their beliefs and consider the implications of these beliefs for classroom practice and students' learning. These activities can also be used with students to help them become more aware of their beliefs and begin the process of establishing more healthy perspectives about the nature and learning of mathematics.

The "draw a mathematician" task is based on a similar task that has been well documented in science education literature (cf. Chambers 1983). To begin, teachers are asked to visualize a mathematician at work. Where is the mathematician? What is the mathematician doing? What kinds of tools or materials is the mathematician using? (*Note:* It is important to avoid the use of gendered pronouns in giving the instructions for the task.) After a few minutes of think time, teachers draw the images that came to their minds. Most adults draw pictures of older, white, balding men with pocket protectors and glasses—not a very glamorous image. The mathematicians are usually alone with a chalkboard, a protractor, or a computer, working on stereotypical mathematics problems, such as 2 + 2 = 4. (See fig. 15.1 for a typical drawing.) Who would want to be good at mathematics if this is the image we associate with mathematicians?

Fig. 15.1. A typical drawing of a mathematician by an adult

The discussion that follows the drawing activity is crucially important. We begin by asking how many people drew

- a female mathematician?
- a person of color?
- a mathematician interacting with other people?
- a mathematician outdoors?
- a person you would want to associate with socially?

A discussion of the overwhelming similarities in the teachers' pictures leads to speculation about the experiences that have led them to hold such images. This is followed by a discussion of the subtle messages we are likely to convey to students if this is what we, as teachers, believe mathematicians are like. For example, if a teacher sees a mathematician (someone who is good at mathematics) as a nerdy, socially inept, middle-aged male working with equations in a lonely room, what does this convey about this teacher's view of mathematics? It likely suggests that she sees mathematics as difficult, boring, and done by very smart people. It is plausible to link these beliefs to beliefs about learning—that students are not capable of doing mathematics themselves (because only very bright people do mathematics), that mathematics must be spoon-fed to students (because it is difficult), and that doing mathematics is an individual endeavor. The classroom practices that result from these beliefs lead to students having limited opportunities to engage in the activities of mathematicians, which likely leads to students' beliefs that mirror teachers' beliefs. In this regard, students who are subject to these types of classroom practices may view mathematics as the province of the gifted and so may not be inclined to seek help when they need it nor persevere in solving problems when faced with the slightest obstacle. Additionally, if students view mathematics as an individual activity, they may resist teachers' attempts to engage them in group work and may be reluctant to share their ideas with peers or elicit ideas from others.

We also ask teachers to predict what elementary school children would draw when given the same task. Teachers usually predict, correctly, that young children will have less stereotypical images of mathematicians but that middle and high school students will have similar images to their own. (See fig. 15.2 for drawings done by fourth-grade students.) After some discussion about why children have more favorable images of mathematicians, we conjecture about what happens to children between elementary school and college that produces these dramatic differences. We also make explicit links to the beliefs about mathematics, teaching, and learning that are implied in these pictures. For example, what do we do in mathematics class that leads students to believe that mathematics is a solitary activity, not shared with others? (Interestingly, many people—children and adults—depict a mathematician as a teacher, complete with mortarboard and chalkboard full of equations. Sometimes there is dialogue from the teacher. However, rarely are there ever students depicted in the picture. Some drawings even include empty desks!) What kinds of teaching practices lead students to conclude that mathematics is done indoors with no apparent connection to the real world? What experiences lead students to conclude that mathematicians are mostly white, male, and older? What is behind the depiction of mathematicians as "geeks?" This activity is an excellent way to set the tone for a semester or professional development session because it helps define one of our goals: to think about how to teach mathematics so that children retain the belief that anyone (and everyone) can do mathematics.

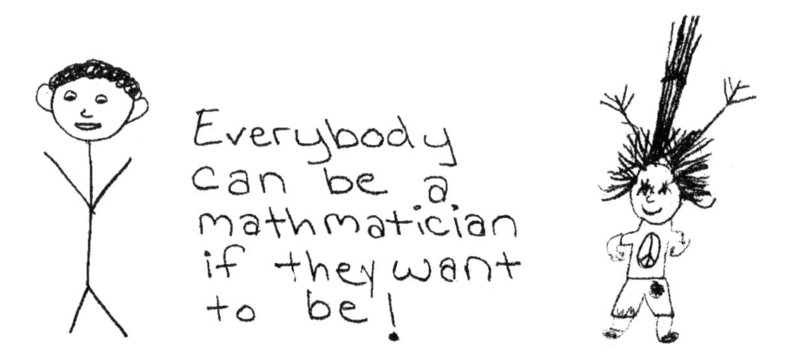

Fig.15.2. Two mathematicians drawn by fourth graders

A second type of task we have used with teachers to help them make their beliefs explicit is a series of open-ended questions. (See Spangler 1992 for the complete set of questions.) Questions include the following.

- If you were playing Password and wanted a friend to guess the word "mathematics," what four one-word clues would you give?
- If you and a friend got different answers to a problem, what would you do?
- Is it possible to get the right answer to a problem but not understand it? Conversely, is it possible to understand a problem but not be able to get a correct answer?
- Would you rather have one method for solving a problem that works all the time or multiple methods that work all the time?
- What other subjects is mathematics most like? Least like?
- How do you know when you have correctly solved a mathematics problem?

These questions can be implemented in several ways—as a journal-writing activity, as a class discussion, or as a small-group or pair "debate." They can be used one at a time or several at a time. We have used them with children, preservice teachers, and in-service teachers.

As with the "draw a mathematician" activity, the activity itself is not as important as the deliberate effort to articulate the beliefs that underlie the responses. For example, the question about what subjects mathematics is most or least like leads to interesting discussions about the level of creativity involved in doing mathematics (comparisons with fine arts), the origin of mathematical ideas (comparisons with history), and room for individual interpretations (comparisons with literature). The point is not to decide whether mathematics is or is not like history; rather, the point is to uncover the beliefs we hold about mathematics and what it means to learn and do mathematics that lead us to say it is or is not like history. Suppose a teacher believes that mathematics and history are alike because both are "old" and static, having been created by dead, white men so that students in future centuries can memorize what they created. What classroom practices are likely to follow from these beliefs? It would be reasonable to envision a classroom characterized by strict adherence to procedures with students practicing worksheet problems. Students in this environment may experience difficulty engaging in novel problem situations where they need to employ multiple strategies to obtain a solution. Because of a lack of exposure to more constructivist practices, students may develop an "unhealthy" conception of mathematics that makes them resistant to problem-solving activities.

In contrast, suppose a teacher sees history and mathematics as constantly changing, still developing fields to which people from a variety of races, cultures, and time periods have contributed and are still contributing. What classroom practices are likely to follow from these beliefs? The learning opportunities provided for the students in these two types of classrooms will likely be very different and may lead to a perpetuation of the teacher's beliefs.

A third activity, which is similar in many ways to the other two, involves offering a choice of similes about mathematics and mathematics learning and asking the teacher to explain why a particular simile was chosen (Cooney, Shealy, and Arvold 1998). Examples of similes for learning mathematics include working on an assembly line, cooking with a recipe, doing a jigsaw puzzle, building a house, watching a movie, picking fruit from a tree, conducting an experiment, or creating a clay sculpture. Once again, there are no right or wrong answers, and the point of the task is not to settle on which activity is most like learning mathematics. Rather, the purpose is to articulate the aspects of learning mathematics and doing a jigsaw puzzle that are alike and different and to posit the way learning opportunities might be structured on the basis of particular similes. For example, if a teacher sees learning mathematics as being like building a house, she might ensure that her students have plenty of opportunities to learn fundamental building blocks of mathematics before moving on to other content that is more conceptually oriented. These students, in turn, may view themselves as unable to solve complex problems before they have been told how to do the component parts. For example, a third grader might claim to be unable to determine how many times her heart has beat in her lifetime because she does not know how to do multidigit multiplication. A teacher who sees learning mathematics as being like watching a movie might be inclined to do mathematics for students at the front of the room on the board, expecting students to watch. If students are exposed to mathematics primarily as a spectator sport, getting them actively engaged in problem solving may be difficult, and they may be unwilling or unprepared to engage in collaborative activity. Rather, they may be inclined to default to the teacher-student or demonstrator-observer roles.

These activities are only some of the methods used to help bring teachers' and students' mathematical beliefs to the forefront, making them more explicit. Through this process teachers have the opportunity to acknowledge their beliefs, beliefs that were previously tacit, and reflect on how they were formed and shaped and the influence they may have had on the instructional decisions they make. It also provides them with the opportunity to confront their own beliefs, the first step in the process of change, and insight into how they can begin nurturing more healthy mathematical beliefs in their students.

Changing Beliefs

We would be remiss if we did not note that changing beliefs is extremely difficult and occurs over long periods of time. Pajares (1992) noted that people are adept at using evidence that would appear contradictory to a belief to support that same belief. This speaks to the power of beliefs. Several researchers (e.g., Gregoire 2003; Middleton 2002; Woolfolk-Hoy, Davis, and Pape in press) have posited that beliefs are changed in the same way that conceptual change is induced—through cognitive dissonance. Thus, they suggest that professional development activities aimed at changing beliefs should be geared toward creating dissonance by offering experiences where the teacher's new understandings from the teacher's perspective conflict with those experienced in their role as a student. Such experiences will help beliefs become more explicit and less tacit, thereby helping teachers confront their beliefs and understandings about mathematics, learning, and pedagogy. For example, let's consider a group of teachers in a professional development project who are given an opportunity to solve the heartbeat problem above. They are then led into a discussion of how they think their students would solve it. Teachers who believe that teaching mathematics is like building a house might be inclined to conclude that their students could not make any headway on this problem. These teachers might then discuss possible ways of presenting the problem to their students so they could attempt to solve it independently. They would then be encouraged to try it and bring back students' work to the next session. Undoubtedly, these teachers will be surprised to see that many of their students can ferret out the structure of the problem: the number of beats per minute times the number of minutes in an hour times the number of hours in a day times the number of days you have been alive. From this experience they would learn a great deal about their students' ability to decipher a problem, their knowledge of relationships among units of time, and their computational or calculator skills. This experience might furnish some cognitive dissonance for the teachers and provoke an awareness and examination of their beliefs about students' learning.

Similarly, students must experience cognitive dissonance in order to begin to change their own beliefs. Take, for example, a student who believes that he is not good at mathematics because he is often unsuccessful with mathematical computations. A possible approach to modifying such a belief would be to expose the student to other types of mathematical activities, such as problem-solving tasks, measurement tasks, or geometry tasks in which he can experience success. Being successful and having his peers see him as successful might cause him to begin to reexamine his beliefs about both the nature of mathematics and himself as a learner of mathematics.

The assumption here is that many of the beliefs teachers hold were developed on the basis of their experiences as students, and so they must be exposed to new positive experiences that force them to confront these previously held beliefs. Green (1971) noted that beliefs that are held on the basis of evidence (e.g., personal experience) are more open to change than those that are held nonevidentially. However, the evidence that leads to a conflict within the belief structure, and thus bringing about change, must be significant and convincing. A passive approach, such as delivering a lecture in a professional development setting, is unlikely to bring about cognitive conflict and a change in teachers' beliefs. As teachers are exposed to new ideas that conflict with their existing beliefs, they need opportunities to test what they are learning in actual classroom settings because such settings can provide new evidence to help solidify the changing beliefs. Pajares noted that "newly acquired beliefs are most vulnerable" (p. 317). Conversely, beliefs that have been held for a long time are most resistant to change. Thus, teachers need sustained and consistent support if they are to undertake a serious examination of their beliefs about mathematics and learning and the classroom practices that result from these beliefs.

Conclusion

As mathematics educators, we seek to establish and maintain successful mathematics learning environments through encouraging and supporting effective classroom practices and teaching strategies. In our efforts to achieve this goal, we must acknowledge the role of teachers and the influence of their beliefs, since they are the primary and often the only medium through which students are exposed to the world of mathematics. Because the type of beliefs these teachers hold will directly affect the conceptions of mathematics that these students develop and the type of learners they become, we cannot bypass the influence of beliefs. Therefore, if our goal is to improve students' learning of mathematics, we must begin the discussion with a focus on teachers since they will ultimately have the greatest impact on the development of future mathematicians, their understanding, and their subsequent achievement.

REFERENCES

Chambers, David W. "Stereotypic Images of the Scientist: The Draw-a-Scientist Test." *Science Education* 67 (April 1983): 255–65.

Cooney, Thomas J. "A Beginning Teacher's View of Problem Solving." *Journal for Research in Mathematics Education* 16 (November 1985): 324–36.

Cooney, Thomas J., Barry E. Shealy, and Bridget Arvold. "Conceptualizing Belief Structures of Preservice Secondary Mathematics Teachers." *Journal for Research in Mathematics Education* 29 (May 1998): 306–33.

Frank, Martha L. "Problem Solving and Mathematical Beliefs." *Arithmetic Teacher* 35 (January 1988): 32–34.

————. "What Myths about Mathematics Are Held and Conveyed by Teachers?" *Arithmetic Teacher* 37 (January 1990): 10–12.

Garofalo, Joe. "Beliefs and Their Influence on Mathematical Performance." *Mathematics Teacher* 82 (October 1989a): 502–5.

————. "Beliefs, Responses, and Mathematics Education: Observations from the Back of the Classroom." *School Science and Mathematics* 89 (October 1989b): 451–55.

Green, Thomas F. *The Activities of Teaching.* New York: McGraw-Hill Book Co., 1971.

Gregoire, Michelle. "Is It a Challenge or a Threat? A Dual Process Model of Teachers' Cognition and Appraisal Processes during Conceptual Change." *Educational Psychology Review* 15 (June 2003): 147–79.

Middleton, Valerie. "Increasing Preservice Teachers' Diversity Beliefs and Commitment." *Urban Review* 34 (December 2002): 343–61.

National Council of Teachers of Mathematics (NCTM). *Curriculum and Evaluation Standards for School Mathematics.* Reston, Va.: NCTM, 1989.

————. *Professional Standards for Teaching Mathematics.* Reston, Va.: NCTM, 1991.

————. *Principles and Standards for School Mathematics.* Reston, Va.: NCTM, 2000.

Nespor, Jan. "The Role of Beliefs in the Practice of Teaching." *Journal of Curriculum Studies* 19 (July–August 1987): 317–28.

Pajares, M. Frank. "Teachers' Beliefs and Educational Research: Cleaning Up a Messy Construct." *Review of Educational Research* 62 (fall 1992): 307–32.

Rimm-Kaufman, Sara, and Brooke Sawyer. "Primary-Grade Teachers' Self Efficacy Beliefs, Attitudes toward Teaching, and Discipline and Teaching Practice Priorities in Relation to the Responsive Classroom Approach." *Elementary School Journal* 104 (March 2004): 321–41.

Romberg, Thomas A., and Thomas P. Carpenter. "Research on Teaching and Learning Mathematics: Two Disciplines of Scientific Inquiry." In *Handbook of Research on Teaching,* 3rd ed., edited by Merlin C. Wittrock, pp. 850–73. New York: Macmillan Publishing Co., 1986.

Schoenfeld, Alan H. "What's All the Fuss about Metacognition?" In *Cognitive Science and Mathematics Education,* edited by Alan H. Schoenfeld, pp. 189–215. Hillsdale, N.J.: Lawrence Erlbaum Associates, 1987.

Skott, Jeppe. "The Emerging Practices of a Novice Teacher: The Roles of His School Mathematics Images." *Journal of Mathematics Teacher Education* 4 (March 2001): 3–28.

Spangler, Denise A. "Assessing Students' Beliefs about Mathematics." *Arithmetic Teacher* 40 (November 1992): 148–52.

Tann, Sarah. "Eliciting Student Teachers' Personal Theories." In *Conceptualizing Reflection in Teacher Development,* edited by James Calderhead and Peter Gates, pp. 53–69. Bristol, Pa.: Falmer Press, 1993.

Thompson, Alba G. "The Relationship of Teachers' Conceptions of Mathematics and Mathematics Teaching to Instructional Practice." *Educational Studies in Mathematics* 15 (May 1984): 105–27.

———. "Teachers' Conceptions of Mathematics and the Teaching of Problem Solving." In *Teaching and Learning Mathematical Problem Solving: Multiple Research Perspectives,* edited by Edward A. Silver, pp. 281–94. Hillsdale, N.J.: Lawrence Erlbaum Associates, 1985.

———. "Teachers' Beliefs and Conceptions: A Synthesis of the Research." In *Handbook of Research on Mathematics Teaching and Learning,* edited by Douglas A. Grouws, pp. 127–46. New York: Macmillan Publishing Co., 1992.

Torff, Bruce, and Edward Warburton. "Assessment of Teachers' Beliefs about Classroom Use of Critical Thinking Activities." *Educational and Psychological Measurement* 65 (February 2005): 155–79.

Woolfolk-Hoy, Anita, Heather Davis, and Stephen Pape. "Teacher Knowledge and Beliefs." In *Handbook of Educational Psychology,* 2nd ed., edited by Pat Alexander and Phillip Winne. Hillsdale, N.J.: Lawrence Erlbaum Associates, in press.

16

Creating an Equitable Learning Environment for Teachers of Grades K–8 Mathematics

Valerie A. DeBellis
Joseph G. Rosenstein

How do teachers learn mathematics? How do students learn mathematics? One answer to those questions is that anyone's learning of mathematics depends on the circumstances in which the learning is taking place. That is, the learning environment plays a crucial role in determining who learns and who does not learn; by its design, any learning environment will either embrace or exclude potential learners.

How do teachers learn about learning environments that encourage mathematical learning for all? Of course, they could be told about such environments. However, unless a teacher has actually been a *learner* in a learning environment that promotes learning for all, she or he may not appreciate the difference that the environment makes for the individual and may not understand how to create an environment that gives everyone a real chance to learn mathematics—what we call in this article an *equitable learning environment.*

Many professional development programs in mathematics focus on mathematical topics with which the participants are, or consider themselves to be, already familiar and on ways that teachers can help students learn that mathematics. Through such programs, the participants may indeed learn new pedagogical techniques and acquire many new activities to use in the classroom. However, such programs will generally not give participants insight about themselves as mathematical problem solvers or as learners of mathematics nor provide them information about how to establish an equitable mathematical learning environment in their own classrooms.

How can teachers become learners of mathematics, and thereby, through their own experience, learn about learning mathematics? How can professional development providers better enable teachers to become learners of mathematics, and thereby to reflect on the mathematical learning process?

Here is the answer that we propose for grades K–8 teachers and that we describe in this article: Expose teachers to *new* mathematical topics, and provide an environment that enables all of them to learn those topics. Amaze them with what they are able to learn, and reflect with them about how they are able to accomplish what they did. Empower them to do the same with their own students.

The fact that the mathematical topics are new to the teachers, as well as the fact that they were never expected to learn them before, enables them to shift from the role of teacher to that of student. This psychological shift is crucial for learning because it enables teachers to let go of their expectation that they already have to "know all the answers." Once this shift takes place, they can begin to experience both the difficulties of the learner when a task is challenging and the exhilaration when a challenging task is accomplished. Like their students, they approach new topics with a certain amount of apprehension; and when they learn in a supportive environment that allows their own apprehensions to dissipate, they learn that their students' apprehensions, although real, also need not be permanent.

What we described above is what we actually do. This article is based on the experience of the authors in developing and conducting extended professional development programs for grades K–8 teachers in discrete mathematics, topics with which grades K–8 teachers are typically unfamiliar. We structured our programs and created a learning environment on the assumption that the participants would become learners, and we found, as discussed subsequently, that *all* participants indeed learned in a few weeks an extraordinary amount of mathematics, more than they could have imagined. We helped them understand the problem-solving process, why they were able to learn difficult content material, and what aspects of the learning environment could be transferred to their own classrooms so they, too, could create a learning environment in which *all* students gain access to important mathematical ideas and experience similar success.

In this article, we discuss both the equitable learning environment that we developed for K–8 teachers and the professional development model in which it was embedded. This ongoing professional development program was the result of a collaborative effort involving a mathematics educator (the first author) and a mathematician (the second author), with the assistance of a number of exceptional K–8 teachers who served in leadership roles in the program.

Although the model of an equitable learning environment discussed in this article was developed in the context of a summer program for practicing teachers, it can also be used, with appropriate adaptations, in undergraduate courses for prospective teachers and in graduate courses for practicing teachers. Indeed, the first author has adapted this model for use in both types of courses. The strategy described in this article can be a powerful tool for bringing about the reforms in mathematics education envisioned by the National Council of Teachers of Mathematics (NCTM) *Standards*—not only through professional development for practicing teachers but also through preservice courses for prospective teachers.

Background

The Program and Its Goals

The teachers were all participants in the Leadership Program in Discrete Mathematics (LP-DM) that was developed and conducted by the authors at Rutgers University with funding from the National Science Foundation (NSF).[1] Although we originally developed a program for high school teachers of mathematically successful students, we soon observed that the participants had also introduced the LP-DM topics both to students in earlier grades and to students whose previous encounters with mathematics were not marked with success (see Kowalczyk 1997; Picker 1997; and Biehl 1997). As a result of the first observation, we created summer institutes that are appropriate for K–8 teachers, and have offered them forty-three times, to more than 1000 teachers, between 1995 and 2005.[2] As a result of the second observation, we realized that the value of discrete mathematics was not only the importance of the content material itself but also the opportunities it provided to achieve the kinds of recommendations that were made in *Curriculum and Evaluation Standards for School Mathematics* (NCTM 1989) and subsequently in *Principles and Standards for School Mathematics* (NCTM 2000).

The introductory article, "Discrete Mathematics in the Schools: An Opportunity to Revitalize School Mathematics" (Rosenstein 1997b), in *Discrete Mathematics in the Schools* (Rosenstein, Franzblau, and Roberts 1997) notes that "in two major ways, discrete mathematics offers an opportunity to revitalize school mathematics":

> Discrete mathematics offers a new start for students. For the student who has been unsuccessful with mathematics, it offers the possibility for success. For the talented student who has lost interest in mathematics, it offers the possibility of challenge.
>
> Discrete mathematics provides an opportunity to focus on how mathematics is taught, on giving teachers new ways of looking at mathematics and new ways of making it accessible to their students. From this perspective, teaching discrete mathematics in the schools is not an end in itself, but a tool for reforming mathematics education.

---

1. The Leadership Program in Discrete Mathematics was sponsored by DIMACS, the Center for Discrete Mathematics and Theoretical Computer Science, and the Rutgers Center for Mathematics, Science, and Computer Education. For additional information about the LP-DM, see Rosenstein and DeBellis (1997).

2. The first twenty-five institutes, from 1995 to 2002, were supported by the National Science Foundation; an additional eighteen institutes were supported with funding from state departments of education and higher education and the Educational Foundation of America and with coordination by LP-DM Associate Director Janice Kowalczyk. LP-DM institutes have been conducted in eleven states (Alabama, Arizona, Indiana, Massachusetts, North Carolina, New Jersey, Ohio, Pennsylvania, Rhode Island, South Dakota, and Virginia).

As a result of reports by participants in the initial years of the LP-DM,[3] we came to understand that discrete mathematics can serve as "a vehicle for giving teachers a new way to think about traditional mathematical topics and a new strategy for engaging their students in the study of mathematics" (DeBellis and Rosenstein 2004).

The Structure of the Program

The LP-DM institute involves fifteen to twenty full days of instruction, including a two-week summer program, a one-week program the following summer, and follow-up sessions during the intervening and following school years. The professional development model that we designed[4] includes four phases, described below, through which we cycle each day. In the first three phases, we expect participants to suspend their role of teacher and take on the role of mathematical learner, so that for a substantial portion of the day, the participants are involved in a variety of *learning* experiences.

Phase 1. Initiate the learning of the mathematics. Each morning, participants are involved in a two-hour content-based workshop on new mathematical topics taught by college faculty. The workshops involve a mixture of whole-group instruction and small-group activity. The pattern that is repeated throughout each workshop involves the introduction of new content material, participants' working on a problem in small groups, and discussion of the problem and the material. This interactive learning environment is supported by a workshop leader (a college mathematics faculty member) and lead teachers (expert K–8 classroom teachers who are LP-DM veterans), who work together to facilitate participants' mathematical learning.

Phase 2. Reinforce the learning of the mathematics. The workshop session is followed by a one-hour study-group session in which participants work in heterogeneous small groups on a set of "homework" problems based on the topic of the workshop, under the watchful mathematical eye of the workshop leader and with appropriate assistance from the lead teachers.

Phase 3. Consolidate the learning of the mathematics. Two important tools for consolidating participants' learning are daily journal writing (see the section titled "Ongoing Assessment and Prompt Intervention") and presentations. Participants jointly present their solutions to homework problems on the following morning, prior to the next workshop. They are encouraged to discuss not only their solutions but also how they thought through the prob-

3. "Participants reported changes in their classrooms, in their students, and in themselves. Their successes taught us that discrete mathematics was not just another piece of the curriculum. Many participants reported success with a variety of students at a variety of levels, demonstrated a new enthusiasm for teaching in new ways, and proselytized among their colleagues and administrators" (Rosenstein 1997b).

4. This model is described in detail in DeBellis and Rosenstein (2004).

lems and what aspects of the problems remain unclear (see the section titled "Reflecting on the Problem-Solving Process").

Phase 4. Implement the learning of mathematics in grades K–8 classrooms. Participants see teacher-tested, age-appropriate classroom demonstrations on the mathematical topics they just learned, as lead teachers model how they brought the LP-DM into their own classrooms. Lead teachers share resources they like, materials they have made, and authentic student work. Following those presentations, participants share their own ideas for introducing discrete mathematics topics to their students, and they develop lessons and units for use in their own classrooms.

The Content of the Program

The content of the LP-DM is discrete mathematics, a collection of topics that NCTM declares should be part of the curriculum at all grade levels (NCTM 2000, p. 31). Although it is difficult to summarize discrete mathematics[5] in a brief statement, *Principles and Standards for School Mathematics* provides a clear guideline by stating, "Three important areas of discrete mathematics are integrated within these standards: combinatorics, iteration and recursion, and vertex-edge graphs." Because many topics of discrete mathematics have few specific prerequisites and are unfamiliar to teachers, they provide an arena in which teachers from a variety of backgrounds can function as learners.

For example, in the first activity of the LP-DM, groups of participants sit on the floor around large maps of the continental United States, in which the interiors of all the states are white. They are asked to color the map by placing a colored chip on each state; bordering states must be colored using different colors, so that, if the states were actually colored, you could tell where one state ends and another begins. They are given chips of many different colors and asked to color the map. After they have colored the map, they are asked to determine how many colors they actually used, and are then asked to color the map using one fewer color. Then they are asked to reduce the number of colors again, and again, and again. Finally, when they have concluded that they cannot reduce the number of colors any further, they are asked to justify their conclusion. Note that the participants are given no suggestion at the outset that their ultimate goal is to color the map using as few colors as possible. Because the participants were allowed to color the map using many colors and then were asked to reduce the number of colors one at a time, the problem is divided into more manageable subproblems that the participants are able to solve.

The focus of the map-coloring activity is not on acquiring information about coloring maps but rather on having the teachers experience for them-

5. For descriptions of discrete mathematics, see Rosenstein (1997), Rosenstein (in press), and the soon-to-be-published volumes on *Navigating through Discrete Mathematics* by Valerie DeBellis, Eric Hart, Margaret Kenney, and Joseph Rosenstein in the NCTM Navigations Series.

selves the challenge of trying to determine the smallest number of colors and to justify their claim to have done so. It is not so much the content of discrete mathematics that makes it a vehicle for reforming mathematics instruction but the opportunity that new content offers to engage people in mathematical thought and activity.[6] The teachers are engaged in the map-coloring activity; they find it inviting, entertaining, and challenging; they are exercising their problem-solving and reasoning skills; they are communicating and developing precision in their mathematical language; and they realize that the same kind of activity, suitably modified, would engage their students, independent of grade level or ability level. Since most K–8 teachers are unfamiliar with discrete mathematics, questions like this are not posed in traditional mathematics classes, eliminating a rich source of problem-solving situations.

Yet it must be noted that although this activity is easy to describe and accessible, it enables us to introduce grades K–8 teachers to the frontiers of mathematics. They discover on their own that the United States map can be colored using four colors but cannot be colored using three colors. After the activity is over, a meaningful discussion of the four-color theorem ensues. The fact that every map can be colored using four-colors was proved in 1976 by Kenneth Appel and Wolfgang Haaken. More than a hundred years passed for the four-color conjecture to turn into the four color theorem, and the method of proof, using computers to verify many hundreds of cases, caused a great stir in the mathematical community and evoked an editorial in the *New York Times*. Although some K–8 teachers may have heard about the four-color theorem and that maps can be colored using four colors, they usually have no idea of why four colors are needed, what the theorem means, why it is interesting, and how it was developed. The introduction of map coloring not only serves as an interesting activity in itself but also provides teachers and students with both an important example of newly discovered mathematics and an entrée into how mathematics develops over time within the community of mathematicians.

This activity is one of several that are included in the first workshop, which uses (*a*) map coloring to introduce the subject of vertex-edge graphs and then discusses the coloring of graphs and their applications. The other four workshops during the first week focus on (*b*) drawing pictures with a single line and Euler paths and circuits, (*c*) finding the cheapest way of linking a number of sites, (*d*) finding the shortest distance between two locations on a map, and (*e*) finding the most efficient route for collecting deposits from all ATM machines in a geographical region (often referred to as the traveling salesman problem or the car-pooling problem).

---

6. If instead discrete mathematics is introduced in the schools as a set of facts to be memorized and strategies to be applied routinely (see Gardiner [1991, p. 12]), then the qualities of discrete mathematics as an arena for problem solving, reasoning, and experimentation are of course destroyed.

The problems in figures 16.1 and 16.2 arise in the workshop dealing with (*c*). Figure 16.1 presents a map of a town[7] in which are located six major sites (Bank, City Hall, Fire Station, Library, Police Station, and School) and twelve roads that link those sites. What is the minimum number of roads that must be paved to guarantee that a person can travel from any of the six sites to any other of the six sites along paved roads? Participants work in pairs to answer this question, and pairs soon conclude that five is the minimum number of roads needed. Although the answer of each pair is five, the sets of roads selected by the different pairs vary considerably, from a linear pattern such as C to P, P to F, F to S, S to L, and L to B to a star pattern such as F to C, F to S, F to L, F to B, and F to P; the problem of which five edges to select has many correct answers. However, the town officials are not satisfied with these answers and want to accomplish their goal more cheaply, using only four roads. Participants work in pairs to develop a justification that the task cannot be accomplished with four roads, that five roads is actually the minimum; providing such justification is a bit more difficult and may involve such problem-solving strategies as "look at a simpler problem," "build a table," and "find a pattern." The discussion leads participants to conclude, more generally, that no matter how many sites are involved, the number of roads needed is one less than the number of sites. This problem thus furnishes an example of mathematical induction that is accessible to upper elementary school students.

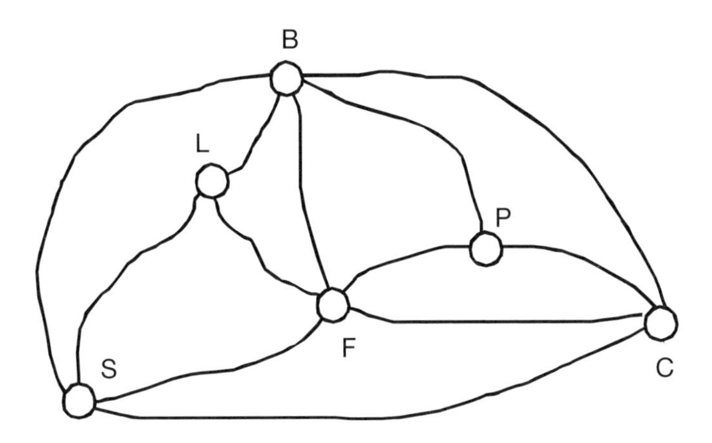

Fig. 16.1

7. This problem, the Muddy City Problem, is based on an activity by Mike Fellows, a computer scientist whose extensive work with children and their teachers is reflected in Casey and Fellows (1993). The town is called "Muddy City" because the roads are not paved, and when it rains, all the roads turn into mud.

In figure 16.2 the number on each of the twelve roads indicates the cost (in hundreds of thousands of dollars) involved in paving that road. Now the problem is, What roads should be paved to minimize the total cost, yet guarantee that one can travel from any of the six sites to any other of the six sites along paved roads? Again, participants work in pairs on this problem and are then asked to report on their solutions. Perhaps not surprisingly, for this problem only one set of roads exists whose total cost is minimal. However, the focus now is on the method for finding the cheapest set of roads, and the participants come up with several different methods. The proposed methods are then discussed in the workshop, and the participants learn that some of their newly created algorithms already have names attached to them, for example, Kruskal's algorithm or Prim's algorithm.

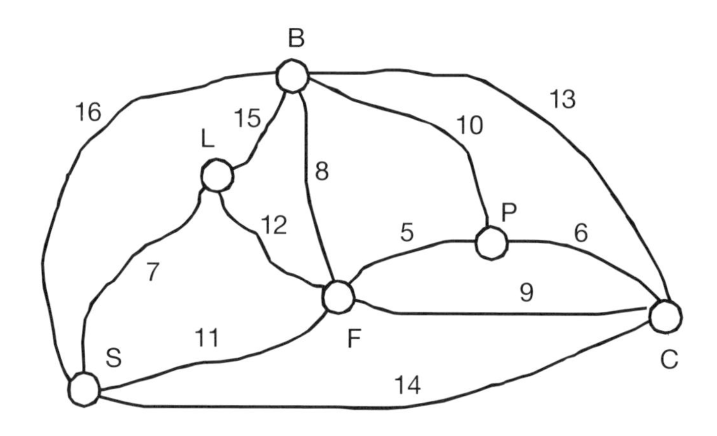

Fig. 16.2

The problem in figure 16.3 arises in the workshop involving (*d*). This vertex-edge graph represents a map that shows the different roads that can be used in traveling from home to school. The problem is, What is the shortest route from home to school, and how can we find it? This graph is usually laid out on the floor, using paper plates to represent the vertices and masking tape to represent the edges. Distances are not supplied in figure 16.3; indeed, participants are asked to place a number of linking cubes on each edge to represent the distance along that edge. (If the map is presented on a transparency, participants are asked to randomly call out numbers between 10 and 30, which are used to represent the distances on the fourteen edges of the vertex-edge graph in figure 16.3.) Participants are encouraged to offer strategies that they might use to find the shortest route, and a list of strategies is generated and discussed, including "as the crow flies," trial and error, start with the shortest edge, examine all possible routes, and tree diagrams. They find that, although such strategies work on small problems like that in figure 16.3, the number of possible routes grows rapidly as the graph becomes more complicated, and so a more sophisticated method is needed.

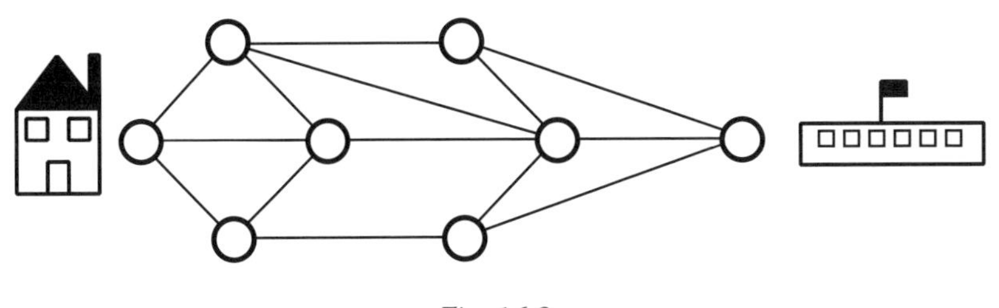

Fig. 16.3

A common feature of all the foregoing problems is that people—whether they are children, teachers, parents, or administrators—can readily understand them, relate them to their own experiences, want to solve them, and quickly become engaged in finding a solution. The problems rapidly bring novice learners into what for them is uncharted mathematical territory and introduce them to the excitement of working on questions that they never before considered. Another common feature of the problems is that initially they do not seem to be "mathematical," so they do not activate in solvers any long-held negative attitudes that they might have about mathematics and their own mathematical abilities.[8]

The question "How do you find the best solution?" introduces the important concept of an algorithm and, reflecting on the solution methods to problems (*c*), (*d*), and (*e*), raises the fundamental idea in computer science of comparing the efficiency of algorithms. Thus, for example, a relatively simple method exists for determining the best solution to the problem in figure 16.2 and all problems like it, one that the teachers can discover and master; what computer scientists call a "greedy algorithm" generates the solution. By contrast, the best solution to the "shortest path" problem in figure 16.3 is not as easy to find; although all participants can learn and use the algorithm themselves, typically only the middle school teachers will be able to introduce the algorithm to their students. And turning to the scenario in the workshop dealing with (*e*), no one in the world yet knows how to efficiently find the best solution to the traveling salesman problem. Indeed, anyone who determines whether an

8. In our experience, many K–8 teachers express negative feelings about mathematics stemming from their difficulties in learning the content as children and young adults, and they identify all of mathematics with the arithmetic that gave them trouble. We suspect that much of their negativism is due to the fact that the material was not presented to them in engaging ways and thus did not appeal to their learning styles.

efficient general solution to the traveling salesman problem exists may earn a million dollar prize offered by the Clay Mathematics Institute (CMI).[9]

This sample of activities from the first week of the LP-DM institute offfers strong evidence for the claim made previously that the participants in the program "learned in a few weeks an extraordinary amount of mathematics, more than they could have imagined."

An Equitable Learning Environment for "All" K–8 Teachers

The use of the word *all* signifies a commitment to equity. The issue of equity is usually discussed in outcomes (such as performance on assessments) or inputs (such as the availability of technology or the training of teachers). However, the focus of this article is on the equitable learning of mathematics, that is, on how we structure our programs to ensure that each participant experiences success in his or her learning of discrete mathematics. We refer to such an environment as an equitable learning environment. Our expectation is that once participants recognize their own success and reflect on the learning environment that facilitates their success, they will better understand what they need to do to help their students achieve similar success.

What are the components of this equitable learning environment?

Engaging activities. We consistently involve participants in the doing of mathematics by providing hands-on, minds-on activities that are both interesting and challenging. The activities are themselves engaging, and the participants are actively engaged in doing the activities. Our main goal is not to convey information—although that too is important—but to involve the participants in *doing* the mathematics. We want participants to understand that mathematics is something that one *does*, not just something that one *memorizes.*

What makes an activity engaging for K–8 teachers? That question can undoubtedly be answered in many ways, but among the components of an engaging activity are that it is inviting, understandable, relevant, interesting, entertaining, and challenging. It has to be engaging on many levels. We can imagine a person asking himself or herself the following questions after being introduced to an activity: Do I understand what I'm supposed to do in this activity? Can I connect it with my own experiences? Is it relevant to everyday life? Am I interested in finding out the answer? Will this endeavor be play as well as work? Can I actually carry out this activity? Is the activity too easy for me; have

9. This is one of seven "Millennium Problems" named by CMI in 2000 to celebrate mathematics in the new millennium. The P vs. NP problem, of which the traveling salesperson problem is a representative, is the only one of the seven Millennium Problems that is accessible to high school students. For a discussion of the P vs. NP problem and the other "Millennium Problems," visit www.claymath.org/millennium/.

I "been there, done that"? The person would go ahead and do the activity if all but the last two questions are answered positively, and would prefer not to do it otherwise.

But making an activity engaging is only part of the story. The real challenge is to develop a *sequence* of activities, each of which is engaging and which together build the teachers' mathematical strength and confidence. This requirement means that the activities are coherent, each building on what comes before, and that they are timely, each occurring when the teachers are prepared to learn its mathematical content. The problems have to be carefully designed and sequenced. This process happens over time, since the initial attempts to develop an activity may be unsuccessful. An engaging activity on a particular theme evolves through several cycles of design, testing, and revision; creating an engaging sequence of activities may also involve experimenting with different ways of ordering the activities. A successful sequence of engaging activities has been achieved when the sequence has been implemented by a variety of instructors, all of whom have been able to engage their participants.

High expectations. We set high but achievable expectations for our participants. The activities and problems that we provide are ones that the participants will indeed find challenging. But we do expect all participants to complete all the activities and solve all the problems. At the beginning of the institute, we discuss and reinforce the idea that a major responsibility of each study group during the morning workshops and study sessions is to make sure that each member of the group understands the material and is prepared to participate in the group's presentation the next day. The lead teachers give special attention to participants who are having difficulties with the material and through appropriate interventions assure that they too will be and feel successful. As a result of our experience with participants in previous cohorts, we know that these participants also will indeed solve the problems we pose. At the end of the institute we reflect on what the participants have learned and what they have achieved. Time and again, participants express a collective "Wow!," amazed at the level of mathematics that they have learned, for none of them could have imagined at the outset what they would attain during the program. We want participants to experience the impact on learning that results from setting high achievable expectations and providing adequate support so that all can achieve them. Achieving high expectations engenders a sense of mathematical competence and accomplishment that helps propel learners further in their study of more difficult mathematics.

Reflecting on the problem-solving process. Paying lip service to "high expectations" is not sufficient. Many teachers espouse high expectations but then give students answers to problems when the students experience frustration. We do not give answers. We expect participants to solve the problems that we pose, and when they find a particular problem frustrating, we offer clues and encouragement so that they will attain the satisfaction that comes from moving from a place of frustration to an "aha!" moment. We want participants to understand that providing answers to problems cheats their students of the satisfaction that comes from solving the problems themselves. To achieve this goal, we introduce participants early in the program to the role of affect in problem solving by discussing with them the emotional states that interact with cognition during mathematical problem solving (DeBellis and Goldin 1997). We reflect on their personal experiences and discuss the emotional dynamics of what it was like for them to try to solve challenging problems. We discuss how they might encourage their students to work through challenging problems, despite the fact that trying to solve problems is often frustrating, and achieve success. In their daily journal writing, participants are encouraged to reflect not only on their solutions to the problems but also on the stages they went through in the process. They are asked, for example, to discuss what confused them, what insight they gained (and how), and what they still do not understand; talking about what one does not know helps one let go of the role of "teacher" and enables one to take on the role of "learner." This experience encourages teachers to reflect on their own problem-solving behaviors and makes explicit that "doing mathematics" is often a series of decision-making steps on what to do next!

Ongoing assessment and prompt intervention. We monitor participants' progress daily and provide quick intervention if needed. We give participants opportunities to describe both their difficulties and their successes in learning the mathematics through daily journal entries that are read and commented on by the lead teachers, who communicate summaries of the journals to the instructor each morning. Such continuous, immediate feedback is important for consolidating learning; often several days are needed to clearly understand some mathematical topic, and the journaling process allows learners to independently discuss previous topics for as long as they need. Participants get individual assistance through the presence and intervention of the lead teachers, in addition to the instructors, and through one-on-one sessions with lead teachers if needed. Although each year several participants need such assistance and, as a result, contemplate leaving the LP-DM, not one teacher left for such reasons during the entire period of the NSF grant; ongoing assessment and prompt intervention were instrumental in maintaining that track record.

A level playing field. We select topics that place all K–8 participants on a mathematically level playing field. As in an ordinary classroom, the backgrounds of the participants vary widely. As in other social settings, some participants are eager to show off what they know, with the unfortunate consequence that feelings of mathematical inadequacy are evoked in the other participants. Discrete mathematics offers the possibility of an equitable learning experience, since few teachers have prior experience with the content. Those topics that some participants will have seen are placed later in the institute program (or, when the program is adapted to a university course, later in the syllabus). In particular, the first week of the institute focuses on topics involving vertex-edge graphs, with which relatively few teachers have had prior experience; not until the second week do we move to topics that may be more familiar, such as combinatorics.

A community of learners. While participating in the program, teachers come to see themselves as part of a community of learners. During much of the day, they work in groups, but the composition of the groups changes depending on the activity being done. As a result, over the course of the program, almost every pair of participants has worked together in a learning situation. For discussions of the mathematical topics and activities, the participants are organized in groups that are heterogeneous by grade level so that teachers at different grade levels can learn from others' perspectives; for implementation sessions, participants are organized in groups that are homogeneous by grade level so that they can collaboratively prepare lessons for their students. Participants are encouraged to propose conjectures and methods of solution and to comment on the solutions proposed by others. They are encouraged to function as a group, discussing possible solution paths respectfully and in a supportive manner even while discarding an unworkable idea. As noted earlier, we expect the members of each group to be responsible for ensuring that all group members understand the activity and can afterward explain to the entire group the solutions to all the problems.

Communication. One important feature of the study groups is that they help participants develop their skills in mathematical discourse in both spoken and written forms of mathematical communication. They come to realize that communication is an essential part of mathematics and mathematics education, and that mathematical language develops over time. They come to understand the important roles that reflection and communication have on the learning process and how precision of language evolves. They experience, often for the first time, that mathematical learning can take place in discourse with other participants (i.e., without the instructor). They learn precision in language as they are expected to take the informal solutions that they have developed in their groups and transform them into written and oral explanations that they will present the next day to their peers, at which time they will be expected to explain and defend their solutions. This experience is new for many teachers, but how can they be expected to discuss precise mathematical language with their students if they have never been called on to generate it? During these sessions, participants come to experience the NCTM Communication Standard (2000, p. 60) as they continue to develop their own understanding of the mathematical topics.

Addressing different learning styles. We consistently address different learning styles in the program—not only through the interactive nature of the instruction but through peer learning and personal reflections as well. We provide a variety of activities that target the kinesthetic learners, the auditory learners, and the visual learners, as well as those who absorb mathematics symbolically. We found that early-grades teachers tend to favor more visual or geometric ways of solving problems and that upper-grades teachers favor more algebraic approaches. By constructing learning groups heterogeneously (across grade levels), we expose each group of teachers to the other group's perspective. Over time, group members learn the value of bringing both representations (geometric and algebraic) to problem-solving situations, since they find that each representation brings different insight into the problem.

Continual self-assessment. We give participants opportunities to offer feedback about their learning environment, and appropriate adjustments are incorporated into the program. Evaluation forms are distributed, completed, reviewed, and discussed with the participants throughout the institute. The evaluation component models and enables real-time feedback and continual self-assessment that create a dynamic learning system in which individual learners and instructional staff members jointly take responsibility for the learning environment and how the program progresses. The mathematical focus of the program is also continually evaluated. For example, as noted previously, we choose problems with care to ensure that the activities are engaging, challenging, and solvable; since new activities are introduced regularly, these requirements mandate ongoing evaluation of the content of the program.

A nurturing environment. As a result of all the components discussed above, we offer a safe, nurturing, supportive, nonthreatening, and respectful environment that helps build mathematical self-esteem and rewards mathematical risk-taking. It is safe and nonthreatening in that not having all the answers is acceptable, and in that all participants feel free to ask questions and discuss their misunderstandings. The environment is nurturing and supportive in that everyone's progress is acknowledged and everyone's achievements are recognized. The environment is respectful in that each person is respected for his or her contributions to the program. All participants and staff members contribute to all these characteristics of the learning environment.

Impact of the Equitable Learning Environment on Participants

In this environment, the participants not only learn discrete mathematics but, through regular reflections on what we do in the LP-DM, also learn about how the learning environment fosters their own learning of new mathematical content.

They discuss engaging students in mathematical learning and discovery after they themselves have become engaged in doing mathematics.

They discuss problem solving in the context of their own problem-solving behaviors. After working on some challenging problems and overcoming their difficulties, they are ready to learn about problem solving. (Teachers who teach what is referred to as "problem solving" may have never had a real problem-solving experience; what often is called problem solving is using the pattern in a worked-out example to do similar problems. Problem solving occurs precisely when one cannot pattern the solution of a given problem on the solution of another problem seen moments before.)

Participants learn about high expectations for students when they see that they have fulfilled higher expectations than they would have imagined. Readers should note that although the participants in the program are self-selected, they come to the LP-DM with low mathematical self-esteem all too common in the elementary school teacher population. Their initial reaction is "What am I doing here?" and disbelief in the possibility of their success. But when they find that they can indeed solve such problems as the map-coloring problem, their perspective changes from "I can't do it" to "We did it!," as reported by Philadelphia middle school teacher Jane Field in a videotaped interview shortly after her group solved the map-coloring problem. Success in mathematics breeds more success in mathematics.

Participants learn about the value of working in groups from their experience in learning mathematics through conversations with their colleagues. Although many of the teachers had previously been using some sort of group activities in their own classrooms, by being a *learner* in a group, they experience firsthand how groups composed of participants of varying abilities can contribute to each individual's learning. They come to understand how mathematics evolves within a community of learners and are better able to facilitate such a community in their own classrooms.

In short, through participation in the LP-DM, participants learn a great deal about how they acquire mathematical knowledge, and how they can transfer that learning to a better understanding of how students learn. An essential aspect of the process is that during the LP-DM, they themselves are learners of mathematics and are able to reflect on the learning environment that facilitated their own learning. A second essential aspect of the process is that they learned about an equitable learning environment by participating in one that was modeled for them by the workshop leaders and lead teachers.

The impact described here of the program on its participants is reflected in evaluation reports prepared by Lesley University's Program and Evaluation Research Group (PERG) on the five years of the LP-DM that were funded by NSF.[10]

---

10. The following are three excerpts from the PERG report: (1) The Leadership Program in Discrete Mathematics has been successful in effecting change in teachers' attitudes. Teachers reported that they were better prepared to teach students, that they appreciated the opportunity to interact with colleagues, that they were treated with respect as learners, and that they gained confidence in their mathematics knowledge. (2) Teachers further reported that these attitudinal

Conclusions

Mathematical training for prospective K–8 teachers has often been minimal; at best, they may have taken two undergraduate mathematics courses that focus on mathematical topics to be taught in elementary schools. (This situation has been changing in the past few years, since an important national report recommended at least three courses for elementary school teachers and at least seven courses for middle school teachers [CBMS 2001, p. 8], and since many states have been developing certification requirements for middle school teachers of mathematics.) In addition, prospective K–8 teachers often view mathematics exclusively as a body of knowledge, as a set of facts and procedures; they perceive that their job, when they get to be teachers, will be to transmit those facts and procedures to their students. This perspective should not be surprising, since they have likely had the same experience in learning mathematics. Many attribute their lack of success in mathematics to their inability to remember all the required facts, formulas, and techniques. In addition, many of these undergraduate students report experiencing math anxiety, and they try to avoid the mathematics rather than embrace it. Yet these students will become the teachers of mathematics for our children.

We believe that college classrooms and professional development programs for K–8 teachers should change these perspectives of mathematics and mathematics learning by establishing engaging learning environments in which all teachers can and do learn mathematics.

We assume that all K–8 teachers, both prospective and practicing, are capable of learning complex mathematical ideas. We assume that our responsibility (whether as faculty members teaching undergraduate students or as professional development providers training established teachers) is to hold teachers to high, achievable mathematical expectations in a supportive, equitable learning environment, so that they experience success in learning complex mathematical ideas and so that they can build the confidence they need to transfer those expectations to their own classrooms.

Enabling K–8 teachers to become successful learners of challenging mathematics is an important step in their professional development. It helps them view mathematics more through problem solving and reasoning, not exclusively through memorizing facts, formulas, and techniques. It empowers them to initiate mathematical explorations, to engage students in classroom dialogue

changes affected their classroom behavior. They reported incorporating DM topics, increasing the use of such instructional strategies as cooperative learning, and embedding mathematics across curriculum topics. (3) Teachers' reactions to the content and instructional strategies presented in the LP were overwhelmingly positive. Their exposure to DM positively affected their attitudes about mathematics, which led to changes in their classrooms. According to teacher reports, those changes in classroom activities in turn led to positive reactions on the part of their students.

and discovery, and to help students work on challenging problems. When K–8 teachers feel empowered mathematically, they can expect their students to feel the same. If they can meet high expectations, their students can do the same. They can now say, with conviction, "If I could do it, then they can, too!"

REFERENCES

Biehl, L. Charles. "Discrete Mathematics: A Fresh Start for Secondary Students." In *Discrete Mathematics in the Schools,* vol. 36 of DIMACS Series in Discrete Mathematics and Theoretical Computer Science, edited by Joseph G. Rosenstein, Deborah Franzblau, and Fred Roberts, pp. 317–22. Providence, R.I.: American Mathematical Society and National Council of Teachers of Mathematics, 1997.

Casey, Nancy, and Michael Fellows. *This Is MegaMathematics! Stories and Activities for Mathematical Thinking, Problem-Solving, and Communication: The Los Alamos Workbook.* Los Alamos, N.Mex.: Los Alamos National Laboratory, 1993.

Conference Board of the Mathematical Sciences (CBMS). *The Mathematical Education of Teachers—Part I.* Washington, D.C.: Mathematical Association of America in cooperation with American Mathematical Society, 2001.

DeBellis, Valerie A., and Gerald A. Goldin. "The Affective Domain in Mathematical Problem-Solving." In *Proceedings of the 21st Conference of the International Group for the Psychology of Mathematics Education,* vol. 2, edited by Erkki Pehkonen, pp. 209–16. Helsinki, Finland: University of Helsinki, Lahti Research and Training Centre, 1997.

DeBellis, Valerie A., and Joseph G. Rosenstein. "Discrete Mathematics in Primary and Secondary Schools in the United States." *International Reviews on Mathematical Education* (Zentralblatt für Didaktik der Mathematik) 36 (2). Electronic-Only Publication. ISSN 1615-679X, 2004.

Gardiner, Anthony D. "A Cautionary Note." In *Discrete Mathematics across the Curriculum, K–12: 1991 Yearbook,* edited by Margaret J. Kenney, pp. 10–17. Reston, Va.: National Council of Teachers of Mathematics, 1991.

Kowalczyk, Janice C. "Fibonacci Reflections—It's Elementary!" In *Discrete Mathematics in the Schools,* vol. 36 of DIMACS Series in Discrete Mathematics and Theoretical Computer Science, edited by Joseph G. Rosenstein, Deborah Franzblau, and Fred Roberts, pp. 25–34. Providence, R.I.: American Mathematical Society and National Council of Teachers of Mathematics, 1997.

National Council of Teachers of Mathematics (NCTM). *Curriculum and Evaluation Standards for School Mathematics.* Reston, Va.: NCTM, 1989.

———. *Principles and Standards for School Mathematics.* Reston, Va.: NCTM, 2000.

Picker, Susan H. "Using Discrete Mathematics to Give Remedial Students a Second Chance." In *Discrete Mathematics in the Schools,* vol. 36 of DIMACS Series in Discrete Mathematics and Theoretical Computer Science, edited by Joseph G. Rosenstein, Deborah Franzblau, and Fred Roberts, pp. 35–41. Providence, R.I.: American Mathematical Society and National Council of Teachers of Mathematics, 1997.

Rosenstein, Joseph G. "A Comprehensive View of Discrete Mathematics: Chapter 14 of the New Jersey Mathematics Curriculum Framework." In *Discrete Mathematics in the Schools,* vol. 36 of DIMACS Series in Discrete Mathematics and Theoretical Computer Science, edited by Joseph G. Rosenstein, Deborah Franzblau, and Fred Roberts, pp. 133–84. Providence, R.I.: American Mathematical Society and National Council of Teachers of Mathematics, 1997a.

———. "Discrete Mathematics in the Schools: An Opportunity to Revitalize School Mathematics." In *Discrete Mathematics in the Schools,* vol. 36 of DIMACS Series in Discrete Mathematics and Theoretical Computer Science, edited by Joseph G. Rosenstein, Deborah Franzblau, and Fred Roberts, pp. xxiii–xxx. Providence, R.I.: American Mathematical Society and National Council of Teachers of Mathematics, 1997b.

———. "Discrete Mathematics in 21st Century Education; An Opportunity to Retreat from the Rush to Calculus." In *Foundations for the Future in Mathematics Education,* edited by Richard Lesh, Eric Hamilton, and James Kaput. Hillsdale, N.J.: Lawrence Erlbaum Associates, in press.

Rosenstein, Joseph G., and Valerie A. DeBellis. "The Leadership Program in Discrete Mathematics." In *Discrete Mathematics in the Schools,* vol. 36 of DIMACS Series in Discrete Mathematics and Theoretical Computer Science, edited by Joseph G. Rosenstein, Deborah Franzblau, and Fred Roberts, pp. 415–31. Providence, R.I.: American Mathematical Society and National Council of Teachers of Mathematics, 1997.

Rosenstein, Joseph G., Deborah Franzblau, and Fred Roberts, eds. *Discrete Mathematics in the Schools.* Vol. 36, DIMACS Series in Discrete Mathematics and Theoretical Computer Science. Providence, R.I.: American Mathematical Society and National Council of Teachers of Mathematics, 1997.

17

Prospective Teachers' Use of Concrete Representations to Construct an Understanding of Addition and Subtraction Algorithms

John F. McAdam

As we begin the twenty-first century, current calls for reform, whether viewed through *Principles and Standards for School Mathematics* (National Council of Teachers of Mathematics [NCTM] 2000) or through *Adding It Up: Helping Children Learn Mathematics* (Kilpatrick, Swafford, and Findell 2001) are reminders of the importance of understanding mathematics generally and algorithms more specifically. I assert that for prospective teachers to understand mathematics, they need to be actively engaged in constructing their own knowledge of basic mathematics, including practical knowledge of standard algorithms. Yet, most prospective teachers are products of classrooms that have emphasized a traditional approach of "teaching by telling," individual work, memorization, and quick recall of mathematical facts and procedures (Ball 1988, 1990). Many prospective teachers may lack the mathematics experiences and basic conceptual understandings that are promoted by the current reform, constructivist learning theory, and "experiences-as-learners" (National Science Foundation 2002, p. 37). The National Science Foundation's 2002 monograph, *Professional Development That Supports School Mathematics Reform,* echoes the 1989 *Everybody Counts* report from the National Research Council (NRC) that "few teachers have had the experience of constructing for themselves any of the mathematics that they are asked to teach" (NRC 1989, p. 66).

The work presented in this article is part of a larger, more extensive study of the experiences of prospective teachers in an elementary mathematics methodology course as they worked collaboratively with concrete materials to construct deeper meanings of algorithms beyond a procedural understanding (McAdam 2000). For purposes of this article, procedural knowledge, also referred to as instrumental understanding (Skemp 1987), refers to mathematics with an emphasis on the rules, memorization, and symbolic representation. Conceptual knowledge or relational understanding (Skemp 1987) emphasizes relationships or connections among representations including concrete-visual models.

Making Connections

Historically, the idea of developing deeper understanding as well as connections between conceptual and procedural knowledge is not a new phenomenon (Brownell 1935; Dienes 1960, 1961, 1963; Freudenthal 1973; Polya 1985). Freudenthal (1973) implies that knowing mathematics and developing mathematically is not a struggle between conceptual knowledge and procedural knowledge (p. 44):

> It is not fair to confront algorithmic and conceptual mathematics with one another as though one is the lofty tower from which you may look down on the other, and we certainly cannot identify this opposition with that between new and old.

It is the complementary relationship between conceptual and procedural knowledge that allows learners to make meaningful connections and to move toward more sophisticated mathematics. Freudenthal stresses that mathematics should be taught as a lived experience in which the process is understood. If prospective teachers are to understand and apply mathematics, they need to experience as learners instructional activities that engage them in the process of "mathematizing"—that is, mathematics situated in a problem-solving context (Freudenthal 1973; Becker and Selter 1996). This focus on mathematization and conceptual understanding of mathematics includes understanding algorithms beyond mere procedures.

Instructional practices in reference to the teaching and learning of algorithms need to emphasize the relationship between conceptual and procedural knowledge. However, as reported in *Adding It Up* (Kilpatrick, Swafford, and Findell 2001), teachers frequently view mathematics as memorized facts and procedures. Many of the prospective teachers participating in this study may have little practice providing justification, reasoning, and proof with their school experiences of learning mathematics. Pedagogical practices need to include multiple methods for developing representations and proofs of algorithms that are explicit in mathematics content, leading to conceptual understanding (Ball, Lubienski, and Mewborn 2001; Kilpatrick, Swafford, and Findell 2001; Shulman 1987). Consistent with this assumption is Ma's (1999) argument that teachers need to have a "profound understanding of fundamental mathematics." This deeper understanding of algorithms facilitates procedural fluency, especially as algorithms are used to support problem solving at higher levels. Teachers at all levels, prospective and practicing, need to understand the underlying reasoning and justification of algorithms if they are to assist children's learning and use of algorithms strategically (Kilpatrick, Swafford, and Findell 2001).

Payne's (1975) summary of a number of research studies related to rational numbers noted that effective use of concrete objects to develop algorithms was an aid to improved achievement and retention for learners. Almost two decades later, Moser (1992) also summarized research findings and noted that the teaching of algorithms with the use of manipulatives, including base-ten blocks, led to better understanding of the mathematics. "Prospective teachers should learn mathematics in a manner that encourages active engagement with mathematical ideas" (NRC 1989, p. 66). When prospective teachers understand and learn algorithms, they establish a stronger fundamental base to continue their study of higher or more advanced mathematics (Freudenthal 1973; Kilpatrick, Swafford, and Findell 2001). However, mathematics education has been caught in "A Cycle of Failure" (Cipra 1992, p. 3), one that stresses mathematics as memorization and quick recall rather than understanding.

Prospective Teachers' Prior Knowledge and Understanding

Research studies on the effective use of concrete materials to develop algorithms suggest that the teaching and learning of mathematics emphasize an understanding of algorithms that moves beyond procedural knowledge (Freudenthal 1973; Moser 1992; Kilpatrick, Swafford, and Findell 2001; Payne 1975). These researchers argue that the use of physical materials as tools in learning mathematics must connect with mental manipulation, resulting in relational understandings. This argument not only applies to children's learning but can apply to teachers' learning as well. The literature reveals that prospective teachers must experience as students the teaching and learning practices that they will use as teachers, since "teachers teach as they were taught" (NRC 1989, p. 6). Ball (1988) suggests that although prospective teachers wish to teach mathematics for understanding, their formative school experiences as students of mathematics may not have provided them with either the conceptual understanding or the instructional model necessary to provide more effective mathematics instruction.

Participants in the study reported in this article related a sense of learned helplessness in their mathematics experiences, reflecting on traditional instructional methods that isolated them as students and narrowly focused their attention on worksheets. "Working straight from ditto sheets, classmates were not allowed to talk things out or use manipulatives to aid in the visualization process of solving the given problem," said Linda. Some of the prospective teachers expressed regret about their school mathematics experiences as lacking in meaning and context and as leading to a sense of personal inadequacy and failure. For example, Torey wrote, "I learned to add, subtract, etc. numbers using algorithms and it wasn't until much later that I understood what 'borrowing' or 'carrying' was. It was something I knew how to do but it never occurred to me to ask why or what was going on." Procedural knowledge dominated their early school experience with mathematics. Now, as part of their preservice development and experiences, prospective teachers were required to work collaboratively using manipulatives (base-ten blocks) to make connections between their conceptual and procedural knowledge of mathematics.

For purposes of this article, the following vignettes and discussion focus on prospective teachers' use of base-ten blocks to understand addition and subtraction algorithms. The twenty-seven prospective teachers that participated in this study experienced a range of cognitive challenges as they worked to connect their prior procedural knowledge of algorithms with conceptual understanding of concrete-visual representations while using base-ten blocks. While manipulating these concrete materials, prospective teachers were required to move back and forth between concrete and symbolic representations to explore the meaning of number, place value, and mathematical operations through whole-number algorithms. Constructing algorithms with base-ten blocks provided the context and opportunity for conceptual growth and development.

During class sessions, participants manipulated the base-ten blocks, exploring, modeling, drawing, and making written notes of their process. For example, they effortlessly represented numerals such as "23" with two longs and three small cubes. Also, they readily made connections with place-value concepts using base-ten blocks and 10-for-1 trading rules. The connection between concrete, physical representations (2 longs and 3 units) and symbolic representation (23 or 2 tens and 3 ones) was readily demonstrated. These prospective teachers easily and successfully made the connection from concrete, physical models to symbolic representations using base-ten blocks. However, some of the prospective teachers' exploration of addition and subtraction algorithms with whole numbers while using base-ten blocks revealed insights into their experiences as they worked to make meaningful connections.

The following vignette is an example of a class problem-solving activity that required the participants to work in small groups or in pairs to represent an addition algorithm, using base-ten blocks as well as showing the pencil-and-paper algorithm.

> Jenny wants to buy a 39-cent slice of cake and a 59-cent cup of coffee. How much will they cost altogether?

These participants accurately represented the pencil-and-paper algorithm with base-ten blocks. In many groups, one partner chose 3 longs and 9 small cubes to represent the first addend, whereas the other partner moved 5 longs and 9 small cubes into position to model the second addend, as depicted in figure 17.1.

After representing each addend with base-ten blocks, prospective teachers quietly put together the units (small cubes), traded 10 units for 1 long, and moved that 1 long into the grouping of the 3 longs plus 5 longs for a total of 9 longs. Their manipulation of the base-ten blocks to arrive at the answer (sum) paralleled the pencil-and-paper notation of the addition algorithm. Each of the addends was represented with the base-ten blocks, the blocks were combined,

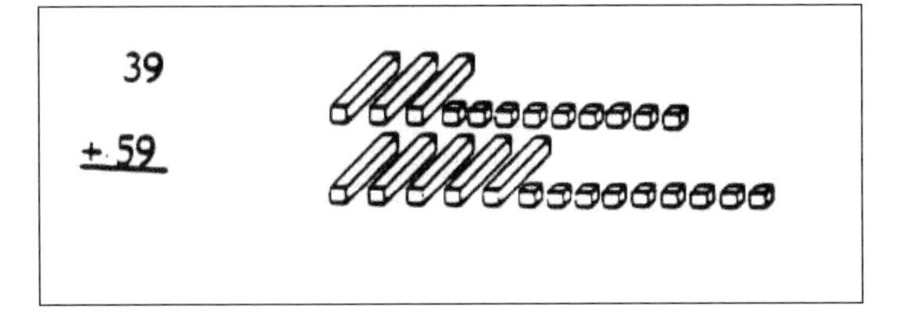

Fig. 17.1. Pencil-and-paper addition algorithm modeled with base-ten-blocks (longs and units)

a 10-for-1 trade was made, and the sum was represented with 9 longs and 8 units. Then they represented their solution with the base-ten blocks as portrayed in figure 17.2.

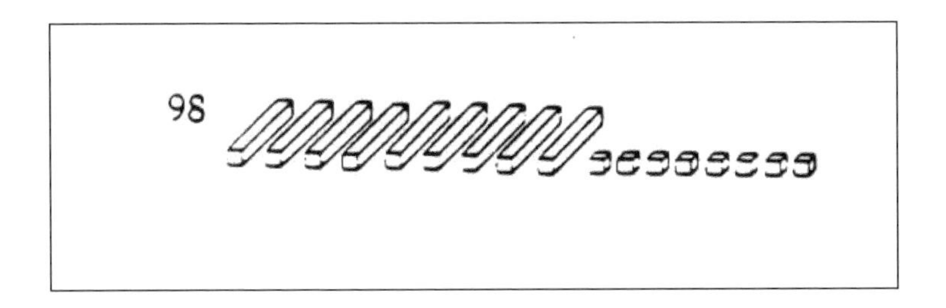

Fig. 17.2. The solution (the sum) to the addition algorithm modeled with 9 longs and 8 units

The transformation from pencil-and-paper symbolic representation to a concrete conceptual model was easily achieved. Prospective teachers' prior procedural knowledge "scaffolded" their construction of a concrete-visual model. However, they experienced confusion when attempting to represent a basic subtraction algorithm with base-ten blocks. For example, prospective teachers were asked to execute the following subtraction problem while using base-ten blocks:

> There were 82 boxes of apples at the fruit stand. At the end of the day, 53 boxes remained. How many boxes were sold during the day?

As with the addition algorithm, setting up the pencil-and-paper algorithm for this subtraction problem was routine. Nearly all the groups of prospective teachers used base- ten blocks to represent the subtraction word problem, as illustrated in figure 17.3. Unlike the addition algorithm, constructing and executing the subtraction algorithm created some confusion for several prospective teachers. Representing the minuend (82) as well as the subtrahend

(53) resulted in a conflict between procedural and conceptual understanding of the problem. Whereas joining base-ten blocks together to represent addition seemed an intuitive process, some of them viewed subtraction with base-ten blocks as counterintuitive. The take-away and the comparison interpretations of subtraction confused them. At that time, these were interpretations of subtraction that they had not studied. For example, Danielle and Christine posed the question, "Should we represent both numbers [meaning the minuend (82) and the subtrahend (53)]?" Torey stated, "Just take away the other number" [meaning simply remove the subtrahend (53) from the base-ten blocks representing the whole, or minuend (82)]. Despite satisfactory manipulation of base-ten blocks to represent the subtraction algorithm, this group was unable to dispel their sense of confusion or disequilibrium.

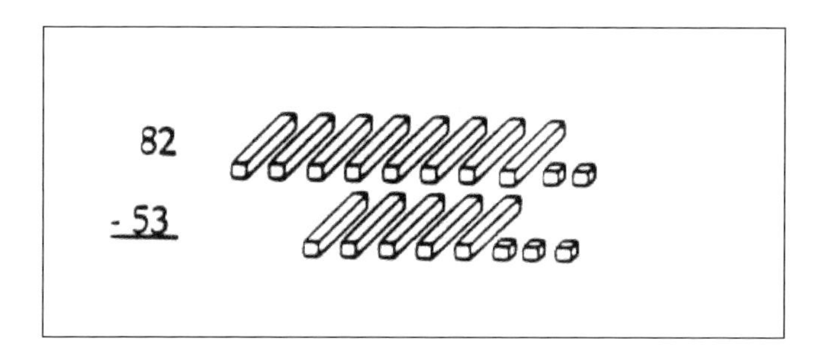

Fig. 17.3. The subtraction algorithm modeled with longs and units

A majority of these prospective teachers characterized subtraction as "working backward." As Sheila wrote, "In subtraction you must work backwards. First [you] must decide if [you] need to regroup and then trade a long in for ten units." Although these same prospective teachers would have no difficulty completing the algorithm 82 – 53 procedurally with pencil and paper, they were stymied on how to mirror the subtraction algorithm using base-ten blocks when the initial step required them to trade 1 long for 10 ones. Although the prospective teachers were confident about regrouping from the tens' column when doing the pencil-and-paper algorithm, they were less likely to know how to start the process conceptually with base-ten blocks.

In a subsequent activity, prospective teachers were very successful in constructing concrete and visual representations of addition and subtraction algorithms with base-ten blocks. Despite this, several of them continued to represent both the minuend and the subtrahend when modeling the subtraction algorithm. Of these models, only Emily's group drew a visual representation of the subtraction algorithm with base-ten blocks, presenting the subtraction algorithm as a comparison interpretation (see fig. 17.4). In the illustration of figure 17.4, Emily sketched three panels. In the first panel, we see the traditional symbolic pencil-and-paper subtraction problem 685 – 204. Also in the first panel is a visual representation of the problem. The numeral 685 is represented with 6 flats, 8 longs, and 5 units. The numeral 204 is represented with the 2 flats and 4 units. Since there are two distinct sets visible in the first panel, this visual might suggest an addition problem. Arrows drawn from the visual model to the pencil-and-paper algorithm clearly indicate, however, that the illustration represents a subtraction problem.

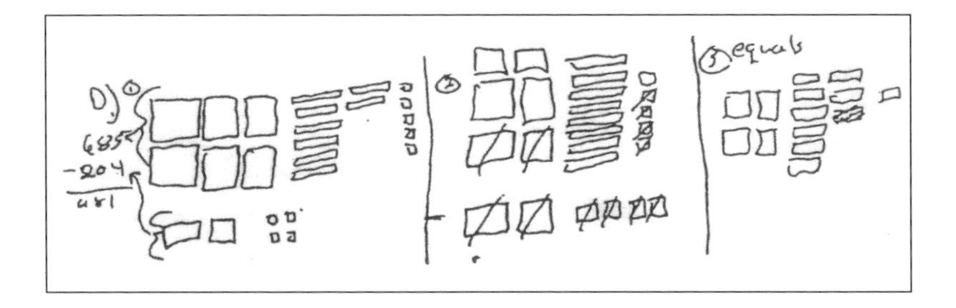

Fig. 17.4. Comparison interpretation of subtract

The illustration in the second panel of figure 17.4 represents a comparison interpretation of subtraction. A comparison relationship involves two distinct sets and the difference between them. Here a comparison is made between two sets of numbers as represented by the two distinct sets of flats, longs, and small cubes. The learner compares the 4 units to the 5 units, leaving a difference of 1 unit and also compares 2 flats to the 6 flats, leaving a difference of 4 flats. Since there are 0 longs and 8 longs, the difference remains as is, or 8 longs. The 4 flats, 8 longs, and 1 unit without a diagonal line through them illustrate the difference, or how many more there are after comparing the two sets. The answer is 481, and the illustration of a comparison interpretation of subtraction as depicted in the second panel is complete.

The illustration in the third panel of figure 17.4 shows 4 flats, 8 longs, and 1 unit. Also written in the third panel is the word *equals*, indicating that the 4 flats, 8 longs, and 1 unit represent, or equal, the answer. The illustration in the third panel is not necessary for a comparison interpretation and is more appropriate for an illustration in a take-away interpretation of subtraction. In a take-away interpretation of subtraction, the 2 flats and 4 units would be taken away from the 6 flats, 8 longs, and 5 units, leaving what appears in the third panel. Although the illustrations in panels 1 and 2 indicate a comparison interpretation, the illustration in the third panel suggests a take-away interpretation of subtraction.

The illustration by Cara's group models the problem 685 – 204, which includes both the minuend (685) and the subtrahend (204) as depicted in figure

17.5. Yet, the interpretation of this visual representation was of a take-away interpretation of subtraction. In figure 17.5, we see 685 (the whole) and "take away" 204 (a part), which equals 481 (the difference). Since this group refers to the take-away interpretation of subtraction, the middle piece in figure 17.5 is not necessary. The 204 could have been taken away from the whole, as more accurately demonstrated in figure 17.6.

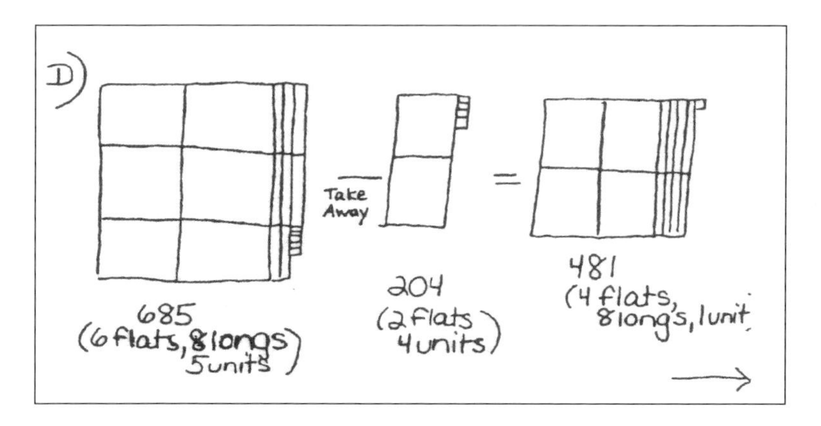

Fig. 17.5. A combination of the comparison and take-away interpretations

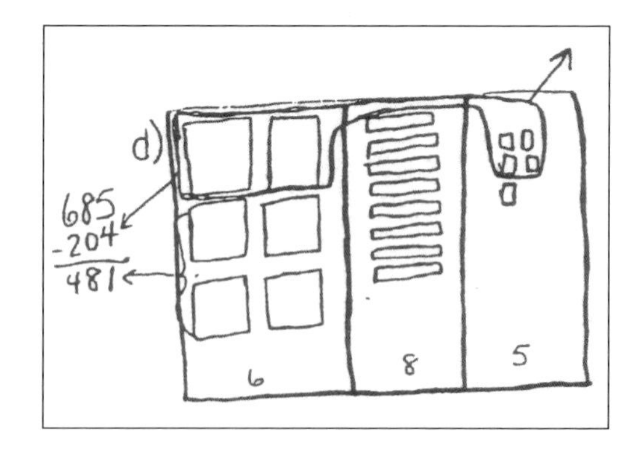

Fig. 17.6. Take-away interpretation of subtraction

Figure 17.6 is representative of the work of Colleen's group, who more accurately illustrated 685 minus 204 as a take-away interpretation of subtraction. The 6 flats, 8 longs, and 5 units (the minuend) are drawn as if the base-ten blocks they represent were on a place-value till. The blocks representing the 204 (the subtrahend) are circled in black and an arrow indicates that they are being taken away. This example not only accurately illustrates the pictorial model but also makes the appropriate connections to the traditional pencil-and-paper symbols that have been included in the drawing. The difference of 481 can be seen in both pictorial and symbolic forms. Those who used this illustration accurately represented all the aspects of a take-away interpretation of subtraction.

When participants were asked to write statements to explain what type of difficulty elementary school children might experience when performing addition and subtraction algorithms with base-ten blocks, they always selected subtraction as the more difficult operation. Emily wrote, "Subtraction uses two numbers but there may only be one concrete object [meaning the minuend] in front of them. This causes them to visualize the number they are subtracting." This explanation helps to explain the confusion between an interpretation of subtraction as a comparison model and a take-away model. When she states, "Subtraction uses two numbers," she is suggesting a comparison relationship from which to start the operation. The statement "but there may only be one concrete object in front of them," however, suggests a take-away interpretation. This is a mixing of two different interpretations of subtraction. Although the subtraction problem may be basic, the experience of relearning an operation in an unfamiliar format (i.e., base-ten blocks) was challenging for many of the participants. Furthermore, some of the prospective teachers experienced difficulty in making a connection between the pencil-and-paper subtraction algorithm and the concrete model. Therefore, it may be inferred that their prior procedural knowledge of subtraction influenced or inhibited the conceptualization process. Working with base-ten blocks allowed prospective teachers to analyze procedures used to solve addition and subtraction problems and to connect concrete actions to symbolic representations while also seeing relationships among the numbers. Torey captured the sentiments of her colleagues in the following statement:

> We found that subtraction was more difficult to represent with Base 10 blocks, especially when the problem involved regrouping. Subtraction was more difficult because it required you to "take [a]way" the second number of blocks [the subtrahend].... Thus, the second smaller number [the subtrahend] is hidden within the larger number [the minuend]. Unlike addition, where both numbers [the addends] are present in front of you in blocks, in subtraction only one of the numbers [the minuend] is actually present in front of you in blocks, and you must abstract the other number [the subtrahend] from the larger number of blocks.... Subtraction requires the students to think more abstractly in order to realize that the smaller number that is being taken away is within the larger number represented by the blocks.

Most of the prospective teachers concurred that addition with base-ten blocks—even addition when regrouping was necessary—was more concrete, more visual, and easier to conceptualize than subtraction with base-ten blocks.

They had an easier time connecting their prior procedural knowledge with conceptual models involving addition while using these manipulatives. They also indicated that in addition "the need for trading is clear," since regrouping occurs after the action of combining base-ten blocks. This process is consistent and compatible with a traditional pencil-and-paper symbolic, rule-bound approach to addition. There is little or no perceived conflict between the traditional symbolic approach of adding and the use of manipulatives such as base-ten blocks. For many of them there seems to be an intuitive connection between the procedural and the conceptual approaches for addition.

Unlike the intuitive flow that the different groups seemed to experience when using base-ten blocks to solve addition problems, subtraction problems thrust them into a state of disequilibrium. Participants wrote that subtraction with base-ten blocks was less visual, more abstract, more mental, more complex, and required you to work backwards. They viewed the "smaller number"—the subtrahend—as "hidden" within the minuend when using base-ten blocks. For example, Linda and Danielle emphasized the visual nature of addition where "both the numbers"—the addends—were concretely represented with base-ten blocks (i.e., "It is much easier to have everything in front of your eyes in a concrete model"). In subtraction, Cara as well as some of her peers noted that one must "imagine" or "visualize what is being taken away" and that students need to "work more in their heads as opposed to [using] the concrete base 10 blocks." Overall, participants in this study had more difficulty making connections between base-ten blocks and the pencil-and-paper subtraction algorithm.

A majority of the prospective teachers experienced a block or confusion when trying to develop a conceptual model with base-ten blocks for subtraction. Their prior procedural knowledge did not complement or match the conceptual model. When they looked at the concrete-visual model they had constructed with the base-ten blocks, they were confused because they had included a concrete representation of the subtrahend (53), as illustrated in figure 17.3. In effect, their representations with base-ten blocks looked exactly like the addition model they had constructed earlier. This time, however, the base-ten blocks representing 53 were a source of confusion. Their way of thinking about the subtraction problem as a take-away model did not match the concrete, visual model that was represented in front of them. The concrete model that they created was a representation of a comparison interpretation of subtraction. Although the subtraction problem was basic, the confusion about the conceptual model in front of them was an indicator of the lack of experience that many of these prospective teachers had with developing conceptual understanding with concrete materials. The lack of compatibility between prior procedural knowledge and the conceptual knowledge required to model the subtraction problem successfully was a source of conflict for most of the prospective teachers.

It is important for prospective teachers to simulate the concrete learning experiences that children will have in their classrooms. Having experiences as learners is an essential element of teachers' learning because many children during their early acquisition stages of learning mathematics may not have the influence of prior procedural knowledge to apply to solving algorithms. Shared learning experiences of constructing meaning of algorithms with base-ten blocks provide prospective teachers with similar, if not exactly the same, types of conflicts and challenges that younger students will encounter. Through shared experiences, prospective teachers will have firsthand insights into the struggles that younger students will encounter. Consequently, prospective teachers will be better prepared to "scaffold" their students with meaningful "hints" (Polya 1985).

Conclusion

The purpose of this article is to share some of the experiences learned from a larger study of a group of twenty-seven prospective elementary school teachers as they completed a mathematics methodology course. The findings of this study are consistent with the literature, which suggests that many prospective teachers do not have well-developed conceptual understanding of basic school mathematics, including standard algorithms. Even as we enter the twenty-first century, problems with mathematics education continue to be of concern. The keys to addressing issues related to mathematics education are preservice teacher education and professional development for practicing teachers. Teachers need to broaden their knowledge of mathematics and develop a deeper understanding of the mathematics they will teach. This includes making sense of basic algorithms. Making sense of algorithms means providing teachers with hands-on activities that assist them in the process of making connections between conceptual and procedural knowledge.

Teachers' knowledge and active experiences with mathematics are keys to improving mathematics education, in particular the teaching of mathematics to, and the learning of mathematics by, children. As simple as it may seem, teachers need to experience the mathematics in a manner similar to how they are expected to teach it. If teachers are to make connections between conceptual and procedural knowledge, they must construct meaning for themselves of mathematics concepts, including basic algorithms.

> It is important that [learners] be fully engaged in any computation they perform. Students often utilize standard written algorithms without thinking about the calculation at hand or the numbers involved. In fact, standard written methods were designed so that one did not have to think. (Sparrow and Swan 2001, p. 8)

However, the power of mathematics is in the "doing" of mathematics. Doing mathematics means actively engaging both prospective and practicing teachers in the NCTM Process Standards of problem solving, reasoning and proof, and representations. Or, as defined by *Adding It Up* (Kilpatrick, Swafford, and Findell 2001), teachers need mathematical proficiency, which includes the five strands of conceptual understanding, procedural fluency, strategic competence, adaptive reasoning, and productive disposition.

Although the use of base-ten blocks could be considered helpful in assisting the learner's performance, they did not guarantee that teachers would be able to make connections between conceptual understandings and procedural knowledge of addition and subtraction algorithms. Conceptual development and the ability to make connections to symbolic representations were observed to be an ongoing process, relating to experiences over time. The process of conceptual development needs to continue throughout preservice teaching and beginning teaching experiences. Also, student-teaching placements need to reflect what schools of education are doing to prepare preservice teachers. Further, teachers need to continue their mathematics development as they enter their first few years of teaching through professional development programs, which bridge conceptual and procedural knowledge. This continued support and development in the field strengthen the linkage between colleges or universities and local-area elementary schools (Johnston 1997). Professional development programs not only assist practicing teachers in the area of school mathematics but also continue to support recent teaching graduates.

Specifically, in the instance of practicing teacher development, teachers need to learn how to work with concrete materials so that they can model for their students these same skills and approaches. Professional development for teachers ought to give them the opportunity to experience working with concrete models, reflecting on that work, and relating the work to children's learning so that they can recognize the potential of such models for learning and understanding algorithms, as well as the potential of failure of the models (Albert, Mayotte, and Phelan 2004). The failure of using concrete models to learn and understand algorithms is not proof of its ineffectiveness as an instructional technique, but rather that it takes time, structure, and practice to make the use of these models effective. Teachers need to participate in hands-on and minds-on learning activities. Active engagement both physically and mentally with the models is a requirement if learners at any level are to develop relational understanding or connections between conceptual and procedural knowledge. Teachers need experiences doing mathematics so they will come to see mathematics as making sense. If we are to change the cycle of failure associated with the teaching and learning of mathematics, prospective and practicing teachers need to view themselves not as passive memorizers of facts and procedures but as active doers of mathematics, confident and self-assured.

REFERENCES

Albert, Lillie R., Gail Mayotte, and Cynthia Phelan. "The Talk of Scaffolding: Communication That Brings Adult Learners to Deeper Levels of Mathematical Understanding." In *Proceedings of the Twenty-sixth Annual Meeting of the American Chapter of the International Group for the Psychology of Mathematics Education*, edited by Douglas E. McDougal and John A. Ross, pp. 1137–38. Toronto, Ont.: OISE/UT, 2004.

Ball, Deborah Loewenberg. "Unlearning to Teach Mathematics." *For the Learning of Mathematics: An International Journal of Mathematics Education* 8 (February 1988): 40–48.

————. "The Mathematical Understandings That Prospective Teachers Bring to Teacher Education." *Elementary School Journal* 90 (March 1990): 449–66.

Ball, Deborah Loewenberg, Sarah Theule Lubienski, and Denise Spangler Mewborn. "Research on Teaching Mathematics: The Unsolved Problem of Teachers' Mathematical Knowledge." In *Handbook of Research on Teaching*, edited by Virginia Richardson, pp. 433–56. Washington, D.C.: American Educational Research Association, 2001.

Becker, Jerry P., and Christoph Selter. "Elementary School Practices." In *International Handbook of Mathematics Education: Part 1*, edited by Alan J. Bishop, Ken Clements, Christine Keitel, Jeremy Kilpatrick, and Colette Laborde, pp. 511–64. Dordrecht, Netherlands: Kluwer Academic Publishers, 1996.

Brownell, William A. "Psychological Considerations in the Learning and the Teaching of Arithmetic." In *The Teaching of Arithmetic*, Tenth Yearbook of the National Council of Teachers of Mathematics, edited by W. D. Reeve, pp. 1–31. New York: Bureau of Publications, Teachers College, Columbia University, 1935.

Cipra, Barry. *On the Mathematical Preparation of Elementary School Teachers*. Chicago: University of Chicago, 1992.

Dienes, Zoltan P. *Building Up Mathematics*. London: Hutchinson, 1960.

————. "On Abstraction and Generalization." *Harvard Educational Review* 31 (summer 1961): 281–301.

————. *An Experimental Study of Mathematics-Learning*. London: Hutchinson & Co., 1963.

Freudenthal, Hans. *Mathematics as an Educational Task*. Dordrecht, Netherlands: D. Reidel Publishing Co., 1973.

Johnston, Marilyn. "One Telling of Our History." In *Contradictions in Collaboration: New Thinking on School/University Partnership*, edited by Marilyn Johnston, pp. 21–47. New York: Teachers College Press, 1997.

Kilpatrick, Jeremy, Jane Swafford, and Bradford Findell, eds. *Adding It Up: Helping Children Learn Mathematics.* Washington, D.C.: National Academy Press, 2001.

Ma, Liping. *Knowing and Teaching Elementary Mathematics: Teachers' Understanding of Fundamental Mathematics in China and in the United States.* Hillsdale, N.J.: Lawrence Erlbaum Associates, 1999.

McAdam, John F. "Examining Prospective Teachers' Conceptual Understandings of Algorithms through Activity-Centered Collaboration with Base-Ten Blocks." (Doctoral diss., Boston College, 2000.) *UMI Dissertation Services* (2000): 9961612.

Moser, James M. "Arithmetic Operations on Whole Numbers: Addition and Subtraction." In *Teaching Mathematics in Grades K–8: Research-Based Methods,* edited by Thomas R. Post, pp. 111–45. Boston: Allyn & Bacon, 1992.

National Council of Teachers of Mathematics (NCTM). *Principles and Standards for School Mathematics.* Reston, Va.: NCTM, 2000.

National Research Council. *Everybody Counts: A Report to the Nation on the Future of Mathematics Education.* Washington, D.C.: National Research Council, 1989.

National Science Foundation. *Professional Development That Supports School Mathematics Reform.* Washington, D.C.: National Science Foundation, 2002.

Payne, Joseph. "Review of Research on Fractions." In *Number and Measurement: Papers from a Research Workshop,* edited by Richard A. Lesh, pp. 145–88. Columbus, Ohio: ERIC, 1975.

Polya, George. *How to Solve It: A New Aspect of Mathematical Method.* 2nd ed. Princeton, N. J.: Princeton University Press, 1985.

Shulman, Lee S. "Knowledge and Teaching: Foundations of the New Reform." *Harvard Educational Review* 57 (spring 1987): 1–20.

Skemp, Richard R. *The Psychology of Learning Mathematics.* Expanded American ed. Hillsdale, N.J.: Lawrence Erlbaum Associates, 1987.

Sparrow, Len, and Paul Swan. *Learning Math with Calculators: Activities for Grades 3–8.* Sausalito, Calif.: Math Solutions Publications, 2001.

18

Making Sense of Decimal Fraction Algorithms Using Base-Ten Blocks

Lillie R. Albert
John F. McAdam

MUCH has been written about the need to involve learners as active participants in the learning process. An emphasis on an active approach to learning is particularly needed in elementary school mathematics. Our assertion is that many of our prospective teachers continue to be products of a traditional approach to mathematics that emphasizes procedural knowledge, memorization, quick recall, and symbolic representations as opposed to the open-ended, higher-level pedagogy espoused by the National Council of Teachers of Mathematics (NCTM 1989, 2000). Therefore, it should be of little surprise when prospective teachers encounter tension when they participate in active experiential learning that focuses on developing a deeper understanding of mathematics. The argument is that teacher educators need to understand more fully the challenges and development of prospective teachers as active learners, especially when working with concrete representations.

This article focuses on the experiences of prospective teachers as learners in an elementary mathematics methods course as they learn to solve decimal fraction algorithms using base-ten blocks. In the first section, we present a brief description of how the array model, a rectangular model of the product, is employed to challenge their understanding of algorithms of the multiplication of decimal fractions. This section includes a classroom example to illustrate how the prospective teachers' knowledge of place value regarding whole numbers served as a scaffold for their learning and thinking about decimals. Next, we present a vignette of prospective teachers learning to solve decimal fraction algorithms. In this section, we present information to exemplify some of the essential components and underlying principles for teaching and learning decimal fractions, such as the use of precise mathematical language.

Algorithms for Multiplying Decimal Fractions

Traditionally, practice with paper-and-pencil algorithms has represented the core of elementary school mathematics (Ginsburg 1989; Bennett and Nelson 2004). Base-ten blocks can assist students in the visualization of algorithms while they are developing a deeper understanding of the procedures that mirror the symbolic representation of these algorithms. From our work and research with prospective teachers, we have found that working with base-ten blocks and decimal fractions created a state of disequilibrium. We believe that prospective teachers bring a history of learning about mathematics based on years of formal and informal instruction. Many of their earlier school experiences have been traditional with a focus on algorithms as calculations, resulting in a procedural understanding of mathematics. We believe that prospective teachers need to experience algorithms in ways that help them develop a conceptual understanding of the content by connecting concrete representations to symbolic ones. In this manner, the prospective teachers' understanding of the underlying principles inherent in the algorithms may be enhanced.

We employed the use of an array, sometimes referred to as an area model, because it is a powerful tool for representing multiplication across several dimensions of mathematics, including fractions, decimal fractions, and algebra (Hatfield, Edwards, and Bitter 2000; Kilpatrick, Swafford, and Findell 2001). Another reason for using an array is because of real-world connections, which include cans shelved at a grocery store, boxes of film in a camera shop, or rows of musicians in a marching band. An important aspect of using an array model to represent multiplication algorithms is that learners can move beyond believing and defining multiplication as *just* repeated addition. Furthermore, when learners explore solving decimal fraction algorithms in which the product is a rectangular representation, they may come to realize that they are examining a part-whole interpretation as well. Because base-ten blocks assist in the development of spatial understanding, they are excellent tools that allow the learners to construct geometric models of multiplication algorithms. They also are consistent with the NCTM (1989) *Curriculum and Evaluation Standards*, which notes that "[a]rea models are especially helpful in visualizing numerical ideas from a geometric point of view" (p. 88).

The Number and Operations Standard states that the "focus should be on developing students' conceptual understanding of fractions and decimals—what they are, how they are represented, and how they are related to whole numbers" (NCTM 2000, p. 152). The prospective teachers' knowledge of place value regarding whole numbers served as a scaffold for their learning and thinking about decimals. For example, this happened in two different ways: applying place-value concepts and using concrete materials to model monetary values. We reviewed place value for whole numbers and worked toward understanding that the decimal system is an expansion of the place-value system we use for naming whole numbers. For instance, considering the numeral 5555.555, we asked:

Instructor: Using place-value ideas of whole numbers, what do we know about the fourth 5 to the left of the decimal point?

Anita: It stands for 5 thousands.

Instructor: The third 5 to the left of the decimal point represents what value?

Lynn: It is hundreds, and the second 5 would be tens.

Instructor: Wonderful, I like your thinking. What does the first 5 to the left of the decimal point represent and what inference can we make?

Nathan: Well, I would say that the first 5 represents ones and then conclude that each time we move one place to the right, the value of the digit becomes smaller.

Instructor: Great. Now, we know that by extending the whole number system, you obtain a set of numerals for representing rational numbers. What are the values of the first 5 and second 5 to the right of the decimal point?

Arthur: Tenths and hundredths.

Instructor: When you say tenths and hundredths, what do these names make you think of regarding their values?

Arthur: Fractions, as 1/10 or 1/100.

Instructor: Are you saying we can call decimal numbers decimal fractions? If so, why?

Erin: Yes. It has something to do with tens.

Arthur: Yes, ah, like Erin said, they have something to do with tens. With whole numbers, as the position of the number expands to the left, the number becomes larger. I guess with fractions and decimals, as the number moves toward the right, it becomes smaller.

Working through this exercise helped scaffold (or gradually increase) the prospective teachers' understanding that to be consistent with our base-ten system for naming whole numbers, the "positions of the digits to the left of the decimal point represent place values that are increasing powers of 10 [$1, 10, 10^2, 10^3, \ldots$]" (Bennett and Nelson 2004, p. 332). Thus, they make the connection that the first 5 to the right of the decimal point represents one-tenth of the 5, concluding that the "positions to the right of the decimal point represent place values that are decreasing powers of 10 [$10^{-1}, 10^{-2}, 10^{-3}, \ldots$], or reciprocals of powers of 10 [$1/10, 1/10^2, 1/10^3, \ldots$]" (p. 332).

In earlier lessons when transitioning to developing decimal concepts and operations, it was important to assist the prospective teachers with how the new unit could be interpreted using base-ten blocks. Therefore, an initial exercise required them to demonstrate money values (e.g., $1.63), using a flat to represent a unit (1), six longs to represent the tenths (6/10), and three small cubes to represent the hundredths (3/100). Learning about our monetary system supports place-value concepts because it corresponds to our base-ten system. Although the designation of the flat as the unit was the benchmark for the activities that follow, prospective teachers were involved in other activities in which different blocks represented the unit, for example, a large cube as the unit or a long as the unit.

After reviewing some of the basic ideas regarding decimal fractions that help prospective teachers understand that algorithms for multiplying decimals are parallel extensions of the whole number algorithms, we explored activities in which they use the base-ten blocks to multiply a whole number by a decimal fraction. Then the prospective teachers performed multiplication algorithms involving tenths and gradually progressed to smaller decimal fractions where flats represented units, longs or rods represented tenths, and small cubes represented hundredths. The idea was to help them understand that "multiplication and division of two numbers will produce the same digits, regardless of the positions of the decimal point. As a result, for most practical purposes, there is no reason to develop new rules for decimal multiplication and division" (Van de Walle and Lovin 2006, p. 107).

An Emerging Model for Decimal Fraction Algorithms

Considering the previous perspective, one of our activities required the prospective teachers to use base-ten blocks to solve for 3.2 × 1.5 and describe an instructional strategy for the multiplication of decimals. The following vignette focuses on one small group, consisting of Anita, Lynn, Nathan, and Rochelle (McAdam 2000). They were observed and videotaped as they built an array model to solve 3.2 × 1.5. Nathan did most of the construction of the model with base-ten blocks, and Rochelle assisted by anticipating and supplying him with the appropriate blocks. Throughout the process, Anita and Lynn observed the construction of the emerging model in which they drew illustrations as well as took notes on the group discussion.

This vignette illustrates that learning to solve decimal fraction algorithms is not an all-or-nothing experience; rather, the learning experience occurs across a spectrum of understanding in which one's reliance on operations with whole numbers converges with one's experience with decimal fractions. In some instances, the use of imprecise language such as "point language" (reading 3.2 as 3 point 2) or "whole number language" (such as counting base-ten longs that represent tenths as 1, 2, 3, 4, 5) can inhibit or even mask the learning of decimal fraction concepts and operations. As we will show, although Anita says, "Oh, okay" in conversation during group collaboration, her written work, which we review, suggests otherwise. Also, Lynn's question, "How are we going to show how we represent decimals?" indicates a lack of full understanding of the connection between the base-ten blocks model and decimal fractions. Although each of the four prospective teachers accurately illustrates the decimal fraction problem, Anita and Lynn experienced difficulty with the symbolic representation. Now, let's look at the vignette as it unfolds.

Nathan appeared reasonably confident manipulating the base-ten blocks while solving the multiplication problem involving decimals. He answered questions that other members asked during the collaborative session. Nathan set up the configuration of four flats as shown in figure 18.1, Step 1. Next, he pulled the bottom flat away from the other three flats, leaving three flats as in Step 2. As he manipulated the blocks, his reasoning can be heard as he ponders aloud, "It's 3 point 2 [3.2] across." Rochelle gave him two longs, which he then placed beside the three flats as illustrated in Step 3. However, Lynn stated the need to start with "1 point 5 [1.5] down," and then, "3 point 2 [3.2] across."

During this somewhat staccato discussion, Rochelle demonstrated active listening as she supplied Nathan with the appropriate base-ten blocks. Rochelle exhibited understanding in regard to which blocks were needed to construct the model. There was nonverbal communication between them—for example, when Rochelle handed Nathan a supply of longs, his reaction was, "We don't need them [the longs]." Then Nathan could be heard to utter, "Okay. We have to have 1 point 5 [1.5] down," and proceeded to take the longs from Rochelle. He continued to build the array as shown in Step 4. After placing 5 longs below each of the three flats, Nathan and Rochelle—simultaneously—reached for the small cubes to complete the rectangular array in Step 5. The discussion was minimal, yet the collaboration between the partners was synchronized. Peer assistance, even though nonverbal, can be active and productive.

After Nathan and Rochelle completed the rectangular array with the small cubes as illustrated in figure 18.1, Step 5, Anita was observed staring pensively at the model. Then she announced, "I have a question. What are these little extra ones?" (She pointed to the ten small cubes.) Nathan responded by pointing to the third column (from left to right) of five horizontal longs (see Step 5) and counting, "This is 1, 2, 3, 4, 5," then pointing to the two longs placed vertically on the far right side, he continues by saying, "6, 7." Then pointing to the group of ten small cubes, Nathan said, "8." "Four eighty [meaning the whole model]. This makes up four eighty." Although Nathan said "four eighty," there appeared no apparent confusion about the base-ten blocks actually representing the decimal 4.80. Again, Nathan pointed to the blocks and stated, "This makes up four eighty," and while pointing to the flats, he repeated, "1, 2, 3." When Nathan paused, Anita pointed and said, "You do 3 [flats] and 2 [longs]. Then [pointing to the three groups of longs] there's 15 all there. Where do you get the little ones?" Nathan replies, "You have to fill in the square [meaning rectangular array]." Once again, he pointed and counted the 3 flats, 10 horizontal longs, 5 more horizontal longs, 2 vertical longs, a column of 5 small cubes, and finally 5 more small cubes. When Nathan counted, it sounded like the following: "1, 2, 3

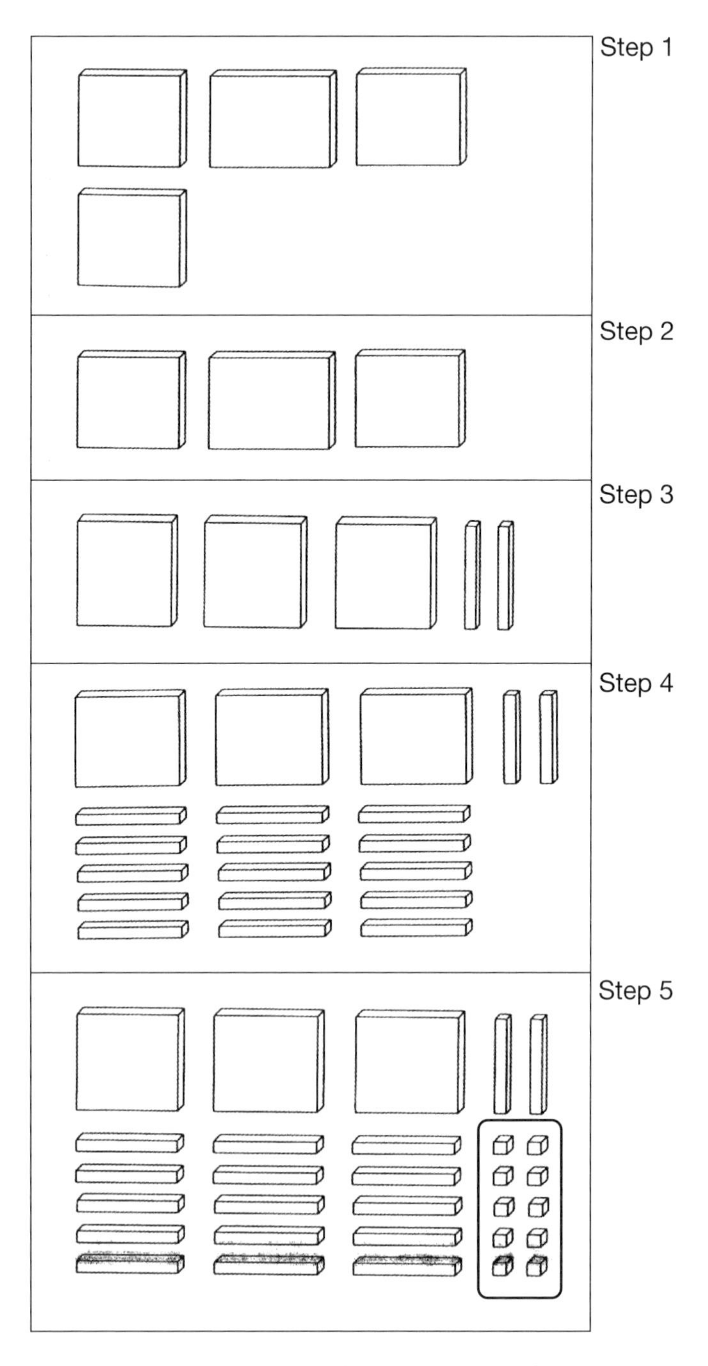

Fig. 18.1. Nathan's group's step-by-step process of 3.2×1.5

[flats], 4 [ten longs mentally regrouped], 450 [five more longs], 460 [first vertical long], 470 [second vertical long], 475 [first column of five small cubes], 480 [second column of five small cubes]." Anita softly said, "Oh, okay." She stared at the model briefly before returning to her paper-and-pencil work.

Anita, although reserved, maintained a connection with the problem and the group throughout the collaborative phase of the work. She made comments and asked questions, which suggested some understanding of the algorithm being modeled. When she says, "Oh, okay," it does give the impression that it is all right. With the base-ten blocks model of 3.2 × 1.5 before them, Lynn asked, "How are we going to show how we represent decimals?" Once again, Nathan responded by pointing, touching, and counting the base-ten blocks. "Because we have four wholes. These [Nathan picks up the two longs] represent tenths. You [Nathan points] have eight tenths. Originally, you had 3 [flats], 17 [longs], 10 [small cubes]. You have 4 wholes and 8 tenths." Nathan even pulled aside the 7 longs (tenths) and 10 small cubes to show the 8 tenths. This was the only time he used the term *tenths* or any precise decimal fraction language during collaborative group discussions, and from all outward appearances, the collaborative problem-solving efforts appeared fruitful. However, it was Lynn's question, which hinted at uncertainty about how this model actually represented decimal fractions.

A review of the written explanations for each of the four papers supported the notion that these prospective teachers, although knowledgeable about using the rectangular array model with base-ten blocks to solve a multiplication algorithm involving whole numbers, continued to be challenged conceptually when using base-ten blocks to represent the multiplication of decimals. Although all four prospective teachers accurately illustrated 3.2 times 1.5 as an array model (see fig. 18.1), two of the four prospective teachers were not totally accurate in their symbolic interpretations. Some of the numeric responses and explanations accompanying the standard paper-and-pencil algorithms suggested confusion when conceptualizing and explaining multiplication with decimal fractions.

As can be seen in figure 18.2, all the prospective teachers' drawings accurately illustrated the array interpretation for the decimal multiplication problem. However, pictorial representations of the multiplication algorithm highlighted the fact that Lynn and Anita made errors. Lynn noted in line B that 0.15 is *5 tenths groups of 3 tenths,* when it should have been *5 tenths groups of 3 wholes*; furthermore, *5 tenths groups of 3 tenths* does not accurately match the pictorial representation labeled B in her drawing. Anita noted that 0.20 is "5 tenths groups of 3 (.5 × .30)" and 1.50 is "1 group of 2 tenths." Perhaps Anita mixed up these explanations because they do not match the illustration rendered. However, these errors again call attention to Lynn's question, "How are we going to show how we represent decimals?" Her query indicated the tension and struggle she experienced as she tried to connect the array model with decimal concepts. Although Lynn and Anita used precise mathematics language in their paper-and-pencil responses, ultimately there were mismatches between the symbolic and the pictorial representations.

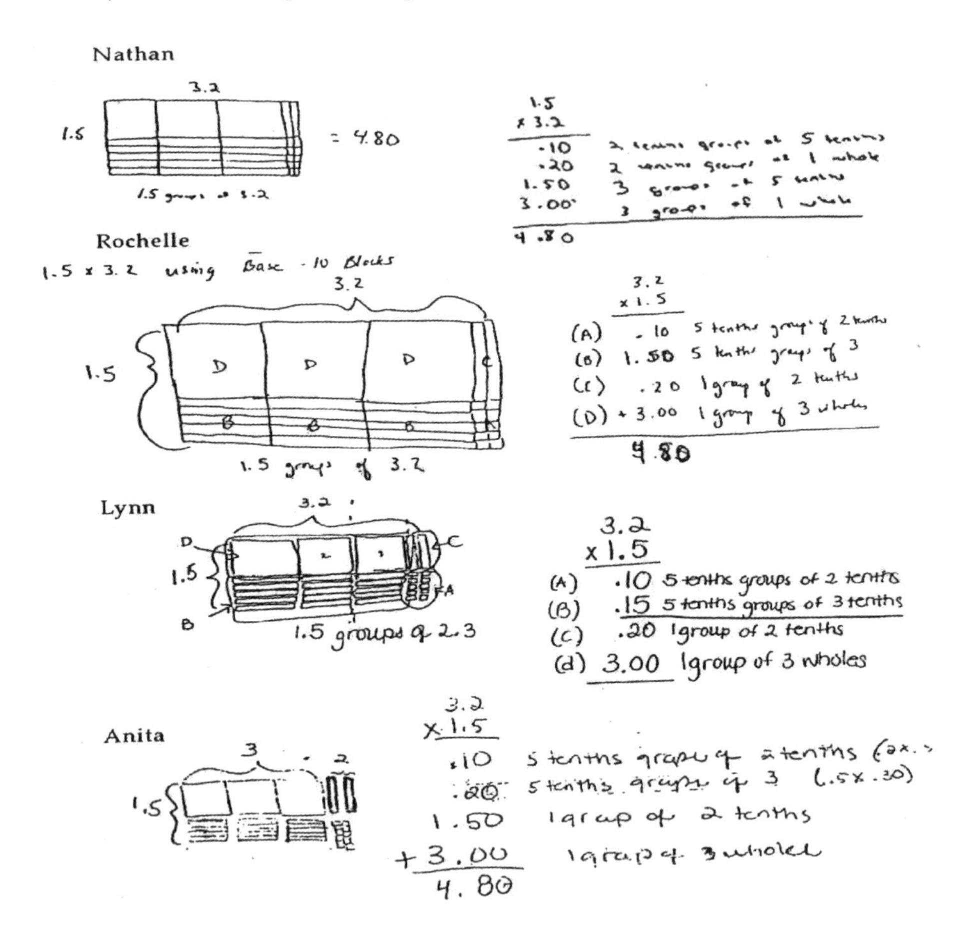

Fig. 18.2. Individual illustrations of a rectangular array model representing 3.2 × 1.5

Resolving Cognitive Conflict

After reviewing the videotapes as well as reflecting on our observations of the prospective teachers' collaborative efforts in solving the decimal fraction algorithm discussed in the previous section, we found a pattern of confusion similar to the experience of Anita and Lynn. Initially, it was assumed that participants actively engaged in the investigation to cognitively extend their knowledge of the multiplication of whole numbers to the multiplication of decimal fractions using an array model. Furthermore, we assumed that they were actively engaged in the process because they asked questions, expressed viewpoints, challenged others' ideas, and asked for alternative explanations, which we believed facilitated an understanding of how the array model might help learners make sense of decimal fraction algorithms. What they might have been experiencing, especially Anita and Lynn, was that their engagement in the task was marginal because they did not have access to the material (Cohen 1994). Access to the material, in this instance the base-ten blocks, was integral to developing an understanding of how to construct an array model to represent decimal fraction algorithms. Another striking component that created conflict among the participants was the lack of the use of precise language for decimal fractions. This aspect will be addressed in the next section.

Central to the prospective teachers' development as conceptual learners was their active involvement and manipulation of concrete materials. To solve this problem of marginal engagement with concrete materials and to assist Anita and Lynn as well as others in the class, we provided follow-up activities whereby they worked in pairs with a more knowledgeable peer to solve similar algorithms in which precise language for decimal fractions was encouraged. First, the more knowledgeable peer would play the major role in manipulating the blocks as well as directing the conversation; then for the second algorithm, the other member (e.g., Anita or Lynn) would manipulate the blocks, directing the conversation, asking questions, and furnishing explanations of her thinking. The third aspect of the paired activity required each individual to draw an illustration that included an explanation that clearly communicated what was done to solve the algorithm. Each person needed to show the paper-and-pencil algorithms with the partial products represented. What emerged was that manipulating the base-ten blocks and constructing an illustration to represent specific algorithms enhanced Anita's and Lynn's understanding of decimal fraction algorithms. The manipulation of the base-ten blocks and the paired collaboration served as tools for resolving the cognitive dissonance or disequilibrium that occurred earlier during the teaching and learning cycle of using an array model to develop an understanding of decimal fraction algorithms. Resolving cognitive conflict through the use of concrete models, collaboration with more knowledgeable peers, or a combination of both (Tudge 1992; Vygotsky 1978) is dependent on the active engagement of the individual learner and having access to the material. Learning mathematics in collaborative groups using concrete materials means students *must* manipulate the physical materials while engaged in the social process of learning (Albert 2000; Vygotsky 1978).

Precise Mathematics Language

The lack of precise mathematics language that accompanies the illustrations is an indicator of the difficulty these prospective teachers experienced using base-ten blocks to represent decimal fraction algorithms. As noted earlier, the language used was not specific to decimal fractions and could be characterized as "whole number" language. Group members referenced the multiplication factors as "3 point 2" and "1 point 5," rather than "3 and 2 tenths" and "1 and 5 tenths." The use of precise language for decimal fractions, which is conceptually oriented, can help prospective teachers make mathematical connections with common fractions and develop a conceptual understanding of decimal representations. Anita exhibited confusion when she made reference to a section of the base-ten blocks model as "0.10 or 10 units." The term *unit* traditionally refers to the ones place, with 10 units being equivalent to 10 ones and 0.10 equivalent to either one-tenth or ten-hundredths. Although Anita seemed to have some basic understanding of the use of base-ten blocks for performing multiplication operations with decimal fractions, she exhibited a lack of clarity in her explanations.

The NCTM (2000) *Principles and Standards* document advances the idea that communication and the language of mathematics that emphasizes the use of precise language and representational models (e.g., a physical model or a mental image) should further conceptual understanding of decimal fractions. When prospective teachers use "whole number language" or refer to decimals as "3 point 2," images of decimal fractions are not readily apparent. Owens and Super (1993) caution that although decimal fractions and whole numbers look similar, research on learning decimals suggests that there is a lack of conceptual understanding. The use of decimal fraction language is mathematical language. It is formal, precise, explicit, and conceptually oriented. Owens and Super advocate the use of conceptually oriented language when "saying" decimals, and they recommend that meaningful decimal fraction language, such as "3 and 2 tenths," be used rather than the "point" reference as described previously. When the teachers were put into pairs to solve additional problems, the more knowledgeable peers and the instructor "scaffolded"(or guided and supported) prospective teachers to be explicit in the use of precise language when referring to decimals.

Not all the prospective teachers experienced this problem. For example, Nathan did seem to have a clear understanding of the multiplication of decimal fractions as judged by his illustration, paper-and-pencil algorithm, and brief explanation:

> To instruct a student, I would use the array model. Each flat is 1, a rod is 1 tenth, and a [small] cube is 1 hundredth. After the array is built I would go through and count them. By combining them you would get 3 wholes, 17 tenths, and 10 hundredths which would be 4 wholes and 8 tenths or 4.8. This will address what students might need to know, place value and decimals. It breaks the numbers up into wholes, tenths, and hundredths and they can see that you place the decimal point between the whole and the tenths.

Although Nathan's oral language was characteristic of whole numbers, this was not so in his written narrative. Nathan used specific decimal fraction language and communicated a good conceptual understanding of decimal multiplication using base-ten blocks on his written response.

Other collaborative groups used the language of decimal fractions while constructing and interpreting the array model. For example, members of another group working on the same problem were observed questioning one another about place value represented by the base-ten blocks. One member was overheard to say, "The longs represent the tenths, right.... The flats represent wholes, and 10 longs is a whole." For some groups, cognitive conflict during group work created opportunities for inquiry, which challenged participants to better explain the process of the multiplication of decimal fractions.

Conclusions and Implications

When constructing rectangular arrays using base-ten blocks, prospective teachers were able to build on their knowledge of the multiplication of whole numbers to illustrate the multiplication of decimal fractions. These physical representations of multiplication—with whole numbers or decimal numbers—follow the same rules, patterns, and discourse parameters that promote conceptual understanding. In spite of this feature, some prospective teachers experienced tension and conflict as they struggled to transition from using base-ten blocks with whole number operations to operations with decimal fractions. This aspect is not surprising, since "the steps of the conventional algorithms, particularly for multidigit multiplication and division, are often every bit as mystifying to teachers as they are to children" (Conference Board of the Mathematical Sciences 2001, p. 67). Precise language was viewed as a key to making sense of decimal concepts (Owens and Super 1993). Oral language can complement the physical model when used to describe partial products, highlighting the place value within the representational model of an array. Specific language assisted in relating decimal fractions to the array model while connecting place-value meaning to the partial products within the multiplication problem. Meaningful language identified patterns within the representational model, which helped to connect number sense, place-value understanding, and, ultimately, the meaning of the decimal point within the symbolic representation of the decimal fraction algorithm.

Mathematics courses need to assist prospective teachers in developing quantitative reasoning through mathematical modeling and problem solving. It is a matter of *transforming and internalizing* the learning that is derived by using models to understand basic mathematics concepts as well as sophisticated ones. More of the same traditional mathematics pedagogy will not suffice. Although base-ten blocks may be considered important tools for thought, they do not guarantee that prospective teachers will be able to make connections between the process and the procedures modeled to their symbolic representations. It is essential that mathematics communication, both oral and written language, be an integral part of the classroom discourse involving the trajectory from concrete to pictorial to symbolic representations.

We believe that effective use of manipulatives is dependent on skilled and knowledgeable teacher educators who combine content with appropriate pedagogy (Sowell 1989). Kozulin (1998) states that materials that support nontraditional activities move the focus of learning from an emphasis on product to an emphasis on process. Teacher educators need to provide activities and encourage interactions in their methods courses and professional development experiences in which teachers manipulate, discuss, draw, and write about what they are learning and understanding with reference to the content. Such pedagogy will assist the prospective teachers in developing content and pedagogical knowledge that will, it is hoped, influence and shape their classroom practices. An implication of this work is that learning activities with manipulatives must involve the manipulation of physical objects that moves learners toward metacognitive engagement. Unless prospective teachers are using manipulatives in a cognitive manner that involves crucial reflection to establish a meaningful concrete representation, the mathematical concept inherent within the concrete model may not be abstracted.

REFERENCES

Albert, Lillie R. "Outside-In—Inside-Out: Seventh-Grade Students' Mathematical Thought Processes." *Educational Studies in Mathematics* 41 (February 2000): 109–42.

Bennett, Albert B., and L. Ted Nelson. *Mathematics for Elementary Teachers: A Conceptual Approach.* Boston: McGraw-Hill Higher Education, 2004.

Cohen, Elizabeth. *Design Groupwork: Strategies for the Heterogeneous Classroom.* New York: Teachers College Press, 1994.

Conference Board of the Mathematical Sciences. *The Mathematical Education of Teachers.* Providence, R.I: American Mathematical Society, 2001.

Ginsburg, Herbert P. *Children's Arithmetic: How They Learn It and How You Teach It.* Austin, Tex.: Pro Ed, 1989.

Hatfield, Mary, Nancy Edwards, and Gary Bitter. *Mathematics Methods for Elementary and Middle School Teachers.* 3rd ed. Boston: Allyn & Bacon, 2000.

Kilpatrick, Jeremy, Jane Swafford, and Bradford Findell, eds. *Adding It Up: Helping Children Learn Mathematics.* Washington, D.C.: National Academy Press, 2001.

Kozulin, Alex. *Psychological Tools: A Sociocultural Approach to Education.* Cambridge, Mass.: Harvard University Press, 1998.

McAdam, John F. "Examining Prospective Teachers' Conceptual Understandings of

Algorithms through Activity-Centered Collaboration with Base Ten Blocks." Doctoral diss., Boston College, 2000. *UMI Dissertation Services* (2000): 9961612.

National Council of Teachers of Mathematics (NCTM). *Curriculum and Evaluation Standards for School Mathematics.* Reston, Va.: NCTM, 1989.

———. *Principles and Standards for School Mathematics.* Reston, Va.: NCTM, 2000.

Owens, Douglas, and Douglas Super. "Teaching and Learning Decimal Fractions." In *Research Ideas for the Classroom: Middle Grades Mathematics,* edited by Douglas Owens, pp. 137–58. New York: Macmillan Publishing Co., 1993.

Sowell, Evelyn J. "Effects of Manipulative Materials in Mathematics Instruction." *Journal for Research in Mathematics Education* 20 (November 1989): 498–505.

Tudge, Jonathan. "Vygotsky, the Zone of Proximal Development, and Peer Collaboration: Implications for Classroom Practices." In *Vygotsky and Education: Instructional Implications and Applications of Sociohistorical Psychology,* edited by Luis C. Moll, pp. 155–72. New York: Cambridge University Press, 1992.

Van de Walle, John, and LouAnn Lovin. *Teaching Student-Centered Mathematics: Volume Three, Grades* 5–8. Boston: Allyn & Bacon Professional Books, 2006.

Vygotsky, Lev S. *Mind in Society: The Development of Higher Psychological Processes.* Cambridge, Mass.: Harvard University Press, 1978.

PART FIVE

Reflections on Mathematics Teaching and Learning

Marilyn E. Strutchens

One of the most important goals for teachers is for them to become reflective about their practices. Reflecting on their teaching enables teachers to think about the type of questions that they ask their students to help them make sense of mathematics and about the type of tasks that they use to help their students develop mathematical understanding. Reflection can provide teachers with meaningful insights that help them assist their students with learning mathematics well; it can also cause the teacher to think about his or her own mathematics learning. Moreover, when teachers reflect on their own learning, they are better able to empathize with their students and question them in ways that help them to understand mathematics. In this section, authors share their insights from reflecting on their mathematics learning in the midst of instructing their students, and they also share the importance of requiring students to reflect on their learning experiences.

Liljedahl, Rolka, and Rösken discuss the deep-seated beliefs that preservice teachers have about mathematics that often run counter to contemporary research on what constitutes good practice. In their article they present research on a mathematics teaching methods course specifically designed to integrate learning outcomes with belief outcomes. In particular, they used reflective journals to examine how a group of preservice elementary school teachers' beliefs about mathematics and the teaching and learning of mathematics evolved as a result of being enrolled in the methods course.

Inspired by a question posed during a course, entitled Numbers and Numerical Thinking, that she taught to sixteen middle school teachers, Fernandez presents an article in which she reflects on the relationship between her understandings of mathematics and the pedagogical strategies she implemented in responding to her teachers' open questions. Her dual roles as a mathematics teacher and a teacher educator were instrumental in guiding these reflections.

While Mau was participating in a study in which she was being videotaped by two of her colleagues, who were researching what students were doing to make sense of mathematics in Mau's mathematics classroom, she found herself worrying about "looking lost," or what the students might call "clueless." That is, she worried that when a student offered her or his thinking, she might not know what to do with it and that her (hopefully temporary) intellectual surprise may lead students to think she did not know what she was doing. As a result of her reflecting on how she was helping her students to make sense of the mathematics, Mau wrote an article in which she describes three learning experiences—one for herself and two for her students.

In the final article, Chazan, Sword, Badertscher, Conklin, Graybeal, Hutchison, Marshall, and Smith share an experience of teaching and learning at the doctoral level. Two instructors, one with a Ph.D. in mathematics and one with an Ed.D. in mathematics education, strove to create an environment in which doctoral students in mathematics education could develop their own mathematical questions and explore them. This particular course was one part of a larger effort in the Mid-Atlantic Center for Mathematics Teaching and Learning to reenvision doctoral education. The authors describe the structure of the course in conjunction with the reflections of the learners. Their goal is to portray images of what mathematical exploration can look like and how it can feel to educators who participate in such an experience.

19

Affecting Affect: The Reeducation of Preservice Teachers' Beliefs about Mathematics and Mathematics Learning and Teaching

Peter Liljedahl
Katrin Rolka
Bettina Rösken

EFFECTIVE mathematics teaching is the result of a complex coordination of specific knowledge and specific beliefs. Too often, however, the emphasis in teacher education programs is placed on the infusion of content knowledge, pedagogy, and pedagogical content knowledge, with only a cursory treatment of the beliefs that, for better or for worse, will govern the eventual application of what has been acquired within those programs. In this article we present the results of a study that examines the effectiveness of a methods course specifically designed to change the mathematical beliefs of a group of preservice teachers. The evolving beliefs are documented in reflective journals and analyzed according to established categories describing mathematical beliefs.

The Study

The primary goal of teacher education programs is that their enrollees learn how to teach. Often visible in such programs are explicit attempts to facilitate the learning of *content knowledge, pedagogy knowledge,* and *pedagogical content knowledge* (Schulman 1986). A problem arises, however, in that on completion of such a program, many novice mathematics teachers revert to a method of teaching that is more reflective of their own experiences as students than of their experiences as prospective teachers. Why do they do so?

The short answer to this question is *beliefs*—beliefs about mathematics and about the teaching and learning of mathematics that prospective teachers hold coming into their teacher education programs. That is, "prospective elementary teachers do not come to teacher education feeling unprepared for teaching" (Feiman-Nemser et al. 1987). "Long before they enroll in their first education course or math methods course, they have developed a web of interconnected ideas about mathematics, about teaching and learning mathematics, and about schools" (Ball 1988). Those ideas are more than just feelings or fleeting notions about mathematics and mathematics teaching. During their time as students of mathematics, they first formulated, and then concretized, deep-seated beliefs about mathematics and what it means to learn and teach mathematics. Those beliefs often form the foundation on which they will eventually build their own practice as teachers of mathematics (cf. Fosnot 1989; Skott 2001). Unfortunately, those deep-seated beliefs often run counter to contemporary research on what constitutes good practice.

Teacher education programs need to attend to this fact and to focus on reshaping the beliefs and correcting the misconceptions that could impede effective teaching in mathematics (Green 1971). More specifically, teacher education programs need to explicitly facilitate the reshaping of beliefs about mathematics as part of the learning of mathematics. Even more specifically, mathematics methods courses, as part of teacher education programs, need to explicitly facilitate the reshaping of beliefs about the teaching and learning of mathematics *as part of* students' learning about the teaching and learning of mathematics.

In this article we presented research on a mathematics method course specifically designed to integrate learning outcomes with belief outcomes. In particular, we use reflective journaling to examine how a group of preservice elementary school teachers' beliefs about mathematics and the teaching and learning of mathematics evolve as a result of being enrolled in the aforementioned method course.

Mathematical Beliefs

Dionne (1984) suggests that beliefs are composed of three basic components called the traditional perspective, the formalist perspective, and the constructivist perspective. Similarly, Ernest (1991) describes three philosophies of mathematics called instrumentalist, Platonist, and problem solving, whereas Törner and Grigutsch (1994) name the three components as toolbox aspect, system aspect, and process aspect. All these different notions correspond more or less with one another. In this article we employ the three components defined by Törner and Grigutsch (1994). In the "toolbox aspect," mathematics is seen as a set of rules, formulas, skills, and procedures, whereas mathematical activity means calculating as well as using rules, procedures, and formulas. In the "system aspect," mathematics is characterized by logic, rigorous proofs, exact definitions, and a precise mathematical language, and doing mathematics consists of producing accurate proofs as well as using a precise and rigorous language. In the "process aspect," mathematics is considered a constructive process in which relations between different notions and sentences play an important role. Here the mathematical activity involves creative steps, such as generating rules and formulas, thereby inventing or reinventing the mathematics. Besides these standard perspectives on mathematical beliefs, a fourth essential component is the usefulness, or utility, of mathematics (Grigutsch, Raatz, and Törner 1997).

Changing Mathematical Beliefs

Robust beliefs are difficult to change (Op't Eynde, de Corte, and Verschaffel 2001). However, an abundance of research purports to produce changes in preservice teachers of mathematics. Prominent in this research is an approach by which preservice teachers' beliefs are challenged (Feiman-Nemser et al. 1987). Because beliefs that preservice teachers possess are often *implicitly* constructed from personal experiences as learners of mathematics (Green 1971), challenging those beliefs helps individuals make *explicit* the basis of their beliefs, and as such, makes them vulnerable to scrutiny and critique.

Another prominent method for producing change in preservice teachers is by involving them as learners of mathematics and mathematics pedagogy, usually immersed in a constructivist environment (Ball 1988; Feiman-Nemser and Featherstone 1992). Such an approach produces change through two distinct, but related, processes. First, it involves the prospective teachers as learners in mathematical experiences that may be completely absent from their prior learning encounters. Second, it models for them *teaching* methodologies that are more conducive to facilitating the aforementioned *learning* encounters.

A third method for producing changes in belief structures has emerged out of the work of one of the authors, in which preservice teachers' experiences with mathematical discovery have been shown to have a profound, and immediate, transformative effect on their beliefs regarding the nature of mathematics, as well as their beliefs regarding the teaching and learning of mathematics (Liljedahl 2005).

Reflective Journaling

Journal writing in mathematics education has a long and diverse history of use. Journaling helps students reflect on and learn mathematical concepts (Chapman 1996; Ciochine and Polivka 1997; Dougherty 1996). It has been shown to be an effective tool for facilitating reflection among students (Mewborn 1999) as well as an effective communicative tool between students and teachers (Burns and Silbey 2001). More relevant to this study, journaling has become an accepted method for qualitative researchers to gain insights into their participants' thinking (Mewborn 1999; Miller 1992). In particular, reflective journals have been shown to be a very good method for soliciting responses pertaining to beliefs, even when such responses are not explicitly asked for (Koirala 2002; Liljedahl 2005).

Methodology

The Participants

The participants in this study are preservice elementary school teachers enrolled in a Designs for Learning Elementary Mathematics course for which the first author was the instructor. The course met for thirteen weeks with weekly four-hour classes. This particular offering of the course enrolled thirty-nine students, the vast majority of whom were extremely fearful of having to take mathematics and even more so of having to teach mathematics. Their fear resided, most often, within participants' negative beliefs and attitudes about their ability to learn and do mathematics. At the same time, however, as apprehensive and fearful of mathematics as these students were, they were extremely open to, and appreciative of, any ideas that could help them become better mathematics teachers.

The Course

All three of the previously mentioned approaches for producing change in beliefs were combined in the design and teaching of the methods course. During the course the participants were immersed in a constructivist learning environment. That is, ideas pertaining to the course content were extensively modeled and experienced prior to any explicit discussion. For example, group work was discussed in the later part of the course, after the class had worked in groups for an extended period of time. The same approach was also taken with assessment, theories about learning, numeracy, enrichment, the use of literature in teaching mathematics, manipulatives, and so on—extensive experience always preceded theory and discussion.

Problem solving was used extensively in the course as a way to introduce concepts in mathematics, mathematics teaching, and mathematics learning. Problems were assigned to be worked on in class, as homework, and as projects. Each participant worked on the assigned problems within the context of a group, but the groups were not rigid, and as the weeks passed the class became a very fluid and cohesive entity that tended to work on problems as a collective whole. As a result, the students had many opportunities to make mathematical discoveries while working on the problems. In fact, all thirty-nine participants demonstrated evidence of mathematical discoveries in their problem-solving journals, and twenty-nine (74 percent) of the participants detailed in their reflective journals the positive effect that those discoveries had on them.

Assessment was used primarily as a formative tool to familiarize participants with contemporary theories of teaching in general, and of assessment in particular. Problem-solving journals, as well as reflective journals, were kept and submitted at a number of junctures within the course. The evaluation of participants' journals evolved throughout the course in correspondence with their evolving understanding of assessment that was fostered by the course design. That is, whereas students' early submissions were evaluated for completeness, later submissions were evaluated using sophisticated rubrics that were coconstructed with the students. A final project required the participants to write a "play" (Calderhead and Robson 1991) about a situation in which a student was struggling to understand a mathematical concept but was hampered by some misconception. This project required the students to demonstrate what they had learned about the teaching and learning of mathematics in the context of what they believed about the teaching and learning of mathematics.

All thirty-nine participants successfully completed the course.

The Data

As mentioned, throughout the course the participants kept a reflective journal in which they were asked to respond to assigned prompts. The prompts varied from invitations to think about assessment to instructions to comment on curriculum. One set of prompts, in particular, was used to assess each participant's beliefs about mathematics, and about teaching and learning mathematics (i.e., What is mathematics? What does it mean to learn mathematics? What does it mean to teach mathematics?). These prompts were assigned in the first, seventh, and final week of the course. The data for this study come from the participants' responses in the first and last week of the course.

The Analysis

The three authors independently coded the data according to each of the four previously mentioned aspects of beliefs: *toolbox, utility, system,* and *process*. We compared the results of the independent codings, discussed the discrepancies, and recoded pertinent entries. This process (Huberman and Miles 1994) led to a more elaborate understanding of the framework, as well as a more consistent coding of the data.

In what follows we use excerpts from the participants' journals to exemplify our shared understanding of each aspect of beliefs with respect to mathematics as well as the teaching and learning of mathematics.

Beliefs about Mathematics

Toolbox Aspect

"My first impression is that math is numbers, quantities, units. In math there is always one right answer.... Math is about ... memorizing formulas that yield the right answer."

"When first pondering the question, 'What is mathematics?' I initially thought that mathematics is about numbers and rules. It is something that you just do and will do well as long as you follow the rules or principles that were created by some magical man thousands of years ago."

System Aspect

"Mathematics is the science of pattern and structure. It uses number sense and mathematical concepts to develop a flexible understanding of the world around us."

"Mathematics is a universal language. It is the study of numbers, proportions, relationships, patterns and sequences. Becoming literate in this language is important in order to understand space and time; to develop logical thinking and reasoning...."

Utility Aspect

"Math is all around us. We live in a quantitative society. On any given day, we may be required to use math to tell us how far over the speed limit we are driving, how much money we have left in our bank account to pay our loans, how many more university credits we need to graduate, how much [P]rozac we need to take to get through the day or simply how many people in this world matter to us."

Process Aspect

"For me, math has truly transformed from being a skill or procedure that can be used merely for efficiency to being imbedded within a process of meaning-making that goes on inside the individual, a construction of understanding that we make up."

Beliefs about Teaching and Learning Mathematics

Toolbox Aspect

"For me math is a puzzle to figure out. All of the questions or problems we were given in school had a solution that I just needed to apply a formula or rule to and the answer would be clear."

System Aspect

"To learn mathematics is to learn how numbers are used to represent concepts and matter, as well as show relationships and solve problems."

"Learning math means understanding patterns, quantities, shapes.... To teach mathematics is to teach fundamental number concepts...."

Utility Aspect

"We also teach mathematics that is related to everyday life, for example our system of currency and how to measure how tall we are."

"[We teach mathematics] to enable students to function successfully in our world. It is such an integral part of everything in and of our world, the more they know, the more choices in life they'll have."

Process Aspect

"The other thing that stands out is the difference between formally teaching students, and actually facilitating learning. By being a facilitator of the learning process, we are able to choose situations, activities and problems for the students to work on either individually or in groups, and through this approach, students are able to ... try different ideas, and develop strategies."

"To teach mathematics means to guide/facilitate critical thinking and problem-solving skills. It means to question and challenge students to think in different ways. To teach math means to create a problem-solving culture within the classroom that encourages and supports students' reasoning."

The quotations presented above represent only a small portion of all the students' journal entries. A wide range of statements supporting each category can be found in the data. The reader should note, however, that not all excerpts are as easily categorized. We follow Dionne's (1984) suggestion that mathematical beliefs constitute a mixture of the above-mentioned aspects; accordingly, clear classification cannot always be made. As a result, many journal entries were coded for more than one aspect. For example, in the following journal entry, the system aspect is intertwined with the utility aspect in beliefs about mathematics:

- "I think it [mathematics] has to do with the complex relationships between numbers and the symbols we use to make sense of the world among us. More and more I see maths as a system put in place to help us better ... make sense of the world around us. Maths allows us to group things, to calculate, to categorize. It's a great way to bring order from chaos."

In addition, the data were checked for comments that are indicative of rhetoric, that is, comments that are hollow echoes of conventional beliefs about mathematics and the teaching and learning of mathematics. Examples of journal entries that were flagged as rhetoric are the following very succinct responses to "What is mathematics?"

- "Math is the study of numbers and patterns and the relationship between them."
- "Mathematics is the study of numbers."

A similar response, followed up with comments that indicate that the beliefs have been internalized, is not flagged as rhetoric.

- "Math is a language that helps individuals reason, problem solve, and

> distinguish relationships. In order to do these activities, we need an understanding of the basics of the language, such as symbol meaning, number values, number relationships, and basic skill counting."

In all, comments from five participants were deemed to be rhetoric, and as such were excluded from the aggregated data.

Results and Discussion

The coded data were aggregated to produce a holistic picture of the evolving beliefs of the class as a whole. The results of this aggregation are displayed in table 19.1.

Table 19.1
Aggregation of Coded Data

	Mathematics Before	Mathematics After	Teaching Before	Teaching After
Toolbox	13	2	7	3
System	22	19	18	8
Utility	10	9	26	1
Process	0	12	6	30

The charts in figures 19.1 and 19.2 show the distribution of beliefs about mathematics at the beginning and at the end of the course. The most obvious change is the degree to which a process aspect of mathematics has been introduced into the collective beliefs of the class. In referring to table 19.1, the process aspect appears to have displaced the toolbox aspect of beliefs about mathematics. However, careful analysis of the disaggregated data reveals a more complex view of changing beliefs. For some participants changes involved the addition of a belief aspect, for others it involved the dismissal of an aspect, and for still others it involved the replacement of one aspect with another. The net effect, however, remains a shift within the class away from the toolbox belief of mathematics and toward the process belief of mathematics.

Figures 19.3 and 19.4 show the distribution of beliefs about teaching and learning mathematics at the beginning and at the end of the course. These figures show a significant shift in beliefs about the teaching and learning of mathematics toward a process aspect. Here, however, the figures are very representative of the changes that occur at the individual level. Most of the change in the participants' beliefs about the teaching and learning of mathematics can be encapsulated as a change from a system aspect or a utility aspect to a process aspect.

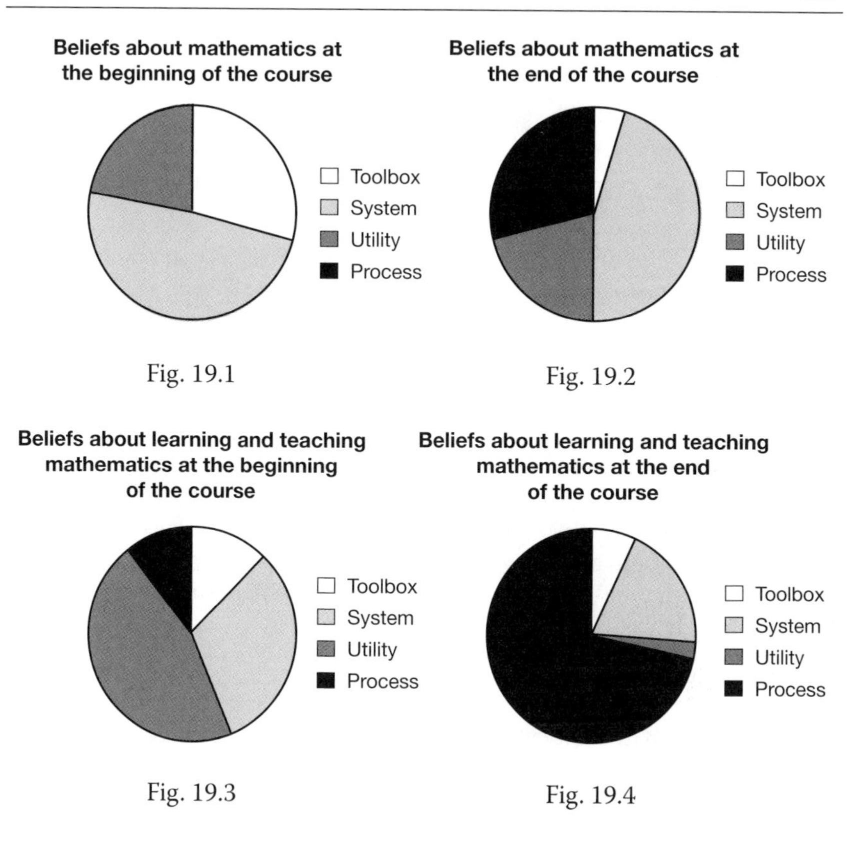

Fig. 19.1

Fig. 19.2

Fig. 19.3

Fig. 19.4

Conclusion

The role of beliefs in the learning of mathematics, as well as in the learning of what it means to teach and learn mathematics, is nowadays widely acknowledged as important (Leder, Pehkonen, and Törner 2002). According to Schoenfeld (1998), beliefs can be interpreted as "mental constructs that represent the codification of people's experiences and understandings" (p. 19). In this article we have shown how a mathematics methods course, designed to deliver both understandings and experiences, affected the beliefs of the students enrolled in the course. Most dramatically, the participants' beliefs about the teaching and learning of mathematics evolved away from a system or utility aspect and toward a process aspect. This evolution can be interpreted as *unlearning* in the process of *learning* to teach mathematics better (Ball 1988).

The observed changes can be attributed, at least in part, to the emphasis placed on problem solving in this course.

- "To teach mathematics means to guide/facilitate critical thinking and problem-solving skills. It means to question and challenge students to think in different ways. To teach math means to create a problem-solving culture within the classroom that encourages and supports students' reasoning."

Through problem solving the participants both experienced progressive mathematical *learning* and had modeled for them progressive mathematics *teaching* (Ball 1988; Feiman-Nemser and Featherstone 1992). As an outcome, they became both *enquiring learners* and *enquiring teachers* (Fosnot 1989). The emphasis on problem solving also afforded the participants the opportunities to experienc mathematical discoveries (Liljedahl 2005).

- "I have never before worked for so long or so hard on a problem. It just seemed like we would never get it. In the end the hard work was worth it though. When we finally broke through and saw the answer, it felt so great. It felt great knowing that we had solved it ... WE had solved it. I know now that it wouldn't have felt nearly as great if Peter [the course instructor and author] told us how to do it. It also wouldn't have felt as great if the problem we had been working on hadn't been as difficult. I definitely want my students to feel this way."

Their experiences with discoveries provided a powerful reference from which the preservice teachers could think about learning mathematics, and what it means to think mathematically and be mathematical in a classroom. Finally, the constant emphasis on reflection about what mathematics is, and what it means to teach and learn mathematics, forced the participants to scrutinize their own beliefs about those issues and to recast or reformulate them in the context of their recent experiences (Feiman-Nemser et al. 1987).

- "For me, math has truly transformed from being a skill or procedure that can be used merely for efficiency to being imbedded within a process of meaning-making that goes on inside the individual, a construction of understanding that we make up."

The results of the research presented in this article show some drastic changes between discrete aspects of beliefs. Further research is required to more closely examine the continuous nature of that change, and to more closely examine the possible continuous nature of the aspects. Also, although the research presented here does not follow the preservice teachers into their teaching, it can be seen from the nature of the participants' responses that beliefs have implications for practice. Further research is currently under way with this same group of participants to examine whether the changes in beliefs reported here are sustained as they enter into their early years of practice.

REFERENCES

Ball, Deborah L. "Unlearning to Teach Mathematics." *For the Learning of Mathematics* 8(1) (1988): 40–48.

Burns, Marilyn, and Robyn Silbey. "Math Journals Boost Real Learning." *Instructor* 110 (7) (2001): 18–20.

Calderhead, James, and Maurice Robson. "Images of Teaching: Student Teachers' Early Conceptions of Classroom Practice." *Teaching and Teacher Education* 7 (1) (1991): 1–8.

Chapman, Kathleen P. "Journals: Pathways to Thinking in Second-Year Algebra." *Mathematics Teacher* 89 (October 1996): 588–90.

Ciochine, John G., and Grace Polivka. "The Missing Link? Writing in Mathematics Class!" *Mathematics Teaching in the Middle School* 2 (March–April 1997): 316–20.

Dionne, Jean J. "The Perception of Mathematics among Elementary School Teachers." In *Proceedings of the Sixth Conference of the North American Chapter of the International Group for the Psychology of Mathematics Education*, edited by James M. Moser, pp. 223–28. Madison, Wis.: University of Wisconsin—Madison, 1984.

Dougherty, Barbara J. "The Write Way: A Look at Journal Writing in First-Year Algebra." *Mathematics Teacher* 89 (October 1996): 556–60.

Ernest, Paul. Th*e Philosophy of Mathematics Education*. Hampshire, U.K.: Falmer Press, 1991.

Feiman-Nemser, Sharon, and Helen Featherstone. "The Student, the Teacher, and the Moon." In *Exploring Teaching: Reinventing an Introductory Course*, edited by Sharon Feiman-Nemser and Helen Featherstone, pp. 59–85. New York: Teachers College Press, 1992.

Feiman-Nemser, Sharon, G. Williamson McDiarmid, Susan L. Melnick, and Michelle Parker. "Changing Beginning Teachers' Conceptions: A Description of an Introductory Teacher Education Course." Paper presented at the annual meeting of the American Educational Research Association, Washington, D.C., April 1987.

Fosnot, Catherine Twomey. *Enquiring Teachers, Enquiring Learners: A Constructivist Approach for Teaching*. New York: Teachers College Press, 1989.

Green, Thomas F. *The Activities of Teaching*. New York: McGraw-Hill Book Co., 1971.

Grigutsch, Stefan, Ulrich Raatz, and Günter Törner. "Einstellungen gegenüber Mathematik bei Mathematiklehrern." *Journal für Mathematikdidaktik* 19 (1) (1997): 3–45.

Huberman, A. Michael, and Matthew B. Miles. "Data Management and Analysis Methods." In *Handbook of Qualitative Research*, edited by Norman K. Denzin and Yvonna S. Lincoln, pp. 428–44. Thousand Oaks, Calif.: Sage Publications, 1994.

Koirala, Hari P. "Facilitating Student Learning through Math Journals." In *Proceedings of the Twenty-sixth Annual Meeting of the International Group for the Psychology of Mathematics Education*, vol. 3, edited by Anne D. Cockburn and Elena Nardi, pp. 217–24. Norwich, U.K.: 2002.

Leder, Gilah C., Erkki Pehkonen, and Günter Törner, eds. *Beliefs: A Hidden Variable in Mathematics Education?* Dordrecht, Netherlands: Kluwer Academic Publishers, 2002.

Liljedahl, Peter. "Aha!: The Effect and Affect of Mathematical Discovery on Undergraduate Mathematics Students." *International Journal of Mathematical Education in Science and Technology* 36 (2–3) (2005): 219–36.

Mewborn, Denise S. "Reflective Thinking among Preservice Elementary Mathematics Teachers." *Journal for Research in Mathematics Education* 30 (May 1999): 316–41.

Miller, L. Diane. "Teacher Benefit from Using Impromptu Writing in Algebra Classes." *Journal of Mathematics Teacher Education* 23 (4) (1992): 329–40.

Op't Eynde, Peter, Erik de Corte, and Lieven Verschaffel. "Problem Solving in the Mathematics Classroom: A Socio-constructivist Account of the Role of Students' Emotions." In *Proceedings of the Twenty-fifth Annual Conference of the International Group for the Psychology of Mathematics Education*, vol. 4, edited by Marja van den Heuvel-Panhuizen, pp. 25–32. Utrecht, Netherlands: PME, 2001.

Schoenfeld, Alan. "Toward a Theory of Teaching-in-Context." *Issues in Education* 4 (1) (1998): 1–94.

Schulman, Lee S. "Those Who Understand: Knowledge Growth in Teaching." *Educational Researcher* 15 (2) (1986): 4–14.

Skott, Jeppe. "The Emerging Practices of Novice Teachers: The Roles of His School Mathematics Images." *Journal of Mathematics Teacher Education* 4 (1) (2001): 3–28.

Törner, Günter, and Stefan Grigutsch. "Mathematische Weltbilder bei Studienanfängern—eine Erhebung." *Journal für Mathematikdidaktik* 15 (3/4) (1994): 211–52.

20

Learning from the "Unknown" in Mathematics Teacher Education: One Teacher Educator's Reflections

Eileen Fernández

THIS article was inspired by a question posed during a Numbers and Numerical Thinking course that I taught to sixteen middle school teachers. Three weeks into the semester, I taught the Egyptian algorithm for multiplication, which works as follows. To multiply 5 by 11, make a two-column table, placing a 1 and either of your factors (5 or 11) in the column headings (see fig. 20.1 for steps in the algorithm). Double the 1 until you obtain powers of 2 that sum to the unused factor (1 + 4 sum to 5). Double the second-column numbers and select those corresponding to the highlighted first-column powers of 2. The sum of these second-column numbers gives the sought-after product ($5 \bullet 11 = 11 + 44 = 55$).

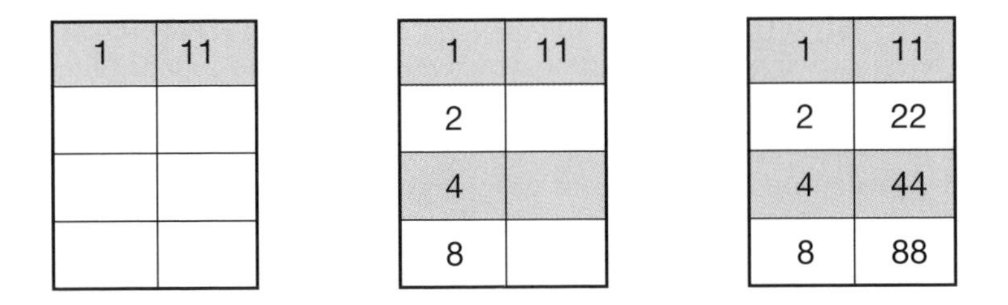

1	11

1	11
2	
4	
8	

1	11
2	22
4	44
8	88

Fig. 20.1. Steps in the Egyptian algorithm for multiplication

I thought this algorithm would be fun and give the teachers material for their own classrooms. Also, I intended to use it to motivate them to come up with ideas about our base-ten positional system (e.g., How do we know there will always be powers of 2 that add to either factor?). But as the teachers experimented with different products in their groups, one teacher, Dana, posed the following questions: "What's so special about doubling? Will this work if we *triple* instead of *double*?"

Learning from the Unknown in Teacher Education

Dana's questions were not the first classroom questions for which I did not have immediate answers. In my mathematics classes, students also pose such questions. As a teacher, the experience of exploring the unknown with students contributes to my lifelong learning of mathematics, since it provides opportunities to learn new mathematics alongside my students. Because my students and I take turns in positions of "confusion" and "knowing," the mathematics is learned from a diversity of perspectives.

Open questions posed by *teachers* also contribute to my learning of mathematics, but they add a new dimension to it. When *students* who are *teachers* take turns in positions of "confusion" and "knowing," it is natural to wonder how the experience will manifest itself in their classrooms. As I investigate the unknown with teachers, what are they learning from me about how they can teach mathematics? In this article, I reflect on the relationship between my understandings of mathematics and the pedagogical strategies I implemented in responding to my teachers' open questions. My dual roles as a mathematics teacher and a teacher educator were instrumental in guiding these reflections.

Background, the Unknown, and Constructivism

The sixteen teachers in my class had teaching experiences ranging between one and thirty years, with an average of eleven years. Seven teachers characterized their schools as urban, five as suburban, and four as blue collar. One teacher taught fourth and fifth grade, whereas the remainder taught some combination of sixth- to eighth-grade mathematics. The numbers course was part of a professional development program, funded by the state of New Jersey and intended to enhance the content knowledge of its participating teachers. Influenced by *Principles and Standards for School Mathematics* (National Council of Teachers of Mathematics [NCTM] 2000), this program's goals for content enhancement included fostering experiences in which learners actively create their own knowledge and reach their own understandings of problem situations. Some readers will recognize the tenets of constructivism in this goal (Lerman 1989).

In an earlier course entitled Algebra and Algebraic Thinking, the teachers and I studied algebraic, geometric, and verbal representations and used different kinds of reasoning to delve into these representations and their connections. Although the Algebra and Algebraic Thinking course provided opportunities to challenge the teachers' knowledge, it generally felt like my own goals and understandings directed our lessons. When Dana posed her questions in the numbers course, I recognized an opportunity for a learning experience guided by constructivist principles. Having the open question come from a teacher enhances its integrity, since my own immediate understandings cannot come into play in resolving it. Confrey (1990, p. 108) notes that a constructivist educator's most basic skill is to learn "to approach a foreign or unexpected response with a genuine interest in learning its character, its origins, its story and its implications." With my interest heightened, I decided to use Dana's questions to sharpen my skills at enabling the teachers' understandings to guide our discussions.

Much to my surprise, Dana's questions started a momentum of open questions that challenged my skill sharpening in ways I hadn't expected. Since I kept a journal on our lessons for the professional development program, the mathematics generated by Dana's questions and my readings on constructivism began to give these reflections a new focus. Effectively, I found myself subjecting the tenets of constructivism to my experiences exploring mathematical unknowns in a professional development setting. This process enabled me to examine my practice and constructivism interactively and to enhance my understanding of this interaction in a teacher education setting. All the recollections recounted in this article are taken from my readings, journal, class notes, scrap paper, and one videotaped lesson. I also used three teachers' notebooks to validate my journal excerpts from the perspective of class notes taken.

Old Strategies and New Conflicts

Dana's questions came at the end of our lesson, so I asked the teachers to think about them for next time. The strategy of asking students to think about questions at home, which I had used before, provides an unconstrained time and privacy for investigation that is not offered by a classroom setting. My hope is that students will come to class eager to share findings (and thereby guide the discussion) after such investigations. At our next meeting, the teachers demonstrated how tripling does not work in the algorithm. (Try to write 5 or 11 as a sum of powers of 3, using each power of 3 only once.)

The question of *why* tripling doesn't work, or what is special about doubling, remained open. I had not had time to think about it, and when I questioned the teachers, they did not volunteer anything. I considered asking, "How might we go about investigating this question?" but I sensed it was time for a change in topics. We turned to the algorithm, exploring new definitions (*linear combinations, coefficients*) and techniques for ensuring that every natural number can be written as a linear combination of powers of 2. For example, the highest power of 2 in 11 is 2^3 with a remainder of 3, so $11 = 2^3 + 3$. The highest power of 2 in 3 is 2^1 with a remainder of 1. Thus, $11 = 2^3 + 2^1 + 1 = 1 \cdot 2^3 + 0 \cdot 2^2 + 1 \cdot 2^1 + 1 \cdot 2^0$. These coefficients give the *binary representation* for 11: 1011_2. (Throughout this article, I adhere to the language of "combinations" and "coefficients," since that was the terminology preferred by my teachers.)

We struggled with these definitions and worked on examples in groups and as a class. The topic of exponents, covered in our Algebra and Algebraic Thinking course, had been challenging and, I reflected, was contributing to difficulties in the current lesson. After writing several numbers as linear combinations of powers of 2, I asked about the coefficients showing up in our work. As the teachers identified 0 and 1, I experienced an "aha!" regarding doubling versus tripling. In my journal, I wrote:

> "selecting" and "not selecting" a power of 2 to represent a factor in the algorithm respectively assigns a coefficient of 1 or 0 in the linear combination for that number. For example, $5 = 1 + 4 = 1 \bullet 2^0 + 0 \bullet 2^1 + 1 \bullet 2^2$. We select 1 or 2^0 ($1 \bullet 2^0$), we don't select 2 or 2^1 ($0 \bullet 2^1$) and we select 4 or 2^2 ($1 \bullet 2^2$).... Because writing numbers as a linear combination of powers of 3 requires three coefficients (0, 1, and 2), the actions of selecting and not selecting fail to provide a third coefficient. So tripling won't work in the algorithm.

I remember thinking, "Should I share my insights, lead the teachers to them, or hold back?" I did not consider pretending nothing happened or a segue back into the tripling issue with a question. Instead, I enthusiastically burst out that something had occurred to me regarding our tripling question and that if I continued with my lesson, the teachers might gain the tools to reach this understanding themselves. They encouraged me to continue. We moved on to linear combinations of powers of 3 and the number of coefficients required (three), and a guess about the number of coefficients for powers of 4 (four). Before ending, the teachers asked for a hint on my insight. Feeling the need to honor their request, I advised them to consider a relation between "selecting" and "not selecting" the coefficients 1 and 0, and the number of coefficients needed to write linear combinations with different powers.

Dilemmas Arising from Exploring the Unknown

As illustrated above, sometimes a teacher comes to understand an open question before a student. Such situations are reminders of the unlevel playing field that underlies every teacher-student relationship. That is, if a student reaches an understanding concerning an open question and shares it, this can be viewed as a "negotiation" in communicating ideas. If the teacher comes to an understanding, this is viewed as an "imposition" of ideas. When teaching is viewed as a continuum on which "negotiation" and "imposition" are endpoints (Cobb 1988), a teacher's dilemmas (in the event of an early understanding) are highlighted: should she share her viewpoints, lead students to them, or hold back until the students' understandings can be elicited? The former options can be interpreted as robbing students of the learning processes that lead to *their* understanding. And yet, if a teacher and student are involved in genuine negotiation, shouldn't the teacher have as much freedom to convey her understandings as the students?

As a mathematics teacher, my sensitivity to this situation was intensified by the fact that the question I resolved had been open and posed by a student. I had hoped for the students' understandings to play a bigger role in reaching a (possibly different) resolution. Yet I had been unable to contain myself and wait until I might support that role. As a teacher educator, I recognized that my disclosure was more honest than "holding back." Yet I remained concerned about the fact that these students were *teachers* and worried how telling them I had "figured out an answer" would influence responses to their own students' open questions.

The following week, the "work at home" strategy proved less successful, since my classroom questioning generated nothing new. Eager to keep the topic alive, I guided the teachers through my findings. Despite my use of a question-answer format, I knew I was guiding them toward *my* understandings. Had my admission of a finding influenced their pursuit of knowledge? I couldn't be sure. Three weeks later, while we worked with a linear combination, a teacher, Abby, noted: "If I can subtract, I can get that algorithm to work." Confused, I asked her to elaborate. Abby struggled to formulate and state her conjecture—an important aspect of mathematical reasoning (see NCTM 2000, p. 57). Eventually, she explained that if "allowed to subtract," she thought she could write every natural number using powers of 3 ($5 = 9 - 3 - 1$ and $11 = 9 + 3 - 1$).

Abby's classmates eagerly took up her conjecture, testing whether natural numbers can be written as sums and differences of powers of 3. I remember being skeptical, but I also tested numbers with one group of teachers while others worked together. For fifteen minutes we investigated combinations, all of which worked! We reached another juncture and promised to think about Abby's conjecture for next time. For me, the feeling that the teachers were back on the path toward constructing their own understandings had returned.

At our next meeting, the teachers asked for more time to think about Abby's conjecture. Then one evening, as I wrote "$5 = 1 \bullet 3^1 + 2 \bullet 3^0$" and "$5 = 9 - 3 - 1 = 3^2 - 3^1 - 3^0$," something hit me. Abby's subtraction introduces a third coefficient of -1 into the linear combination. That is, $5 = 3^2 - 3^1 - 3^0 = 1 \bullet 3^2 - 1 \bullet 3^1 - 1 \bullet 3^0$. Abby's conjecture suggests that 0, 1, and 2 can be replaced by 0, 1, and -1 in linear expansions using powers of 3! Later that evening, I explained this to my husband (who is a mathematician), illustrating with $21 = 1 \bullet 3^1 + 2 \bullet 3^2 = 1 \bullet 3^1 - 1 \bullet 3^2 + 1 \bullet 3^3$. As we exchanged ideas, my husband subtracted "$1 \bullet 3^1$" from both sides of $1 \bullet 3^1 + 2 \bullet 3^2 = 1 \bullet 3^1 - 1 \bullet 3^2 + 1 \bullet 3^3$, yielding $2 \bullet 3^2 = -1 \bullet 3^2 + 1 \bullet 3^3$. He conjectured that $2 \bullet 3^n = -1 \bullet 3^n + 1 \bullet 3^{n+1}$, which I verified by adding $1 \bullet 3^n$ to both sides. With this equation, a coefficient of 2 can be replaced by -1 and 1 in linear combinations involving powers of 3, that is,

$$\begin{aligned} 5 &= 1 \bullet 3^1 + 2 \bullet 3^0 \\ &= 1 \bullet 3^1 - 1 \bullet 3^0 + 1 \bullet 3^1 \\ &= 2 \bullet 3^1 - 1 \bullet 3^0 \end{aligned}$$

$$= -1 \cdot 3^1 + 1 \cdot 3^2 - 1 \cdot 3^0$$
$$= 1 \cdot 3^2 - 1 \cdot 3^1 - 1 \cdot 3^0.$$

An Analysis of Dilemmas and Remaining Classes

For me, the experiences above demonstrated an aspect of constructivism called *equilibration* (Piaget 1977) that began to help me understand why I kept finding myself torn between enabling the teachers to create their own (possibly different) understandings of our problem situations or sharing (what felt like) mine. Equilibration describes an interplay in the relationship between the parts and the whole of one's knowledge: as learning unfolds, new, uncertain parts find their place in the whole and the whole reorders itself in the face of these new parts. In the passage above, the unknown of "subtraction" transformed into a coefficient of -1 and found its place in connecting 0, 1, and 2 to 0, 1, and -1. Equating $1 \cdot 3^1 + 2 \cdot 3^2$ with $1 \cdot 3^1 - 1 \cdot 3^2 + 1 \cdot 3^3$ found its place in a general equation that verified this connection.

When the feeling associated with this learning is one of "things coming together," of challenges and capabilities being in balance, an experience called *flow* can be associated with these creative endeavors (Csikszentmihalyi 1975). In this optimal state, the learning process is experienced "as a unified flowing from one moment to the next" (Csikszentmihalyi 1975, p. 36). One characteristic of being in a state of flow includes self-forgetfulness to the point where a subject is unaware of "outside reality" (p. 42). For the teacher educator, this implies a possibility of becoming immersed in, and resolving, a teacher-posed problem regardless of the educator's desire to enable her teachers to reach their own solutions. For me, the influence of this learning was never certain. When I "burst out" my discovery, I questioned whether it would have been better to wait. Did my learning compromise my classroom questioning? Did it influence how the teachers tackled the tripling question at home? What role did my hints, or just the fact that I gave hints, play? For answers to these questions, I could only speculate.

Our next meeting brought us to week 11 (of 14) of the semester. I asked the teachers whether they "had come up with anything" on Abby's conjecture. Did they think it was true? False? How could we investigate it? I purposely asked open questions, concerned that my discovery would unduly influence a more tailored question. No discussion ensued. In reflecting back, I wondered whether enabling the teachers to brainstorm in groups might have generated discussion. At the time, I contemplated the constructivist tenet that no one can ever really *know* anything inside another's perception (Jaworski 1994). Had I imagined our last lesson's momentum? The teachers' notebooks I used to write this article suggested otherwise, containing numerous examples of numbers written as linear combinations of powers of 3 using subtraction. I worried about time and the possibility that *no* understandings would be shared concerning Abby's conjecture.

I decided to ask, "How can subtraction be written using multiplication?" I also asked the teachers to pick numbers to expand using subtraction but to rewrite the subtraction as "multiplication by −1." What did this imply about possibilities for coefficients? Having discovered 0, 1, and −1, the teachers responded with a homework task: test whether 1 to 100 can be written using powers of 3 if subtraction is allowed. I added that the teachers should equate representations using traditional, and Abby's, coefficients. At the end of our next class, four teachers (Debbie, Paula, Pam, and Margot) came to me with calculations. These included the 1-to-100 calculations and attempts to equate different representations. They said they were stuck and showed me calculations like Margot's for the numbers 54 and 25 (see fig. 20.2). In response, I highlighted all equations of the form "$2 \cdot 3^n = 1 \cdot 3^{n+1} - 1 \cdot 3^n$" on their papers (like Margot's 54) and asked, "What do these have in common?" The teachers not only extrapolated the general formula but also cited our Algebra and Algebraic Thinking class, observing that this "pattern was not a proof" and that a mathematical proof was still needed to verify the formula.

$$2 \cdot 3^3 = 1 \cdot 3^4 - 1 \cdot 3^3$$

$$25 = 27 - 3 + 1 = 1 \cdot 3^3 - 1 \cdot 3^1 + \boxed{1 \cdot 3^0}$$

$$25 = 2 \cdot 3^2 + 2 \cdot 3^1 + \boxed{1 \cdot 3^0}$$

Fig. 20.2. Margot's calculations for the numbers 54 and 25

Our Last Class Together: A Joint Understanding

At this juncture, the reader may be wondering about Margot's calculations for 25. After simplifying, the resulting equation is not of the form "$2 \cdot 3^n = 1 \cdot 3^{n+1} - 1 \cdot 3^n$." My own work contained similar calculations, which I opted to ignore, assuming they could be made to fit the pattern. At the end of our last class, one teacher asked whether we could just "ignore the other (such) equations," but the question was not pursued. Perhaps their "teacher's" steering them toward a formula to explain a long, unresolved problem overshadowed this question.

Our last (videotaped) class was devoted to Abby's conjecture. Working in groups, some teachers shared patterns discovered in the 1-to-100 exercise. Others equated representations using traditional, and Abby's, coefficients. For example, $19 = 2 \cdot 3^2 + 1 \cdot 3^0$ (with traditional coefficients) and $1 \cdot 3^3 - 1 \cdot 3^2 + 1 \cdot 3^0$ (with Abby's coefficients). Equating yields $2 \cdot 3^2 = 1 \cdot 3^3 - 1 \cdot 3^2$. As we moved into whole-class mode, we wrote several such connections on the board, as well as the emergent pattern. We then investigated how $2 \cdot 3^n = 1 \cdot 3^{n+1} - 1 \cdot 3^n$ could be used to transform a linear combination with traditional coefficients into one using Abby's coefficients:

$16 = 1 \cdot 3^2 + 2 \cdot 3^1 + 1 \cdot 3^0$	traditional coefficients
$= 1 \cdot 3^2 + (1 \cdot 3^2 - 1 \cdot 3^1) + 1 \cdot 3^0$	letting $n = 1$ in our formula and substituting
$= 2 \cdot 3^2 - 1 \cdot 3^1 + 1 \cdot 3^0$	combining "like terms"
$= (1 \cdot 3^3 - 1 \cdot 3^2) - 1 \cdot 3^1 + 1 \cdot 3^0$	letting $n = 2$ in our formula and substituting
$= 1 \cdot 3^3 - 1 \cdot 3^2 - 1 \cdot 3^1 + 1 \cdot 3^0$	Abby's coefficients

Some teachers noted how our formula "pushes the 2-coefficient" into higher powers of 3 until it is eliminated. And Paula proudly shared a proof of the formula she had obtained, with a little help from us: $1 \cdot 3^{n+1} - 1 \cdot 3^n = 3^n(3^1 - 1) = 3^n \cdot 2 = 2 \cdot 3^n$.

But the question about calculations that do not "fit the pattern" was revived. A teacher, Kara, conveyed that she had been looking for something to "fit all the calculations" and wondered about those that didn't look like "$2 \cdot 3^n = 1 \cdot 3^{n+1} - 1 \cdot 3^n$." Her observation challenged, and engaged, us in the kind of reexamination that is so essential to doing mathematics (see Lampert 1990). She asked us to write representations for 48 and 5 on the board. After equating representations for 48 and for 5 and canceling like terms in the representations for 48, we obtained the following:

$$1 \cdot 3^3 + 2 \cdot 3^2 = 1 \cdot 3^4 - 1 \cdot 3^3 - 1 \cdot 3^2$$

and

$$1 \cdot 3^1 + 2 \cdot 3^0 = 1 \cdot 3^2 - 1 \cdot 3^1 - 1 \cdot 3^0) \quad (*)$$

Kara asked whether these could be expressed in the form $2 \cdot 3^n = 1 \cdot 3^{n+1} - 1 \cdot 3^n$.

During the sixteen-minute, whole-class discussion devoted to Kara's problem, the teachers and I remained *genuinely* confused over how to convert the equations in (*) into the form "$2 \cdot 3^n = 1 \cdot 3^{n+1} - 1 \cdot 3^n$." As the teachers made suggestions, I transcribed ideas. For example, Dave suggested adding "$1 \cdot 3^1$" to both sides of $1 \cdot 3^1 + 2 \cdot 3^0 = 1 \cdot 3^2 - 1 \cdot 3^1 - 1 \cdot 3^0$ ($2 \cdot 3^1 + 2 \cdot 3^0 = 1 \cdot 3^2 - 1 \cdot 3^0$). As we reflected, he turned the discussion toward ensuring that both sides of the equation were equal. Kara came to the board, added "$1 \cdot 3^0$" to both sides of Dave's equation, $2 \cdot 3^1 + 3 \cdot 3^0 = 1 \cdot 3^2$, and resolved that this did not help. Although these suggestions led to dead ends, the teachers were articulating their insights, publicly admitting their limitations, and responding with revisions in their thinking (another important part of the process of doing mathematics) (cf. Lampert 1990).

The discussion continued thusly, until Kara asked us to add "$1 \cdot 3^0$" to both sides of $1 \cdot 3^1 + 2 \cdot 3^0 = 1 \cdot 3^2 - 1 \cdot 3^1 - 1 \cdot 3^0$:

(I wrote) $1 \cdot 3^1 + 3 \cdot 3^0 = 1 \cdot 3^2 - 1 \cdot 3^1$

Kara: Now, now, let's just break this down. So wait, I can rewrite this, the left side, I can rewrite, 3 to the first plus 3 to the first times 3 to the 0. I'm just putting the exponents, getting rid of the coefficients. Cause I'm putting, everything is in terms of, has a coefficient of 1 now. Plus 3 to the first, plus 3 to the first times 3 to the 0. Equals 3 to the first times 3 to the first.

$3^1 + 3^1 \cdot 3^0$

$3^1 \cdot 3^1$

Teacher: Cause that's 3 squared?

Kara: Minus 3 to the first. Okay and then, what's gonna end up happening. . . Actually, I didn't have to do, the part I'm writing. I'm sorry. What you're gonna be left with on the left side is 3 to the first plus 3 to the first, which is equal to 2 times 3 to the first (Jointly)

$3^1 \cdot 3^1 - 3^1$

$3^1 + 3^1$

$2 \cdot 3^1$

Teacher: 2 times 3 to the first (Jointly)

Kara: And then you, back it, back into

$2 \cdot 3^1 = 1 \cdot 3^2 - 1 \cdot 3^1$

Teacher: And then this side becomes, ah! Do you see it?!

A semester's work had generated a shared learning experience between my teachers and myself!

Another Look at Learning from the Unknown in Teacher Education

It is useful to consider how the unknown influenced my teachers' and my own learning of mathematics. We all learned about new coefficients in linear combinations involving powers of 3—a finding with relevance to the domain of mathematics and to our experiences within that domain. Mathematically, our finding raises inquiry-level issues about characteristics of base-three coefficients. For example, Abby's positive and negative coefficients (±1) permit expansions of *all* integers, not just nonnegative ones (see Michalek 2001). On my own, I discovered an alternative method for writing numbers as linear combinations of powers of 3, which, interestingly, connected to work my teachers and I did with series (see Appendix). Also, connections were made to our algebra class and to working with patterns and exponential expressions. With good problems characterized as those for which "the solution method is not known in advance," (NCTM 2000, p. 52), "alternative representations and solution strategies" exist (NCTM 1991, p. 58), and "mathematical ideas interconnect and build on one another" (NCTM 2000, p. 64), there is no doubt about the richness of these teachers' questions and findings.

The unknown also influenced our work with the *process* of doing mathematics. We *all* experienced important facets of problem solving from conjecture to theorem to proof. The teachers' use of patterns, their observations (formulas "pushing" coefficients), and calculations with exponential expressions all acquired a more playful and natural quality. Perhaps the genuine, versus manufactured, need for this material facilitated its finding a place in the teachers' learning (as Piaget's equilibration would predict). "Habits of persistence and curiosity, and confidence in unfamiliar situations" (NCTM 2000, p. 52) also were developed in our lessons. But risk and vulnerability, also, can be associated with exploring open questions (Lampert 1990). These teachers, fourteen of whom had relayed anxiety over studying mathematics a semester earlier, were taking the very risks needed to push their questions from the unknown into the resolved. They were publicly questioning one another and their "teacher," articulating tangents (successful and unsuccessful ones), opening up ideas to revision, and necessitating justifications. The famous mathematician Paul Halmos once said, "The best way to learn is to do—to ask and to do" (Halmos 1974). We could not have learned what we did without the teachers' asking and doing what they did.

In my struggles to foster constructivist learning, this experience gave me a new perspective for analyzing my work with teachers. As a teacher educator, the constant awareness that teachers are learning to teach by watching me introduced an added burden to my struggles. I doubted the role of my learning in my classroom questioning and investigations, worrying that my teachers would learn, "It is the teacher's job to resolve a student's question." But "just as models of learning treat the learner as someone who is attempting to make sense of a teaching encounter, so too, for consistency, should a teacher be treated as someone who is attempting to make sense of that same encounter" (Jaworski 1994, p. 33). This suggests that the teacher, also, is a learner in classroom problem solving, and it enabled me to recognize how our learning was supportive of constructivist principles. For example, we did investigate "every instance where the students deviate(d) from the teacher's expected path" (Jaworksi 1994, p. 33). Also, our roles were genuinely responsive to our confusion and included questioning, explaining, challenging, revising, listening, working alone and together, being confused and in the know. Today, I would even feel comfortable sharing comparable situations with future teachers and eliciting their viewpoints. Accordingly, I learned to place my conflicts constructively within my learning experiences, to appreciate the natural course of learning that arises between my teachers and myself, and to acknowledge that the integrity of our contributions was an important component in advancing our work.

What had the teachers learned? In a videotaped discussion after our last class, I asked the teachers to describe the influence of our experiences on their decisions to pursue students' open questions. Initially, they said they would pursue such discussions, characterizing our classes as "exciting," "teachable moment," "discovery," and "important." Paula told me this was the first time she felt like she was "doing math," and Laura noted the importance of learning that "not everything in the world is figured out yet." As the discussion progressed, however, Abby noted real-life classroom concerns:

> When students are able to develop an idea on their own, they have a tendency to remember it, more readily, and easily. We have a tendency to tell them, "This is what it is" and without an explanation, they don't have a tendency, that that's the thing they're gonna recall. So I think this is an important process. The problem with the process is the time factor and that is a problem particularly on levels that we have state mandatory testing, that we have to cover so many topics in a certain amount of time before a certain day.

Abby's remark reminded us of the complexities in relating professional development to professional practice. In the spirit of "not giving answers," I asked what we could learn from our investigations to address the *time issue*. Some teachers noted that open questions could be explored after class or at home without taking up class time. I mentioned that using shorter portions of consecutive classes for open questions (as I had) also could save time. We discussed how these approaches could more equitably allocate time for both mandatory and unexpected mathematical ideas.

In writing this article, I had the idea that keeping a journal of students' open questions (as I had) might help teachers address the time issue. In such a journal, teachers could document the pros and cons of pursuing such questions (in relation to time), as well as the teacher's and the students' attitudes and feelings. For example, if teachers discover connections between open questions and mandatory mathematics (as I discovered connections to the algebra we studied in my teachers' questions), this could alleviate anxiety about the question of time. Other such discoveries could be made and shared. And because a journal records concrete (versus speculative) experiences, it can help teachers balance the predictable versus the unexpected aspects of classroom problem solving with actual examples.

Concluding Remarks

Influencing professional development is only a beginning. Hunting (2000) reminds us of the need for "models, examples and interpretations" that illustrate the "interactive nature of the teacher's goals and actions" (p. 238) during attempts to foster constructivist learning. It is hoped that communicating one teacher educator's goals and actions, and her teachers' experiences and insights, provides a point of discussion for educators at all levels wishing to explore the unknown with their students.

REFERENCES

Cobb, Paul. "The Tension between Theories of Learning and Instruction." *Educational Psychologist* 23 (spring 1988): 87–103.

Confrey, Jere. "What Constructivism Implies for Teaching." In *Constructivist Views on the Teaching and Learning of Mathematics, Journal for Research in Mathematics Education* Monograph No. 4, edited by Robert B. Davis, Carolyn Maher, and Nel Noddings, pp. 107–22. Reston, Va.: National Council of Teachers of Mathematics, 1990.

Csikszentmihalyi, Mihaly. *Beyond Boredom and Anxiety.* San Francisco: Jossey-Bass Publishers, 1985.

Halmos, Paul. "The Problem of Learning to Teach." Paper presented at the Joint Annual Meeting of the American Mathematical Society and the Mathematical Association of America, San Francisco, 1974.

Hunting, Robert P. "Themes and Issues in Mathematics Teacher Education." In *Radical Constructivism in Action: Building on the Pioneering Work of Ernst von Glasersfeld,* Vol. 15, Studies in Mathematics Education Series, edited by Leslie P. Steffe and Patrick W. Thompson, pp. 231–47. London: Routledge Falmer, 2000.

Jaworski, Barbara. *Investigating Mathematics Teaching: A Constructivist Enquiry.* London: Falmer Press, 1994.

Lampert, Magdalene. "When the Problem Is Not the Question and the Solution Is Not the Answer: Mathematical Knowing and Teaching." *American Educational Research Journal* 27 (spring 1990): 29–63.

Lerman, Stephen. "Constructivism, Mathematics, and Mathematics Education." *Educational Studies in Mathematics* 20 (May 1989): 211–23.

Michalek, Gary. "Writing Numbers in Base 3, the Hard Way." *Mathematics Magazine* 74 (February 2001): 51–53.

National Council of Teachers of Mathematics (NCTM). *Professional Standards for Teaching Mathematics.* Reston, Va.: NCTM, 1991.

———. *Principles and Standards for School Mathematics*. Reston, Va.: NCTM, 2000.

Piaget, Jean. "Problems of Equilibration." In *Topics in Cognitive Development*, Vol. 1, *Equilibration Theory, Research, and Application*, edited by Marilyn H. Appel and Lois S. Goldberg, pp. 3–13. New York: Plenum Press, 1977.

Appendix

In this appendix, I illustrate a method for finding coefficients in linear combinations of powers of 3 when the coefficients (to select from) are given. For the number 5 and coefficients 0, 1, and –1, write:

$$\begin{aligned} 5 &= c_0 + c_1 3^1 + c_2 3^2 + c_3 3^3 + c_4 3^4 + \ldots \\ &= c_0 + 3(c_1 + c_2 3^1 + c_3 3^2 + c_4 3^3 + \ldots) \\ &= c_0 + 3k \end{aligned}$$

Testing the three possibilities for c_0 yields a solution for k: with $c_0 = -1$, $k = 2$. The process can now be repeated for subsequent values of k (in this example, for $k = c_1 + c_2 3^1 + c_3 3^2 + c_4 3^3 + \ldots$) until all coefficients are obtained.

21

Under the Microscope: Looking Closely at One's Own Teaching and Learning

Sue Tinsley Mau

Two of my colleagues are currently researching what students do to make sense in my mathematics classroom. They videotape class sessions daily and copy students' work, including homework, journals, tests, and quizzes. They analyze the more than ninety hours of videotapes using a Reggio Emilio process (Forman and Fyfe 1998; Helm, Beneke, and Steinheimer 1998), reporting their findings back to the students and asking students for their thoughts. We, my colleagues and I, also meet routinely over lunch to discuss issues that either they or I bring to the table. Often the issues that I raise have arisen from something students have done or said in class that I could not quite reconcile.

When we first discussed this research project, I thought the researchers' work would be risk free for me. Their focus was to be the students—what they did to make sense of and learn mathematics; I was not to be the focus. Additionally, I have spent considerable effort in becoming the reform teacher envisioned by many current publications (Lester and Mau 1993; Lester et al. 1994; National Council of Teachers of Mathematics [NCTM] 1989, 2000; Sherin 2002; Steffe and D'Ambrosio 1995). My students work in groups of four to solve problems, and then I help them formulate their thinking into commonly accepted mathematical language and symbols. I attempt to use what Sherin (2002) calls an—

> adaptive style of teaching in which teachers attend to the ideas that students raise in class. Consider, for instance, a typical lesson in which students work on a problem and then explain their solution strategies to the class. The students' solutions cannot be predicted entirely in advance; thus the teacher must listen to the ideas that the students raise in class and then use that information to decide how to proceed. (P. 122)

Although I know that I cannot predict what will happen in class, I am confident that if students share their thinking and if I judiciously use their emerging understandings as foundations for learning, good things will happen with regard to students' learning. The course syllabus states that my goals for my students are their development of "an adult understanding of the elementary school curriculum" and the development of "intellectual mathematical autonomy." Those are lofty goals, consistent with *Principles and Standards for School Mathematics* (NCTM 2000). Why on earth should I feel any risk in this endeavor?

Yet I find myself worrying about "looking lost," or what the students might call being "clueless." That is, I worry that when a student offers her or his thinking, I might not know what to do with it and that my temporary, I hope, intellectual surprise may lead students to think I do not know what I am doing. I well remember the days when I called that feeling "getting comfortable with being uncomfortable." That worry, that discomfort, often accompanies a student's unique question or solution regarding fractions. In this article, I describe three learning experiences—two for my students and one for me.

The Students' Learning—the First Problem

We began division of fractions by saying that division merely asks the question "How many of these in this?" When students see 3/4 ÷ 1/2, the real question is "How many one-halves are in three-fourths?" Students easily answer one and one-half. They can see the answer, or I thought they could. Once they had offered verbal solutions, I asked them to draw a picture that would show how they thought about the problem. They were to use the picture to determine the answer, not draw the picture of the answer after using the algorithm. Most groups of students drew a whole, split it into fourths, colored three of them, and then started marking off the one-half units. I wrote on the board, "There are one and one-half 1/2 in 3/4."

One group, Amber's group, did something different (see fig. 21.1). They made two bars of four color tiles and then marked three tiles in each bar to indicate 3/4. Then, because of the invert-and-multiply rule, they put those two bars together, that is, they *added* those two, to get their answer of 1 1/2. They told me that this tactic made sense because 3/4 times 2 is 1 1/2. They asked whether they were right. I was my normal noncommittal self and did not confirm or deny their answer. I did go home and think about how to push their thinking—what problem could I use that would cause them to think about the magnitude of the divisor, dividend, and quotient?

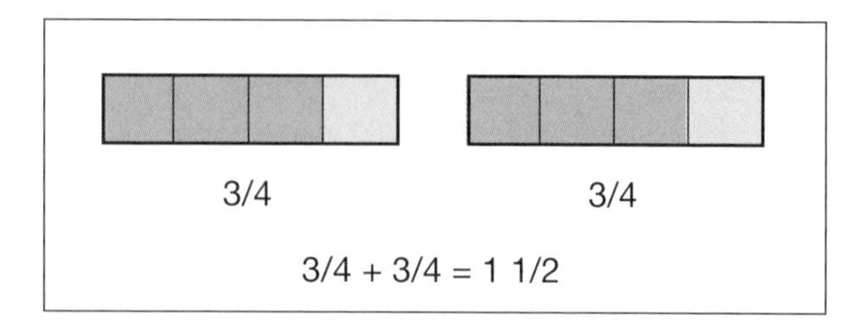

Fig. 21.1. One group's pictorial solution to the problem 3/4 ÷ 1/2

The Students' Learning—the Second Problem

As we continued to work on division of fractions, we established the ground rule that the students could use a linear, set, or region model but could not simply use the algorithm for dividing fractions. I posed the problem 3/4 ÷ 2/3. Before they were allowed to begin the work, I asked them to estimate whether the answer would be greater than or less than 1. Most of the students agreed that it would be greater than 1 because 2/3 is less than 3/4. With that question answered, I told them to begin their group work, and I circulated among the groups to observe what strategies they selected.

Not surprisingly, most groups arrived at a correct solution for correct reasons. One group, though, got an answer of 4 1/3. They listened to others' solutions but did not offer their solution. When we were almost finished, one of the researchers asked for this particular group's solution to be considered. They had produced the picture seen in figure 21.2.

One	2/3	Four 2/3			
Two	2/3		Three	2/3	1/3

Fig. 21.2. One group's representation for the problem 3/4 ÷ 2/3

They had drawn a rectangle and appropriate lines to indicate cutting the rectangle into fourths. They discarded one of the fourths, then cut each of the three remaining fourths into three equal pieces. As they did their division, they colored two of their "thirds" and counted how many times they could make 2/3 in the 3/4. They were able to make four "two-thirds" pieces with one of their "thirds" left over; hence their answer was 4 1/3. They wanted to know why they got a different answer from the others.

An observer can easily see that the group changed the unit when they divided each fourth into thirds, and I did ask them about that aspect of their approach. They were able to see that their "thirds of each fourth" were really twelfths of the whole and could see that they needed to count the pieces differently, but they were still a bit unsatisfied with this explanation. Although the class session ended as we were discussing their solution, I thought they had come to some resolution. I had failed to attend to their 1/3 remainder, however.

During the next class session, I asked whether students had questions about the previous day's work. We reexamined the 3/4 ÷ 2/3 problem using a set model. Students chose to use twelve color tiles for their whole and explained why they chose twelve tiles rather than fifteen, sixteen, or some other number of tiles. Basically, they were finding common denominators and then comparing numerators. Several groups had found it beneficial to make two models, both 3/4 and 2/3, and literally to place the 2/3 on top of the 3/4 to see how many times they could fit 2/3 into 3/4. Most of the class decided it was 1 1/8 times, but the group with the 4 1/3 solution from the previous day could not understand the 1/8 portion of the answer. They could see where the 1 came from, but the remainder of 1/8 was not visible to them.

One group member, Phyllis, asked whether the 1/8 came from comparing the one tile left in the 3/4 with the eight tiles that were not covered by those tiles representing 2/3 and 3/4 (see fig. 21.3, the tiles labeled *r*, *p*, and *q*). Perhaps important to note is the fact that she was using the one tile left from the 3/4 (labeled *p* in fig. 21.3) in both her numerator and denominator; that is, when she counted the eight tiles, she counted the leftover one shaded tile from 3/4 (tile *p*), the three unshaded tiles in 3/4 (tiles *q*), and the four unshaded tiles from the 2/3 (tiles *r*) as part of the eight tiles she was denoting for the denominator.

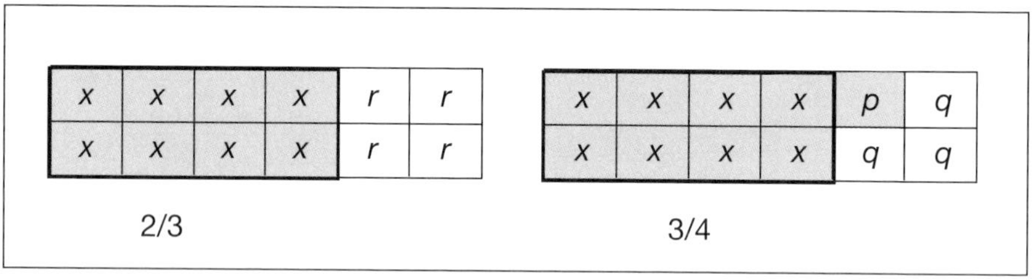

x	x	x	x	r	r
x	x	x	x	r	r

2/3

x	x	x	x	p	q
x	x	x	x	q	q

3/4

Fig. 21.3. The reexamination of the 3/4 ÷ 2/3 problem using a set model

We decided that if her strategy were viable, it should work with all numbers, not just this particular problem. We tried a couple of other problems, and it did not work, so Phyllis agreed that it was just lucky in this particular problem. We also, as a class, tried to help Phyllis see what the others saw—that the leftover tile must be compared with the divisor, not with the unshaded part of the drawing.

In this class and in successive classes, I often mention to students the need for teachers to understand all the subtle nuances of ideas. In this situation, the idea is division and we are merely doing division with nonintegral numbers. To answer these questions arising in fraction division, we need to think specifically about the meaning of division and what that looks like, literally, in a picture. At this point, I began to learn, to renegotiate the meaning of my own mathematics.

My Learning

I have said that I was worried about "looking lost" and that those moments often came as a result of the students' questions about fractions. That feeling of being clueless had been nagging at the fringes of my consciousness for weeks, perhaps months. I seemed to be having a hard time "getting comfortable with being uncomfortable." Those words have taken on new meaning for me.

When Amber asked about the 1 1/2 and showed her model to me, I knew right away what she had done. What to ask Amber was a little less clear to me. When Phyllis and her group said 4 1/3, I knew what they had done and knew what to ask to push their thinking. When Phyllis asked about the 1/8, I really had no conclusive response. I bought precious minutes by commenting that an algorithm that is true should work every time. Even as those words came out of my mouth, I was quickly trying to find numbers that would serve as a counterexample. She pressed for an answer, and I had to say that I had never considered this problem before; I would have to think about it before I could give a firm response. In the meantime, others in the class were making up their own problems and trying them using Phyllis's strategy. Thankfully, we quickly arrived at counterexamples.

Ball (1997) urges us to know our students as individuals, to be sensitive to their emerging understandings, and to listen to their thinking. She also describes the difficulty of using that approach when students' ideas are in formative stages (p. 735):

> Listening across chasms of age, culture, and class, teachers face a problem common to most forms of cross-cultural communication. The problem is one of trying to understand what students mean with their words, pictures, gestures, and tone. Students often do not represent their thinking in ways that match adult forms.

Although my students are prospective teachers, they too use words, pictures, and gestures that do not match more "adult" forms of mathematical communication. On many days I find their communication difficult to follow. That eventuality does not allow me to ignore it or to demand that their words match my expectations. I must know my students.

Davis (1996, 1997) suggests that teachers listen to students to renegotiate their own teacher understanding of mathematics. He describes three kinds of listening: evaluative listening to pronounce the correctness of solutions, interpretive listening to build a model of students' thinking, and hermeneutic listening to rethink one's own understanding of mathematics. I was involved in teaching fractions—part of the elementary school curriculum that I thought was so familiar, so known to me, that I had nothing more to learn. So what new connections, what renegotiations, have I made in my own understanding of fractions as a result of these two problems and listening to my students' mathematics talk?

Rethinking My Mathematical Content Knowledge

In rethinking my mathematical content knowledge, I have become exceptionally aware of the need to go back to the most basic concept and to use that concept as the action on objects. I have given more thought to the issue of comparing magnitudes, the ideas of division, and the physically visible "appearance" of the answer. For example, when dividing integers, the dividend does not change magnitude; it is redistributed. In 48 ÷ 7, we get six groups of seven with six left over, or 6 6/7. The dividend does not change size; it changes "shape." The same thing happens with fractions. When students using set models found common denominators and compared numerators by covering the dividend numerator with the divisor numerator, the dividend did not change size; it changed "shape." The color tiles that represented the fractions were reorganized into the answer. Further, that procedure was mathematically acceptable *because the quantities being compared were composed of individual pieces of the same size.* That is, each piece of each of the fractions was the same size because the students had found common denominators. Doing so is crucial to division's making any sense at all.

From a mathematical stance, I also wonder about the notions of equivalent fractions. When students use a region or linear model for equivalent fractions and show those equivalences by cutting the pieces into successively smaller bits, the size, the *magnitude,* of the whole does not change. When students try to form equivalent fractions with a set model, invariably they alter the number of items in the set. If they want to show that 1/2 is the same as 2/4, they will double the number of pieces in the numerator and the denominator. I keep wondering what distinctions I should be making as my students, whose mathematical knowledge is fragile and in development, make judgments that seem inconsistent to me. I continually question the mathematical meaning of equivalent fractions—do *I* understand it?

I have given new thought to the meaning of the remainder and to students' thinking in determining remainders. In division with integers, students very naturally compare remainders with divisors to determine the fractional part of their answers. Why are students' actions so different in division with fractions? Phyllis's thinking of the "third" left over as a third instead of comparing it with the divisor is not an uncommon perception (Perlwitz 2005). I have seen it occur over and over again when students are required to use models rather than algorithms. I cannot help but wonder about our approach to teaching arithmetic involving fractions. Would we be better off to force students at an early age to build models and to show us, literally, what things they are comparing? Would a set model that uses comparison of two fractions with common denominators force our students to consider the remainders in fraction division? Should I encourage my students, future elementary school teachers, to do just that?

Rethinking My Pedagogy

How have all these experiences influenced my thinking about my teaching? What a thought! I have spent many hours considering the logic of Davis's (1996, 1997) admonition that we listen to students' ideas with a disposition of willingness to rethink our own understanding of mathematics. I have given much thought to Fleener, Carter, and Reeder's (2004) likening of hermeneutic listening to the willingness to play with an idea, and I often wonder how a teacher plays with an idea when he or she has a clear solution path in mind. What is the value of "ignorance" when teaching? Is playing with an idea possible when we already know a solution path and an answer? Or do we end up listening interpretively, so that we might move students closer to our own understanding of mathematics? And, when we live in this world of pedagogical logic problems, how do we get comfortable with being uncomfortable? I do not have the answers right now, and perhaps I never will. However, I am increasingly certain that if I do not think about these issues, I will become the very teacher I do not wish to emulate—the teacher who does not listen at all.

I have given thought to the cognitive development of my students. Vygotsky and Piaget are "alive and well" in my classroom. I see emerging understanding in Amber and in Phyllis, and I see them push themselves and each other to make better sense of mathematics. Each day their questions indicate that they have incorporated the previous day's thinking to the best of their abilities, and now a new question raises its head. Each day, they are a bit better able to articulate their thinking through their words and their pictures. I am amazed at, and curious about, their thinking.

I am also more acutely aware of the notion of saliency for students. Do they recognize the same constructs as important that I recognize? What do they attend to as we develop meaning for arithmetic with fractions, or for any other mathematical idea, for that matter? Do they consciously realize the mathematical concept that we are using, or does it become more of an algorithm for them? Clearly, Phyllis and her group attended to the two-of-three meaning of 2/3; they attended less to the size of the unit. Does a particular question exist that would cause students to attend to the entire situation, not just a part of it?

I think I now more consciously look for indicators of potential confusion in students' mathematics talk. I find myself asking specific questions to try to tease out their thinking as we do this work. In doing so, I hope I am demonstrating a willingness to listen to students' thinking and to play with their ideas. I hope I am demonstrating perseverance and a belief that we will make sense together.

In work with fractions, I find myself attending to remainders in very specific ways. I ask students to explain why their remainder is 1/8 instead of 1/3. I find myself asking students (and some of my mathematics colleagues) to justify whether a set model indicates equivalent fractions (and the mathematics colleagues are not all in agreement about this representation!). I find myself pushing the mathematics talk in the classroom as my best attempt to tease out *their* thinking and generate questions that cause them to reorganize their thinking. Many of those questions have come as a result of my listening to them and playing with their thinking. Sense making is no small task.

Potentially most important, I cannot imagine that all this would have been as important to me had I not been aware of "looking lost" or of feeling clueless. Having the video camera running, even if it is to capture students' work and not my teaching, is a bit daunting. Most days, I am unaware of it until I look up and see that red light glowing in front of me. On the days when I do not have a ready question or response, I worry about looking like I do not know what to do—I worry that I have been clueless.

My colleagues are incredibly trustworthy; they would never use my words *looking lost* or *clueless.* We talk about this trustworthiness as researchers—we have no idea how important it is until we are under that microscope. However, they are trustworthy for more than just their respect and professional stance toward me in their dissemination of their learning. They are trustworthy for rethinking mathematics and mathematics teaching with me and for supporting me as I engage in this pursuit. I want to be clear about this point. They do not patiently listen while I come to their understanding. Rather, they wrestle with these ideas and questions with me; they have no better answers than I do when we start our thinking. Sometimes they offer insight; sometimes I offer insight. Our work is a give-and-take of equals trying to make sense of what appears, at first glance, to be senseless.

I am acutely aware of what they hear, and what I am missing, in my classroom. Their research and their continued discussions of what the students' thinking means have led me to, have supported me in, thinking about the mathematics and about the students' learning of mathematics. They often come back with their own investigations based on either what I have said or what a student has questioned. This continued and sustained rethinking has been both exhilarating and exhausting—it has been a joy that I cannot completely articulate.

Colleagues often wonder how we can build a model of teacher learning—what that model looks like and what should be included so that we capture all the complexities of such learning. I suspect that to do so is impossible. My growth has come about in small steps over years of focused conversation and reading. I remember a graduate school peer telling me that people look closely at themselves until the introspection becomes too painful, and then they back away until they have the strength to look again. This process, she said, is how growth occurs. Would the same be true for teacher learning? Do we look closely until doing so is too painful, and then do we back away, waiting for the strength to look again? I suspect we do. Making progress takes years of small steps and strength from trustworthy others.

Yet another aspect of this lengthy growth is the cost of supporting teachers while they engage in this learning. I had the opportunity to be the "teacher" in this research. The structure of our work at the university often leaves large gaps of time, which we use for professional talk. Teachers of kindergarten through grade 12 almost never have this luxury unless someone brings large amounts of money to buy out teachers' time. Public school corporations generally cannot afford what we did.

So what do I believe are the crucial components of teacher learning, of my learning? First, I needed a disposition of learning. I read current research and consider current philosophical pedagogical questions as I teach. I talk with my peers about the logic of the philosophy and about how that pedagogical stance plays out in practice. We search together for understanding, for sense making. Second, I needed a disposition of being a reflective practitioner. Crucial to that reflective stance is spending time thinking about my students' mathematics talk and about how to use it in my classroom. After that, acting on my understandings is also crucial. Merely reflecting with no change in practice is not enough. I must use my current understandings in my daily teaching if I want to call myself a reflective practitioner. And I must reexamine those new understandings to yield yet more understanding. The process is a constant cycle of getting comfortable with being uncomfortable. It is a constant cycle of feeling lost and finding my way home.

Conclusion

The theoretical ideas from Piaget and Vygotsky are often hard to practice. Try as I might, I often find it difficult to formulate the questions that hit the zone of proximal development and that cause students to rethink and reorganize their understandings. Knowledge development has an ebb and flow to it—students move forward and backward as they reorganize their thinking into a cohesive whole or as they add new information into their existing structure. To further complicate things, students often lack the skills to articulate their thinking clearly. Not surprisingly, teachers may have difficulty situating students' current thoughts in the mathematics continuum. Such is the hard work of teaching.

In this endeavor, listening for the purpose of playing with an idea becomes important. The value of looking lost, of temporary "ignorance," might just be in a teacher's ability to play with an idea with students. To be a collaborator in knowledge development is vital. A collaborative stance sends the message that the teacher respects and values the students' thinking. It also sends the message that teachers do not have to be the possessors of all knowledge, that teachers may not have all the answers at their fingertips every moment of the class. Perhaps an important lesson in this feeling of looking lost is learning to embrace uncertainty rather than fear it. Embracing uncertainty and working with the students, playing with ideas along with them, may be the very learning space we need to move forward.

I am reminded that correct answers do not necessarily indicate correct strategies. We need to draw out students' solutions in different formats. Pictures and actions on objects often say more than the symbols of arithmetic written on a page. Students are adept at using algorithms and then drawing pictures to illustrate the answer; they have more trouble actually moving tiles on the table. Only when we see their thinking through their actions or pictures can we begin to negotiate genuine understanding. And often as we engage in this process, we ourselves begin to play with ideas and enhance our own mathematical understandings. We begin to build new mathematical connections and meanings.

In the process of this work with my colleagues, I find a renewed sense of commitment to my students as thinkers. I find that I must, even when confronting potential evidence to the contrary, believe that the students are thinking and that their answers make sense *to them*. I must respect them as learners and as collaborators in this teaching-learning process. I also find that I have to trust them and trust myself—them, to work and to continue working until we arrive at a solution and understanding, and myself, that I can guide them, regardless of their unexpected answers. This pursuit is quite a task!

I hope I am teaching my students that learning mathematics and rethinking the meaning of mathematics never ends for teachers. I hope I am teaching them that students' thinking is complex and curious. If the day comes that I do not feel that way, my learning—and my teaching—will be over. If that day comes, I should hang up my book bag and go home. Until then, I hope to keep learning.

REFERENCES

Ball, Deborah Loewenberg. "From the General to the Particular: Knowing Our Students as Learners of Mathematics." *Mathematics Teacher* 90 (December 1997): 732–37.

Davis, Brent. *Teaching Mathematics: Toward a Sound Alternative.* New York: Garland, 1996.

————. "Listening for Differences: An Evolving Conception of Mathematics Teaching." *Journal for Research in Mathematics Education* 28 (May 1997): 355–76.

Fleener, M. Jayne, Andy Carter, and Stacy Reeder. "Language Games in the Mathematics Classroom: Teaching a Way of Life." *Journal of Curriculum Studies* 36 (July/August 2004): 445–68.

Forman, George, and Brenda Fyfe. "Negotiated Learning through Design, Documentation, and Discourse." In *The Hundred Languages of Children: The Reggio Emilia*

Approach—Advanced Reflections, 2nd ed., edited by Carolyn Edwards, Lella Gandini, and George Forman, pp. 239–60. Westport, Conn.: Ablex Publishing Corp., 1998.

Helm, Judy Harris, Sallee Beneke, and Kathy Steinheimer. *Windows on Learning: Documenting Young Children's Work.* New York: Teachers College Press, 1998.

Lester, Frank K., Jr., Joanna O. Masingila, Sue Tinsley Mau, Diana V. Lambdin, Vânia Maria Pereira dos Santos, and Anne M. Raymond. "Learning How to Teach via Problem Solving." In *Professional Development for Teachers of Mathematics: 1994 Yearbook,* edited by Douglas B. Aichele, pp. 152–66. Reston, Va.: National Council of Teachers of Mathematics, 1994.

Lester, Frank K., Jr., and Sue Tinsley Mau. "Teaching Mathematics via Problem Solving: A Course for Prospective Elementary Teachers." *For the Learning of Mathematics* 13 (June 1993): 8–11.

National Council of Teachers of Mathematics (NCTM). *Curriculum and Evaluation Standards for School Mathematics.* Reston, Va.: NCTM, 1989.

———. *Principles and Standards for School Mathematics.* Reston, Va.: NCTM, 2000.

Sherin, Miriam Gamoran. "When Teaching Becomes Learning." *Cognition and Instruction* 20 (June 2002): 119–50.

Steffe, Leslie P., and Beatriz S. D'Ambrosio. "Toward a Working Model of Constructivist Teaching: A Reaction to Simon." *Journal for Research in Mathematics Education* 26 (March 1995): 146–59.

22

Learning to Learn Mathematics: Voices of Doctoral Students in Mathematics Education

Daniel Chazan
Sarah Sword
Eden Badertscher
Michael Conklin
Christy Graybeal
Paul Hutchison
Anne Marie Marshall
Toni Smith

> At an early age, students know that when you walk into a mathematics class, the mathematics has already beaten you through the door. It came in with the teacher. It lives within the hard covers of the text. As a student, your job is to sit, listen, and take in the numbers, formulas, equations, and rules. True understanding can seem irrelevant in these places. Long before taking high school calculus, I had stopped listening to my own mathematical curiosity or wondering about where mathematics comes from. There was no place to ask my questions, seek answers to my problems or become an authentic doer of mathematics.
>
> —*Excerpt from a graduate student's mathematical autobiography*

WITH its *Principles and Standards for School Mathematics*, the National Council of Teachers of Mathematics (NCTM) promotes a vision of mathematics classrooms in which students learn mathematics by doing mathematics (NCTM 2000). In this vision, doing mathematics is not just solving exercises or doing problems. It is exploring and developing mathematical understandings. Thus, the vision is in tension with some current practice in which doing mathematics means practicing solution methods introduced by the teacher. As we seek to move in the direction of the *Principles and Standards*, an ongoing challenge for the profession is to figure out how to create classroom communities in which mathematical exploration can be a vehicle for the learning of mathematics. Although there is much folk wisdom and lore about how to create effective mathematics classrooms in which students practice algorithms, a rich parallel discourse about the exploratory mathematics classroom is just beginning to develop.

This challenge of creating classroom communities in which students learn through exploration is many faceted. We have to learn how to craft exploratory problems that are within reach of our students. We have to learn to manage conversations around these problems that will lead to the crystallization of knowledge, not just to conversation. But, even more basically, we need to understand what we mean by mathematical exploration. Few mathematics educators have had the opportunity to learn mathematics in the way advocated by the *Principles and Standards.* More to the point, for most of us, formal mathematics education has not included the opportunity to pose our own problems or to develop our own conjectures. Few of us, perhaps with the exception of those who have pursued doctorate study in mathematics, can point to experiences where we explored mathematics, particularly in a group. As one doctoral candidate in mathematics education wrote,

> I have often wondered how we can expect teachers to teach in a reform inspired manner when they, themselves, and we have never experienced the joy in learning that way. In graduate school, as a future mathematics teacher educator, I needed a space to ask questions about things that I didn't understand and ask about those things that I naturally wondered about; the opportunity to boldly raise my hand, and with a confident voice, ask my own questions, seek answers to my own problems and become an authentic doer of mathematics.

In this article, we share an experience of teaching and learning at the doctoral level. Two instructors, one with a Ph.D. in mathematics and one with an Ed.D. in education, strove to create an environment in which mathematics education doctoral students could develop their own mathematical questions and explore them—in short, to do research on the mathematics of the grades K–12 curriculum. This particular course is one part of a larger effort in the Mid-Atlantic Center for Mathematics Teaching and Learning to reenvision doctoral education.

In this paper, we describe the structure of the course. This description is offered in conjunction with the reflections of the learners, some the doctoral students who were students in the course. In this way, we hope to portray images of what mathematical exploration can look like and how it can feel to educators who participate in such an experience. The perspectives of the participants in this course are our contribution to this yearbook's focus on the learning of mathematics.

The Basic Structure of the Course: An Introduction by the Instructors

We created a mathematics course for doctoral students in mathematics education, all former K–12 teachers. While designing the course, we took as a starting assumption that mathematics educators need to use mathematical knowledge in the course of their professional work. Furthermore, we did not imagine that we could teach all the mathematics for which participants might find a need in their professional lives. Rather than have us choose the mathematics that we deemed most important for them to know (they were too diverse a group to support such a strategy), we worked to design a course that would allow participants to explore their own mathematical interests from the areas of the curriculum relevant to their professional goals. We also wanted students to leave the course with strategies that would support lifelong learning of mathematics, particularly of mathematics related to their professional work. In this way, they might be prepared for a range of mathematical challenges they might find ahead of them. To this end, the course had two components: (1) multiweek in-class explorations around particular mathematical themes and (2) individual course projects.

In-Class Explorations

As the instructors, we chose the topics for the in-class explorations. But we wanted these topics to be explorations. We identified problems we wanted to explore and topics we hoped might come up, but not topics that we would "cover" in particular sessions. We worked to center our explorations on participants' questions and insights in response to the problems we posed. At the same time, we did not exclude formal definitions, theorems, and proofs from the class explorations. As we studied inequalities, sequences, continued fractions, cardinality, measure, and so on, we often needed formal mathematics to proceed. We tried, however, to introduce those formal ideas that the students did not know only in the service of students' explorations, not as material for which the participants would be held accountable. For example, when we began our study of real numbers in the first semester, we asked participants to bring their questions to class. Questions students asked included the following:

- Why do the decimal expansions of some rational numbers repeat and others terminate?
- How do we know that e and π are transcendental? Are there other transcendental numbers? How can we find them?
- Why do people say that $0.\overline{9} = 1$?
- How can there be different sizes of infinity?

In most instances, the participants knew answers to these questions, but often their concern was that their understanding of the answers to these questions was not satisfactory. Our exploration of their dissatisfaction with the answers they knew prompted us to introduce and explore the "Cauchy sequence" definition of a real number.

A Quick Synopsis of One Exploration

To better illustrate the in-class explorations, here is a quick synopsis of one exploration that extended over eight sessions. We began with a *balanced assessment* task for high school students. The problem asks students to design a measure of "squareness": the degree to which a particular rectangle is close to, or far from, being a square. In the discussion of creating strategies for finding such a measure, which included the ratio of adjacent side lengths, we introduced a process of cutting off squares (called "anapharysis" by Fowler [1987] and illustrated in fig. 22.1). This process generates a sequence of numbers, and we explored how to make that sequence into a measure of squareness.

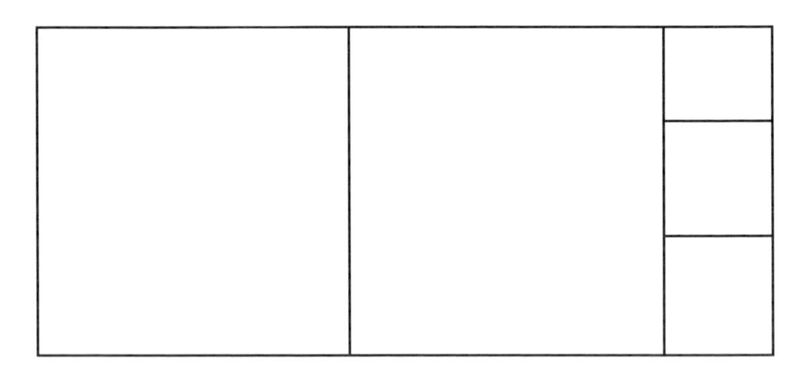

Fig. 22.1. This figure has "squareness" 2, 3—we cut off (or created) two squares and then three squares.

Our exploration of this process led us to consider how to turn a sequence into a rectangle with particular side lengths. This focus led us to view the sequences from right to left and to build up a rectangle from squares, rather than to cut off squares from a given rectangle. A participant came to the board and connected this way of thinking to continued fractions as a way of representing our sequences. His point was that this procedure would enable us to compute without recourse to a diagram.

Inspired by the notion that the decimal representation of a number is a sequence, we began to explore what happens when one has a repeating, but unending, anapharytic sequence. A participant conjectured that terminating sequences would always be rational, and that any rational could be represented by such an "anapharytic" sequence. The question prompted us to consider what kind of numbers are those that result from a repeating sequence. We began to explore particular cases. The original conjecture was embellished to suggest that nonterminating sequences would always produce irrational numbers—but what sorts of irrational numbers? Transcendental? Algebraic?

Members of the class did not fully resolve that question, but we did explore continued fraction representations of particular irrational numbers. This was the springboard into ideas from measure theory and an exploration of the Cantor set.

Our explorations of anapharysis ended with a return to the Greeks. For the Greeks, anapharysis was a way of comparing ratios. One participant read Euclid's algorithm and showed the class that in all our explorations of anapharysis, we had always chosen rectangles whose sides were relatively prime. Relaxing that unintended choice, one can appreciate that anapharysis is also a method for determining greatest common factors.

Sessions involved work on problems, presentation to the class, and then a summary. At the end of each session, each participant wrote in a class journal. We prepared a summary of our understandings of what had been claimed and argued in class. Participants did not have any homework or assignments on the in-class exploration between class sessions, though most found it difficult not to pursue thinking about these topics outside of class, sometimes even in informal groups.

Projects

To complement the in-class explorations, we gave the participants the chance to explore their own questions with class projects. William Thurston's (1994) criticism of the definition-theorem-proof model of communicating mathematics resonated with us: "A clear difficulty with the DTP [definition-theorem-proof] model is that it doesn't explain the source of the questions" (p. 3). We wanted our students to have access to the sources of their own questions.

Generating questions for research was one of the most difficult components of the course, even with resources like Brown and Walter's (1983) book *The Art of Problem Posing.* Outside of class, participants spent the first three weeks of the semester developing questions for a semester-long mathematical research project. They mined their teaching experience. We hoped that from struggling with developing questions, they would develop strategies for asking good mathematical questions and strategies for exploiting those questions for mathematical learning experiences. One student wrote the following excerpt in her weekly journal:

> It's been difficult to come up with possible problems, select one, and to know what to do with it. For me, the hardest part has been not knowing if the problem

> is solvable. When a teacher gives you a problem, you can be pretty sure that it is solvable and that you have the knowledge/ability to do so. When we make up our own problems, we have no idea if it is solvable and what the solution might involve.

In reflecting on the class, the same student described the experience of working on the projects as "both liberating and frustrating":

> The research component of this course allowed the students the time and support necessary to focus on their own questions in a way that none of us had ever experienced before. As a student, this was both liberating and frustrating. It was wonderful to be able to focus on questions that were of interest to us and that were relevant to our own teaching. It was also very challenging because in our past experiences as both students and teachers of mathematics, the questions were always given to us. Developing and working on one's own questions is a very different, and often maddening experience. Through the process of posing and solving our own questions, we experienced some of the ups and downs that mathematicians experience.
>
> In my experiences in other math classes, I had a sense of how to solve a problem simply by looking at when in the course the question was posed; the proof required only the theorems on the previous pages. Because the teacher assigned the questions, I trusted that I had all of the knowledge and skills required to answer the question. In this course, since we were asking our own questions, we did not know if we had the prerequisite skills necessary to solve the problem, or if the problems were even solvable. In math courses, students often look in the back of the book or other textbooks to find the solution to a question assigned by the instructor. Since we had posed our own questions, we could not do this, and the students in this class purposefully avoided known solutions.
>
> Although it may have been naive, at times of exasperation, some of us wondered if we had stumbled upon another great question of mathematics (like trisecting an angle) that might require years of work by the greatest mathematicians to solve or prove unsolvable. In fact, one of the students said that while he was working on his problem, to him, his problem was one of the great problems in mathematics. Our problems may not have been research problems in the sense of being unsolved in the mathematical field, but they were very real to us because we had posed them and did not know if or how anyone else had answered them. This provided an avenue to experience the joy of posing and solving problems that matter to us.

Learning Mathematics Together

We hadn't explicitly connected the research projects with the in-class explorations, but an unexpected result was that the students learned from others' projects and were inspired to extend their work on the basis of what others had presented.

Students regularly made mathematical connections between the two. For example, while enrolled in our class, one student was teaching a mathematics content course for preservice elementary school teachers. Since she was teaching her own students about the arithmetic of numbers in bases other than ten, she decided to focus her project on that subject. In the third week of the course, she presented her topic and preliminary questions to the class. She inspired another student, who had been doing an in-class exploration trying to characterize rational numbers with terminating decimal expansions, to extend his exploration to bases other than ten. He presented his results enthusiastically to the class. Later, we asked him to reflect on what inspired him to publicly present his work. He wrote:

> In some ways, my desire to "go the board" (which in my mind is more of a general phrase for any sort of open-ended public sharing) co-existed with and was co-dependent with my excitement for the course, and ultimately mathematics in general.
>
> Why share? I think of the times I have gone home and shared (what I thought were) interesting mathematical musings with my wife, to be greeted with supportive, though not always interested, nods. In a similar vein, my brother, the writer, is often sought out at family gatherings to get his take on the latest best-seller or Oscar prediction buzz. I don't recall ever being cornered at Thanksgiving to give my 2 cents on the future of two-column proofs in the high school curriculum.
>
> Please do not misunderstand, I do not offer such statements as complaints. I would neither trade my family for a more mathematical-minded bunch nor my experiences in mathematics for more mainstream interests. But what this course offered, more than Thanksgiving dinner at my mom's cousins' house, was the opportunity to truly share. So, I suppose, more than anything, my going to the board was simply taking advantage of this opportunity, the opportunity to both share and be "shared to."

This kind of interaction was accessible to students with diverse mathematical backgrounds. One student wrote this comment:

> Within this group, the majority of students were interested in high school and post-secondary level mathematics. Many students had either a masters' level background in mathematics classes or enough mathematics classes to be certified to teach high school mathematics. I came to the class as a former elementary teacher, still very much interested in young children's mathematical thinking. I had taken "enough" mathematics to be considered a doctoral student in mathematics education but clearly had taken considerably less mathematics courses than my classmates. As the semester went on, my "elementary-ness" evolved from something that was felt as a deficit to something that became a unique perspective to add to our learning.

A diverse group of students were working on the question of why people say that $0.\overline{9} = 1$. The students had written down several arguments, but two students thought that these arguments were not personally meaningful. When they saw the characterization of repeating decimals base n, one of them noticed that in base 4, we can write $0.333333... = 1$, and in base 6, we can write $0.55555... = 1$. "Hey," she said. "This is just an artifact of the base!" In reflecting on this, one student described her own developing sense of mathematical authority:

> I would hypothesize that the students in this class had rarely had the opportunity to ask their own questions about mathematics, let alone raise questions about the validity of formal mathematics. However, what this class enabled us to do was to develop a personal sense of authority in studying mathematics. As our personal authority developed so did our efficacy in analyzing mathematical arguments and judging their validity in reference to our own experiences and beliefs. Over the semester we each to varying degrees developed a sense of not only that our questions mattered, but they were important to consider mathematically. Moreover, we learned to listen to our own intuition and, when this contradicted what was mathematically accepted, to delve deeper into both our own understanding as well as the mathematical argument.
>
> In particular I found it beneficial to understand that I was able to decide whether or not a mathematical argument was valid and/or satisfying. Even when I could follow a proof or logical argument, it was still appropriate for me to say, "Sure, I follow that, but it does not effectively account for or answer this question I still have. So, it is not particularly useful for me, and until an argument effectively deals with these, I may follow the argument, but I will not say it is the right argument."
>
> As the course progressed, there was a greater recognition that we as students of mathematics did not have to "buy in" to an argument that did not make sense personally. Somehow an argument had to be personally satisfying and if it did contradict particular beliefs or experiences, delving into the argument had to help develop new experiences and intuitions. Arguments had to provide understanding and answer questions.

We recognize that this environment was luxurious in the sense that we needed to justify decisions about how we spent our time only to one another. The instructors had not planned to spend class time on explorations of numbers in bases other than ten. We had time to keep the $0.999... = 1$ (base ten) question in the public conversation until students found personally meaningful resolutions. We had time to spend on student-generated explorations in bases other than ten despite the fact that the work was not part of the original syllabus and not a major strand of the course. We had the luxury of allowing student-generated questions and students' thinking to drive the course, even when the direction of the course did not perfectly match the instructors' intent. One of the course participants wrote about this from his dual perspective as a mathematics teacher and a doctoral student:

> Part of what made my insight [about representation in mathematics] so compelling to me was that I hit upon it on my own. It was my insight into doing math. This realization raises a bit of a dilemma for many math teachers. On the one hand there are things we want our students to know. These things might include various content objectives, and they might also include knowledge about the doing of math, like my realization that looking for different representations can enrich my understanding of the content. On the other hand allowing students the freedom to make their own discoveries can lead to really powerful mathematical experiences for them, as was the case for me as a math learner. But giving them such freedom diminishes our ability as the teacher to lead the students to the objectives we want them to get to.
>
> Lots of people have written about this tension in the classroom, the tension between valuing student thinking and the traditional content of math courses. I don't have anything to add from the perspective of the teacher, except maybe to acknowledge I feel that tension as a teacher as well. But my experience as a math learner in this class gives me reason to comment from the learner perspective. For me, being put into a math learning context where my ideas and the ideas of my classmates were frequently the building blocks of the course led to me taking myself more seriously as a math learner. The implicit message in such an approach is that the teacher deems the students capable of engaging rigorously in math. By giving us considerable freedom, the instructors of the course were telling us they took us seriously as mathematical learners, and I think this encouraged us all to take ourselves more seriously.
>
> Having established such an atmosphere in the classroom, what we did felt more relevant somehow. My insight into representation, which came early in the course, had an impact on the math I did that followed it. And I suspect it will have an impact on the math and the teaching I do in the future. Though I don't know how to test this claim, clearly the claim I am making here is taking students seriously as math learners will result in more powerful mathematical experiences for them. Powerful here means insights into math content and the doing of math that will remain with them after they turn in their textbooks. There is no doubt that creating such an atmosphere can be challenging, and that there are trade-offs to be made along the way, but having been on the learner side of the fence I can tell you the potential benefit is worth the struggle.

Final Thoughts

For many of the participants in this classroom community, the potential benefit of having space to explore questions of personal mathematical interest seemed "worth the struggle." This played out mathematically in the second semester of the course while participants were engaged in parallel explorations of cardinality and the measure of real numbers. As a part of these explorations, the instructors wanted to introduce the Cantor set, an uncountable set of real numbers with measure zero. The very existence of this set is counter-intuitive. The set is very difficult to understand if the elements are represented in base ten but far easier to understand when the elements are represented in base three. At that point, working in base three posed no problem for the participants, since they had done a number of explorations themselves. In the second semester, that body of (entirely participant-driven) work supported the engagement in exploration around the Cantor set.

Our wish as both participants and instructors is that we continue to enrich our professional lives with opportunities for mathematical exploration. And isn't that the very beauty of mathematics? That a life even partly centered on the subject is a life in which there is always more to wonder about, more to explore, more to learn? We want to end this article with one last reflection from a participant:

> What seems so central here is that the mathematics we studied mattered to us. Changing roles from teacher to student and from question answerer to question poser was sometimes a scary transition, but this transformation had the potential to awaken the mathematical mind and to allow mathematics to be intrinsically meaningful. There are so many questions that have not been answered yet, and so many that have not even been asked.

REFERENCES

Brown, Steve, and Marion Walter. *The Art of Problem Posing.* Hillsdale, N.J.: Lawrence Erlbaum Associates, 1983.

Fowler, D. H. *The Mathematics of Plato's Academy: A New Reconstruction.* Oxford, England: Clarendon Press, 1987.

National Council of Teachers of Mathematics (NCTM). *Principles and Standards for School Mathematics.* Reston, Va.: NCTM, 2000.

Thurston, William. "On Proof and Progress in Mathematics." *Bulletin of the American Mathematical Society* 30 (April 1994): 161–77.

ADDITIONAL READING

Ball, Deborah Loewenberg, and Hyman Bass. "Making Mathematics Reasonable in School." In *A Research Companion to "Principles and Standards for School Mathematics,"* edited by Jeremy Kilpatrick, W. Gary Martin, and Deborah Schifter, pp. 27–44. Reston, Va.: National Council of Teachers of Mathematics, 2003.

Cuoco, Al, E. Paul Goldenberg, and June Mark. "Habits of Mind: An Organizing Principle for Mathematics Curriculum." *Journal of Mathematical Behavior* 15 (December 1996): 375–403.

Lave, Jean, and Etienne Wenger. *Situated Learning: Legitimate Peripheral Participation.* Cambridge, England: Cambridge University Press, 1990.

Skemp, Richard. "Relational Understanding and Instrumental Understanding." *Mathematics Teaching* 77 (November 1976): 20–26.